AF478648

Computational Fluid and Solid Mechanics

Series Editor

K.J. Bathe
Massachusetts Institute of Technology, Cambridge, MA, USA

For other titles published in this series, go to
http://www.springer.com/series/4449

Tatiana G. Elizarova

Quasi-Gas Dynamic Equations

 Springer

Tatiana G. Elizarova
Russian Academy of Sciences
Institute of Mathematical Modeling
Miusskaya Sq. 4
Moskva
Russia 125047
telizar@yahoo.com

ISSN 1860-482X   e-ISSN 1860-4838
ISBN 978-3-642-00291-5   e-ISBN 978-3-642-00292-2
DOI 10.1007/978-3-642-00292-2
Springer Dordrecht Heidelberg London New York

Library of Congress Control Number: 2009926099

Cover design: deblik, Berlin

Printed on acid-free paper

Springer is part of Springer Science+Business Media (www.springer.com)

Preface

The monograph is devoted to modern mathematical models and numerical methods for solving gas- and fluid-dynamic problems based on them. Two interconnected mathematical models generalizing the Navier–Stokes system are presented; they differ from the Navier–Stokes system by additional dissipative terms with a small parameter as a coefficient. The new models are called the quasi-gas-dynamic and quasi-hydrodynamic equations. Based on these equations, effective finite-difference algorithms for calculating viscous nonstationary flows are constructed and examples of numerical computations are presented. The universality, the efficiency, and the exactness of the algorithms constructed are ensured by the fulfillment of integral conservation laws and the theorem on entropy balance for them.

The book is a course of lectures and is intended for scientists and engineers who deal with constructing numerical algorithms and performing practical calculations of gas and fluid flows and also for students and postgraduate students who specialize in numerical gas and fluid dynamics.

Moscow   Tatiana Elizarova
December 2008

Contents

Introduction

This book is devoted to modern mathematical models and numerical methods for solving gas- and fluid-dynamic problems based on them. Two interconnected mathematical models generalizing the Navier–Stokes system are presented; they differ from the Navier-Stokes system by additional dissipative terms with a small parameter as a coefficient. The new models are called the quasi-gas-dynamic (QGD) and quasi-hydrodynamic (QHD) equations. We present a method for obtaining QGD/QHD equation systems and the theoretical justification of them. Based on these systems, we construct finite-difference algorithms for calculating viscous nonstationary gas and fluid flows. The universality, the efficiency, and the accuracy of the algorithms constructed are ensured by the fulfillment of integral conservation laws and the entropy balance theorem for them. Additional dissipative terms regularize the numerical solutions. Another merit of the algorithms is the simplicity of the numerical implementation that is of especial importance in applications of three-dimensional nonstructured grids and parallel computers.

The description of gas and fluid flows based on the Navier–Stokes equations has a rich history. At present, many commercial program packages using numerical algorithms for solving these equations are created and are successfully applied. Nevertheless, approaches used in them cannot be considered as perfect. At different times, enlarging the possibilities of describing the flows by the Navier–Stokes equations was attempted. However, the proposed models turned out to be essentially more complicated than the classical system and found no application in practical computations.

For the first time, the system of quasi-gas-dynamic equations enlarging the possibilities of the Navier–Stokes system has appeared in the process of investigations performed by a small group of scientific research workers of the M. V. Keldysh Institute of Applied Mathematics of the Russian Academy of Sciences in the 1980s led by Professor Boris N. Chetverushkin.

In the very beginning of these studies, in 1982, the author had a good fortune to enter this work, and the first variant of quasi-gas-dynamic equations was written with her direct participation. These equations differ from the classical gas-dynamic equations by additional terms having the form of the second spatial derivatives. The new models immediately allow one to construct effective algorithms for solving the Euler equation and then those for the Navier–Stokes equations.

Later on, in the works of Yu. V. Sheretov, the quasi-gas-dynamic equations were represented in the form of conservation laws, were studied in detail, and were theoretically justified. Moreover, he constructed a quasi-hydrodynamic system contiguous to these equations. The principal and essential distinction of the QGD/QHD approach from the Navier–Stokes theory is the use of the time-spatial averaging procedure for the definition of the main gas-dynamic quantities: density, velocity, and temperature. An additional smoothing with respect to time is the cause of appearance of additional dissipative summands in the equations, by which the QGD/QHD systems formally differ from the Navier–Stokes system.

In the stationary case, both QGD and QHD systems differ from the Navier–Stokes system and one from the other by divergent terms having asymptotically small orders of smallness $O(\mathrm{Kn}^2)$ as $\mathrm{Kn} \to 0$, where Kn is the Knudsen number. The approximations of the boundary layer for the new equations as well as for the Navier–Stokes equations are the Prandtl equations. The influence of additional terms is inessential for the stationary and quasi-stationary flows for small Knudsen numbers. However, for strongly nonstationary flows and also for the Kn numbers close to unity (for example, $\mathrm{Kn} \sim 0.5$), their contribution becomes essential. Precisely in these classes, one can find the advantages of the new models. In the numerical modelling, the additional terms manifest themselves as effective regularizers.

Each of the two QGD/QHD system obtains its own method for solving the closure problem. The quasi-gas-dynamic equation should be used in modelling the flows of ideal polytropic gas, whereas the quasi-hydrodynamic equations should be used in modelling the flows of gas and fluids with more general state equation.

The book consists of nine chapters and four appendices.

In the first chapter, we formulate the general ideas, which along with the classical gas-dynamic system of equations, the Navier–Stokes system, allows one to construct two new mathematical models for describing the viscous flows – the quasi-gas-dynamic and quasi-hydrodynamic systems.

In the second chapter, we present elements of the kinetic theory, which are necessary for the subsequent presentation, and describe a method for constructing kinetically consistent difference schemes, which have served as a base for first variants of the QGD equations.

In the third chapter, which plays a key role, we present two developed methods for constructing quasi-gas-dynamic equations. The first of them is based on the kinetic model of particle motion and the regularized kinetic equation; the second is based on integral conservation laws written for a small but finite volume. Also, we present here a method for writing the obtained equations in the form of conservation laws, write the entropy balance equation, and trace the connection of the QGD system with the Navier–Stokes equations. One of the results of this chapter is obtaining approximate formulas for the viscosity, bulk viscosity, and heat conductivity coefficients.

In the fourth chapter, we write the QGD equations for an arbitrary coordinate system.

The fifth and largest chapter is devoted to constructing effective finite-difference algorithms for solving the QGD equations for the numerical modelling of gas-dynamic flows by using orthogonal meshes. Here, we present examples of calculation of test problems.

In the sixth chapter, we generalize the proposed algorithms to the case of non-structured two-dimensional grids.

The seventh chapter is devoted to the quasi-hydrodynamic system. Based on these equations, we construct new effective algorithms for calculating the nonstationary flows of viscous incompressible fluid and present examples of modelling two- and three-dimensional flows.

In the eighth and ninth chapters, we construct generalizations of the quasi-gas-dynamic equations for modelling the gas flows with translational and rotational degrees of freedom being not in the equilibrium and the equations for describing the binary mixture of gases not interacting with each other.

In the appendices, we present a detailed deduction of the quasi-gas-dynamic system for the planar one-dimensional flow and also examples of calculation of the rarefied flow in microchannels, the one-dimensional shock wave structure, and the numerical modelling of a laminar-turbulent transition in a gas flow after a backward-facing step.

The present book is based on a course of lectures given by the author for students and postgraduate students of the Physical Department of the M. V. Lomonosov Moscow State University in 2004–2008 and was published in 2005 as a textbook and in 2007 as a book.

I express my sincere gratitude to all my colleagues who participated in the development of the QGD/QHD theory at different times of their collaboration. Their contribution is reflected in the publications which serve as a base for this book. I sincerely express my gratitude to Boris N. Chetverushkin whose ideas served as a base for the QGD approach. I express my deep gratitude to Yuri V. Sheretov for his help in writing the first chapter and a number of sections in Chaps. 2, 3, 5, and 7. I express my gratitude to Aleksander A. Samarskii and Aleksey G. Sveshnikov for constant support of this scientific direction. I also express my gratitude to Aleksey V. Ovchinnikov for assistance in the preparation of the English version of the monograph.

In the English translation of the monograph, misprints and inaccuracies found in the last Russian version were corrected. Moreover, the list of references was supplemented with some recent papers.

Chapter 1
Construction of Gas-Dynamic Equations by Using Conservation Laws

In this chapter, we recall physical principles serving as a base in deducing equations of the classical gas dynamics and the new quasi-gas-dynamic (QGD) and quasi-hydrodynamic (QHD) systems. As the Navier–Stokes equations, the QGD/QHD equations are a consequence of integral conservation laws, have a dissipative character, and can be obtained from the general system of conservation laws. A principal and substantial distinction of the QGD/QHD equations from the Navier–Stokes equations is the use of the time-spatial averaging procedure in order to find the main hydrodynamic quantities—the density, the velocity, and the temperature. The use of spatial averages leads to the Navier–Stokes system. For time-spatial averages, we propose two variants of closing the general system of equations, which lead to QGD/QHD systems. In this chapter, we present the expressions for the vectors of mass flux density, heat flux, and the tensor of viscous stresses for QGD/QHD systems without any deduction. Two methods for constructing these quantities for the QGD system are presented in Chap. 3. We discuss the physical meaning of the vector of mass flux. The presentation of this chapter is mainly based on [84, 181, 184, 190].

1.1 Averaging Procedure

Consider a monoatomic gas consisting of sufficiently large number N of spherical atoms of radius r_0 and mass m_0. In the Euclidean space $\mathbb{R}^3$, we choose a Cartesian coordinate system (x_1, x_2, x_3) and time t.

The motion of each of the atoms can be described by the Newton equations. However, such an approach to modelling gas-dynamic problems turns out to be very far from practice since N is very large. Moreover, there arise problems connecting the determination of initial conditions for the problem and the subsequent averaging of the quantities obtained, which is necessary for the calculation of the quantities being measured—the density, the velocity, and the temperature of the medium.

In the classical hydrodynamics, one uses another approach based on the transition from a large number of separate particles to a continuous medium by using averaging procedures. These procedures can be chosen in different ways.

T.G. Elizarova, *Quasi-Gas Dynamic Equations*, Computational Fluid and Solid Mechanics, DOI 10.1007/978-3-642-00292-2_1,

1.1.1 Spatial Averages

In the Navier–Stokes theory, one uses the so-called instant spatial averages, which are defined as follows.

Let ΔV be a ball of radius R_V centered at a point x (Fig. 1.1). Let an atom belong to the ball whenever its center belongs to this ball. Let $N_{\Delta V}(t)$ be the number of molecules in the volume ΔV at an instant of time t. We define the density, the mean momentum, and the mean energy of the volume unit as follows:

$$\rho(x,t) = \frac{m_0}{\Delta V} N_{\Delta V}(t), \tag{1.1}$$

$$I(x,t) = \rho u = \frac{m_0}{\Delta V} \sum_{i=1}^{N_{\Delta V}(t)} \xi_i(t), \tag{1.2}$$

$$E(x,t) = \rho \left(\frac{u^2}{2} + \varepsilon \right) = \frac{m_0}{\Delta V} \sum_{i=1}^{N_{\Delta V}(t)} \frac{\xi_i^2(t)}{2}, \tag{1.3}$$

where $\xi_i(t)$ is the velocity of the ith particle at the instant of time t and m_0 is its mass. In the presented expressions, $\varepsilon = \varepsilon(x,t)$ is the specific internal energy. Let us introduce the temperature T, which is found from the expression

$$\varepsilon = c_v T, \tag{1.4}$$

where $c_v = c_p - \mathcal{R}$ is the specific heat capacity at constant volume, c_p is the specific heat capacity at constant pressure, $\mathcal{R} = k_B/m_0 = \mathcal{R}_\star/\mathcal{M}$ is the perfect gas

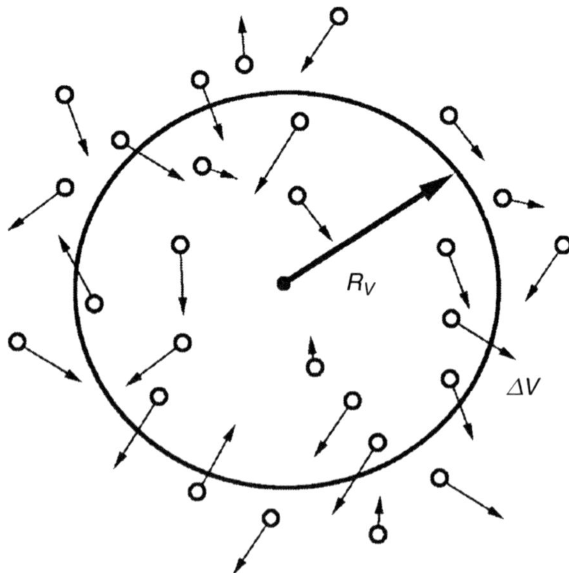

Fig. 1.1 The definition of mean quantities

constant, $k_B = 1.38 \times 10^{-16}$ erg/K is the Boltzmann constant, $\mathcal{R}_\star$ is the universal gas constant, $\mathcal{M}$ is the molar mass of the gas, and $\gamma = c_p/c_t$ is the specific heat ratio.

When the averaging volume ΔV changes, the values of the averages can change. Let us accept the conjecture on the existence of two scales R_{max} and R_{min} such that $R_{max} \gg R_{min}$, and for fixed x and t and for any $R_{max} > R_V > R_{min}$, the values of the above averages are practically constant and independent of ΔV. Then the corresponding averages are called the gas-dynamic quantities—the density, the momentum, and the total energy. Also, assume that all these functions are sufficiently smooth, i.e., they are continuously differentiable as many times as required.

1.1.2 Spatial-Time Averages

Let us define the spatial-time averages, which differ from definitions (1.1), (1.2), and (1.3) by additional smoothing in time. For this purpose, we additionally introduce a certain interval of time Δt and define the density, the momentum, and the energy of volume unit as follows:

$$\rho(x, t) = \frac{m_0}{\Delta V} \frac{1}{\Delta t} \int_t^{t+\Delta t} N_{\Delta V}(t') \, dt', \tag{1.5}$$

$$I(x, t) = \rho u = \frac{m_0}{\Delta V} \frac{1}{\Delta t} \int_t^{t+\Delta t} \left(\sum_{i=1}^{N_{\Delta V}(t')} \xi_i(t') \right) dt', \tag{1.6}$$

$$E(x, t) = \rho \left(\frac{u^2}{2} + \varepsilon \right) = \frac{m_0}{\Delta V} \frac{1}{\Delta t} \int_t^{t+\Delta t} \left(\sum_{i=1}^{N_{\Delta V}(t')} \frac{\xi_i^2(t')}{2} \right) dt'. \tag{1.7}$$

Further, assume that along with two spatial scales R_{max} and R_{min}, there also exist two timescales $\Delta t_{max} \gg \Delta t_{min}$ such that for any $\Delta t_{max} > \Delta t > \Delta t_{min}$, the values of the above averages are practically constant and independent of ΔV and Δt. The characteristic values of these scales are connected with each other as $\Delta t_{max} \sim R_{max}/c$ and $\Delta t_{min} \sim R_{min}/c$, where c is the sound speed determining the speed of mean particle velocity in the gas. Then the corresponding averages can be considered as the gas-dynamic quantities—the density, the momentum, and the total energy. Also, assume that these functions are sufficiently smooth.

The hierarchy of characteristic spatial and timescales for gas flows are discussed, in particular, in [5, 119, 121] in detail.

The introduction of the additional smoothing in time in the definition of the gas-dynamic quantities seems to be natural for many reasons. In experiments, the measurement of all gas-dynamic quantities is performed for a finite time, which automatically leads to smoothing over a certain time interval. The number of particles in a small volume ΔV with transparent boundaries naturally change on times that accounts for particles intersecting its boundary in a chaotic way. For a sufficiently large number of particles in the averaging volume ΔV, the spatial and spatial-time averages can be very close to each other, which corresponds to the ergodic conjecture on the coincidence of the instant spatial and spatial-time averages.

In what follows, we do not identify the instant spatial and spatial-time averages preserving the usual notation for all gas-dynamic quantities. If necessary, we denote the instant spatial averages by the subscript s and the spatial-time averages by the subscript st.

1.1.3 Galileo Transform

Consider two inertial coordinate systems. Let $K^\star$ be a coordinate system moving relative to the initial coordinate system K with a constant speed U (Fig. 1.2). Then the coordinates of a material point $x^\star$ and time $t^\star$ in $K^\star$ are connected with the coordinates x and t in the system K by the relations

$$x^\star = x - U(t - t_0), \tag{1.8}$$
$$t^\star = t. \tag{1.9}$$

At an instant t_0, both systems coincide. Frank called formulas (1.8) and (1.9) the "Galileo transforms" (see [155]). For the first time, the Galileo transform was written for a material point in the Newtonian classical mechanics. In this case, the conjecture on the absoluteness of time (1.9) was substantially used. According to the Galileo principle, two inertial coordinate systems moving one relative to the other with a constant speed are equivalent from the viewpoint of mechanical phenomena occurring in them, i.e., the equations of motion in these systems are invariant.

However, for the study of the invariance of the hydrodynamic equations, formulas (1.8) and (1.9) are insufficient. In addition, it is necessary to know in which way the macroscopic parameters—the density ρ, the hydrodynamic velocity u, and the temperature T—change when passing from K to $K^\star$. The answer to the latter question depends on the averaging procedure used in the definition of these macroparameters.

When using the instant spatial averages, we have the following relations:

$$\rho_s^\star = \rho_s, \quad u_s^\star = u_s - U, \quad T_s^\star = T_s. \tag{1.10}$$

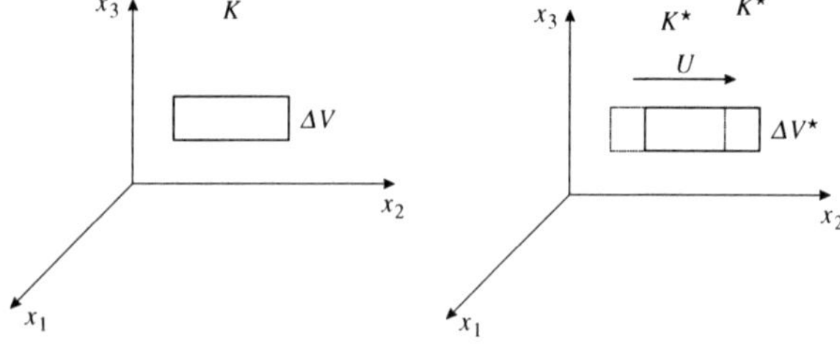

Fig. 1.2 Spatial-time averages and the Galileo transform

The invariance of the Navier–Stokes equations, which are constructed by using the spatial averages, with respect to transformations (1.8) and (1.9) is directly verified.

For the spatial-time averages, the volumes in which the averaging is performed in the fixed coordinate system ΔV and the moving coordinate system $\Delta V^\star$ are different; the volume ΔV fixed for the observer from the coordinate system K seems to be movable for the observer from the coordinate system $K^\star$ and vice versa. Therefore, relations (1.10) hold not exactly but only approximately,

$$\rho_{st}^\star \approx \rho_{st}, \quad u_{st}^\star \approx u_{st} - U, \quad T_{st}^\star \approx T_{st}. \tag{1.11}$$

Therefore, the density, the velocity, and the temperature turn out to be relative, and the invariance with respect to the Galileo transform violates.

A similar situation arises in relativistic hydrodynamics where conjecture (1.9) on the absoluteness of time is not true. Therefore, the equations of relativistic hydrodynamics are also not invariant with respect to the Galileo transform.

For the existence of the invariance in the hydrodynamic equations, the fulfillment of the following two conditions is necessary:

(i) the definition of the hydrodynamic quantities as spatial averages (i.e., the same averaging volume is used in the fixed and movable coordinate systems, which yields $\rho = \rho^\star$, etc.)
(ii) the absoluteness of time (time is the same in the fixed and movable coordinate systems, and hence $t = t^\star$)

The invariance violates if at least one of these two conditions does not hold. For the Navier–Stokes equations both conditions hold. For the equations of relativistic mechanics, the second condition does not hold. For the equations that are constructed by using spatial-time averages, the first condition does not hold. The work [188] of Yu. V. Sheretov is devoted to a detailed discussion on the fulfillment of the Galileo transform for the hydrodynamic equations.

1.1.4 Continuity Equation

The principle of mass conservation or the mass-balance equation is laid as a base of hydrodynamics; it is written as follows:

$$\frac{\partial \rho}{\partial t} + \operatorname{div} j_m = 0, \tag{1.12}$$

where j_m is the vector of mass flux density. This equation is called the continuity equation. The integral form

$$\frac{\partial}{\partial t} \int_V \rho\, dV + \int_\Sigma (j_m \cdot n)\, d\Sigma = 0 \tag{1.13}$$

of the continuity equation means that the change of the mass in a certain closed volume V is defined by the mass flux j_m through its boundary Σ. Note that in the construction of Eq. (1.12) the definitions of the gas-dynamic quantities are not used.

For spatial averages, it is assumed (see, e.g., [173, 184]) that the density of the mass flux is equal to the momentum of the volume unit:

$$j_{ms} = \rho_s u_s. \tag{1.14}$$

In this case, Eq. (1.12) has the form

$$\frac{\partial \rho_s(x, t)}{\partial t} + \mathrm{div}\left(u_s(x, t)\rho_s(x, t)\right) = 0, \tag{1.15}$$

and the Galileo transform holds for the equation of continuity (1.15), as well as for the spatial averages themselves.

Let us show that when using the definitions of spatial-time averages, the equation of continuity of the form (1.15) for these averages does not hold. We integrate Eq. (1.15) over a small interval of time Δt. Then, according to the definition of ρ_{st}, the first term takes the form

$$\frac{1}{\Delta t} \int_t^{t+\Delta t} \frac{\partial \rho_s(x, t')}{\partial t} \, dt' = \frac{\partial}{\partial t} \rho_{st}(x, t).$$

The second term in Eq. (1.15) becomes

$$\frac{1}{\Delta t} \int_t^{t+\Delta t} \mathrm{div}(\rho_s(x, t')u_s(x, t')) \, dt'$$

$$= \mathrm{div}\left(\frac{1}{\Delta t} \int_t^{t+\Delta t} \rho_s(x, t')u_s(x, t') \, dt'\right) \neq \mathrm{div}\left(\rho_{st}(x, t)u_{st}(x, t)\right),$$

since

$$\int_t^{t+\Delta t} \rho_s(x, t')u_s(x, t') \, dt' \neq \int_t^{t+\Delta t} \rho_s(x, t') \, dt' \cdot \int_t^{t+\Delta t} u_s(x, t') \, dt'.$$

Therefore, for spatial-time averages, the mass flux density j_{mst} may not coincide with the momentum of the volume unit $\rho_{st}u_{st}$. This reflects the fact that even for a small time Δt, the instant values of the density and momentum change.

In this case, the expression for the mass flux density can be written in a more general form. Introduce a small addition to the velocity, which is denoted by w_{st}, and write the mass flux density as follows:

$$j_{mst} = \rho_{st}(u_{st} - w_{st}).$$

As was shown above by using qualitative reasons, the Galileo transform does not hold for the gas-dynamic equations based on spatial-time averages.

1.2 Conservation Laws in the Integral Form

In the Euclidean space $\mathbb{R}^3$, we choose an inertial Cartesian coordinate system (x_1, x_2, x_3). Let (e_1, e_2, e_3) be the orthonormal basis of unit vectors corresponding to it and let t be the time. We use the following standard notation for the quantities characterizing the flows of compressible, viscous, heat-conducting medium: $\rho = \rho(x, t)$ is the medium density, $u = u(x, t)$ is the velocity, $p = p(x, t)$ is the pressure, $\varepsilon = \varepsilon(x, t)$ is the specific internal energy, $T = T(x, t)$ is the temperature, and $s = s(x, t)$ is the specific entropy.

Assume that the medium is two-parameter, i.e., among five thermodynamic parameters ρ, p, ε, T, and s, only two parameters are independent and the following state equations are given:

$$p = p(\rho, T), \quad \varepsilon = \varepsilon(\rho, T), \quad s = s(\rho, T). \tag{1.16}$$

Let $F = F(x, t)$ be the mass density of exterior forces. For example, for a fluid in the gravitational field of the Earth we have $F = g$, where g is the free fall acceleration.

In the domain of the flow, let us distinguish a bounded moving volume $V = V(t)$ with smooth surface $\Sigma = \Sigma(t)$ oriented by the exterior unit normal field n (Fig. 1.3). We assume that the volume $V(t)$ arises from the volume $V_0 = V(t_0)$, where t_0 is the initial instant of time, via a continuous deformation stipulated by its movement along trajectories of a certain vector field v. Let us write the well-known Euler–Liouville identity [173]:

$$\frac{d}{dt} \int_V \varphi \, dV = \int_V \left[D\varphi + \varphi \operatorname{div} v \right] dV = \int_V \left[\frac{\partial \varphi}{\partial t} + \operatorname{div}(\varphi v) \right] dV, \tag{1.17}$$

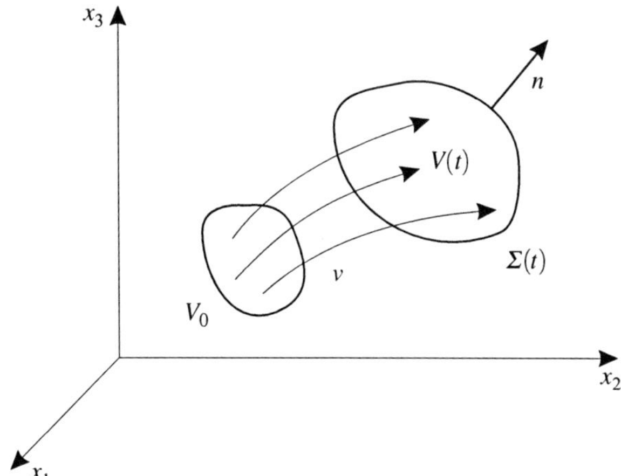

Fig. 1.3 The deduction of conservation equation

where $D = \partial/\partial t + v \cdot \nabla$ is a differential operator, $\varphi = \varphi(x, t)$ is a certain continuously differentiable scalar or vector field, and dV is a volume element in $\mathbb{R}^3$.

At each point x in the domain of the flow, at an instant of time t, let the mass-flow density vector $j_m = j_m(x, t)$ be defined. Let the volume V_0 move along trajectories of the vector field $v = j_m/\rho$. According to conjecture (1.12), this ensures the conservation of mass in the volume under its displacement. Then the Euler–Liouville identity takes the form

$$\frac{d}{dt} \int_V \varphi \, dV = \int_V \left[\frac{\partial \varphi}{\partial t} + \operatorname{div}\left(\frac{\varphi j_m}{\rho} \right) \right] dV. \tag{1.18}$$

Let us present the postulates lying in the base of the construction of the gas-dynamic equations.

As the first postulate, we take the mass conservation law (1.12) and (1.13), which, for the uniformity of the subsequent presentation, we write in the equivalent integral form as follows:

$$\frac{d}{dt} \int_V \rho \, dV = 0. \tag{1.19}$$

The second postulate is the momentum conservation law:

$$\frac{d}{dt} \int_V (\rho u) dV = \int_V \rho F \, dV + \iint_\Sigma (n \cdot P) d\Sigma, \tag{1.20}$$

where $d\Sigma$ is an area element of the surface Σ near the unit vector n. The velocity of varying of the momentum in the volume V is equal to the sum of forces applied to it. The first integral in the right-hand side of Eq. (1.20) is the volume force of the exterior field; the second integral defines the pressure and internal viscous friction forces applied to the surface Σ. The quantity $P = P(x, t)$ is called the internal stress tensor. The symbol $(n \cdot P)$ denotes the contraction (inner product) of the vector n and the rank-2 tensor P with respect to the first index of the tensor P. Similarly, the symbol $(P \cdot n)$ means that the contraction of P and n is performed with respect to the second index of the tensor P. In the case where the tensor P is symmetric, we have $(n \cdot P) = (P \cdot n)$.

The third postulate is the total energy conservation law

$$\frac{d}{dt} \int_V \rho \left(\frac{u^2}{2} + \varepsilon \right) dV = \int_V (j_m \cdot F) dV + \iint_\Sigma (A \cdot n) d\Sigma - \iint_\Sigma (q \cdot n) d\Sigma. \tag{1.21}$$

The first integral in the right-hand side of Eq. (1.21) is equal to the power of exterior mass forces applied to the volume V; the second integral is interpreted as the power of the pressure surface forces and internal viscous friction forces. The last term in Eq. (1.21) describes the influx of energy in unit time through the surface Σ that accounts for the heat transfer processes. The concrete expressions for the vector fields $A = A(x, t)$ and $q = q(x, t)$ are presented below.

The next postulate expresses the angular momentum conservation law:

$$\frac{d}{dt} \int_V [x \times (\rho u)] dV = \int_V [x \times \rho F] dV + \iint_\Sigma [x \times (n \cdot P)] d\Sigma. \tag{1.22}$$

It is represented in the classical form. The internal moments and also the distributed mass and surface pairs are not taken into account. The symbol $\times$ denotes the cross product of two vectors.

The second thermodynamic law being our fifth postulate has the form

$$\frac{d}{dt} \int_V (\rho s) dV = - \iint_\Sigma \frac{(q \cdot n)}{T} d\Sigma + \int_V X \, dV. \tag{1.23}$$

The surface integral in the right-hand side of Eq. (1.23) determines the speed of the entropy variation in the volume V that accounts for the heat flow; it can be positive or negative. The last integral is always nonnegative and yields the increase of the entropy that accounts for internal noninvertible process. The quantity X is called the entropy production.

1.3 Conservation Laws in the Differential Form

To pass from the integral relations (1.19), (1.20), (1.21), (1.22), and (1.23) to the corresponding differential equations, we use the Euler–Liouville formula (1.18) on the differentiation of the integral over a moving material volume. In this case, we assume that all basic macroscopic parameters of the medium are sufficiently smooth functions of the spatial coordinates and time.

Setting $\varphi = \rho$, ρu, $\rho(u^2/2 + \varepsilon)$, $[x \times \rho u]$, and ρs and taking into account the arbitrariness of V, we obtain the differential equations of the mass balance

$$\frac{\partial \rho}{\partial t} + \operatorname{div} j_m = 0, \tag{1.24}$$

of the momentum balance

$$\frac{\partial(\rho u)}{\partial t} + \operatorname{div}(j_m \otimes u) = \rho F + \operatorname{div} P, \tag{1.25}$$

of the total energy balance

$$\frac{\partial}{\partial t} \left[\rho \left(\frac{u^2}{2} + \varepsilon \right) \right] + \operatorname{div} \left[j_m \left(\frac{u^2}{2} + \varepsilon \right) \right] = (j_m \cdot F) + \operatorname{div} A - \operatorname{div} q, \tag{1.26}$$

of the angular momentum balance

$$\frac{\partial}{\partial t}\left[x \times \rho u\right] + \mathrm{div}\left(j_m \otimes \left[x \times u\right]\right) = \left[x \times \rho F\right] + \frac{\partial}{\partial x_i}\left[x \times P_{ij}e_j\right], \qquad (1.27)$$

and of the entropy balance

$$\frac{\partial(\rho s)}{\partial t} + \mathrm{div}(j_m s) = -\mathrm{div}\left(\frac{q}{T}\right) + X. \qquad (1.28)$$

Here, $(j_m \otimes u)$ is the rank-2 tensor obtained as a result of the direct product of the vectors j_m and u. In calculating the divergence of the rank-2 tensor, the contraction is performed with respect to the first index. In Eq. (1.27), the symbol P_{ij} denotes the portrait of the tensor P in the basis (e_1, e_2, e_3). We perform the summation with respect to repeated indices i and j.

Let us show that the obtained system of Eqs. (1.24), (1.25), and (1.26) is dissipative. Let all quantities entering this system be defined. Assume that the gas flow takes place in a closed vessel V_0 with rigid wall Σ_0 which does not conduct heat. Let us add to system (1.24), (1.25), and (1.26) the initial conditions

$$\rho\big|_{t=0} = \rho_0, \quad u\big|_{t=0} = u_0, \quad T\big|_{t=0} = T_0, \quad x \in V_0, \qquad (1.29)$$

and the boundary conditions

$$u\big|_{\Sigma_0} = 0, \quad (j_m \cdot n)\big|_{\Sigma_0} = 0, \quad (q \cdot n)\big|_{\Sigma_0} = 0, \quad t \geq 0. \qquad (1.30)$$

Here, $\rho_0 = \rho_0(x) > 0$, $u_0 = u_0(x)$, and $T_0 = T_0(x) > 0$ are given values of the density, velocity, and temperature, respectively, at the instant of time $t = 0$. The first of conditions (1.30) means that the gas adheres to the walls of the vessel, the second condition ensures the absence of the mass flux through the boundary, and the third condition ensured the vanishing of the normal component of the heat flow on Σ_0. Integrating Eq. (1.28) over the volume V_0 and taking Eqs. (1.29) and (1.30) into account, we obtain the inequality

$$\frac{dS(t)}{dt} \geq 0 \qquad (1.31)$$

for the total thermodynamic entropy

$$S(t) = \int_{V_0} \rho s \, dx.$$

It follows from Eq. (1.31) that the quantity $S(t)$ is a nondecreasing function of time.

System (1.24), (1.25), (1.26), (1.27), and (1.28) is not closed. It is necessary to represent the quantities j_m, P, q, A, and X as functions of the macroscopic parameters of the gas and their derivatives. The closing problem can be solved by using different methods.

1.4 Euler and Navier–Stokes Equations

First, we present the classical approach in which for the definition of the hydrodynamic quantities the instant spatial averages are used (see, e.g., [134, 144, 173]). In this case, the vector of mass flux density j_m at any point (x, t) coincides with the mean momentum of the volume unit ρu, and the first closing relation has the form

$$j_m = \rho u. \tag{1.32}$$

Further, one introduces the idea on the pressure forces and the internal viscous friction that instantly act on the surface of the material volume. The law of motion of the latter is chosen as that in the rigid-body mechanics. This assumption is called the solidification principle. The angular momentum balance equation (1.27) is a consequence of the momentum conservation law (1.25) under the symmetry condition of the stress tensor P. In the Newton medium theory, the following expression is used for $P = P_{\mathrm{NS}}$:

$$P_{\mathrm{NS}} = \Pi_{\mathrm{NS}} - pI, \tag{1.33}$$

where

$$\Pi_{\mathrm{NS}} = \mu\left[(\nabla \otimes u) + (\nabla \otimes u)^T - \frac{2}{3}I \operatorname{div} u\right] + \zeta I \operatorname{div} u \tag{1.34}$$

is the rank-2 tensor called the Navier–Stokes tensor of viscous stresses and I is the rank-2 identity invariant tensor. The superscript T denotes transposition.

The heat flow $q = q_{\mathrm{NS}}$ is given in accordance with the Fourier law:

$$q = -\varkappa \nabla T. \tag{1.35}$$

For ideal monoatomic gases and small Knudsen numbers, conjectures (1.34) and (1.35) are justified by a kinetic calculation.

The work of surface pressure forces and the internal viscous friction per unit time is calculated by the same formula as in the rigid-body mechanics, namely,

$$A = (P_{\mathrm{NS}} \cdot u) = (\Pi_{\mathrm{NS}} \cdot u) - pu. \tag{1.36}$$

It is assumed that the specific thermodynamic entropy satisfies the differential Gibbs identity

$$T\,ds = d\varepsilon + p\,d\left(\frac{1}{\rho}\right). \tag{1.37}$$

The entropy balance equation (1.28) can be obtained by using the Gibbs identity, the conservation laws of mass, momentum, and entropy (1.25) and (1.26). In this case, the entropy production $X = X_{\mathrm{NS}}$ has the form

$$X = \varkappa \left(\frac{\nabla T}{T}\right)^2 + \frac{(\varPi_{\mathrm{NS}} : \varPi_{\mathrm{NS}})}{2\mu T} = \varkappa \left(\frac{\nabla T}{T}\right)^2 + \frac{\varPhi}{T}, \tag{1.38}$$

where

$$(\varPi_{\mathrm{NS}} : \varPi_{\mathrm{NS}}) = \sum_{i,j=1}^{3} (\varPi_{\mathrm{NS}})_{ij}(\varPi_{\mathrm{NS}})_{ij}$$

is the double inner product of two equal tensors. The quantity $\varPhi$ is called the dissipative function; its value defines the energy dissipation that accounts for the viscous friction forces. Note that the right-hand side of relation (1.38) is nonnegative. The substitution of expressions (1.32), (1.33), (1.34), (1.35), and (1.36) in Eqs. (1.24), (1.25), and (1.26) yields the classical Navier–Stokes system for the viscous compressible heat-conducting medium:

$$\frac{\partial \rho}{\partial t} + \mathrm{div}\, \rho u = 0, \tag{1.39}$$

for the momentum

$$\frac{\partial(\rho u)}{\partial t} + \mathrm{div}(\rho u \otimes u) + \nabla p = \rho F + \mathrm{div}\, \varPi_{\mathrm{NS}}, \tag{1.40}$$

and the total energy

$$\frac{\partial}{\partial t}\left[\rho\left(\frac{u^2}{2} + \varepsilon\right)\right] + \mathrm{div}\left[\rho u\left(\frac{u^2}{2} + \varepsilon + \frac{p}{\rho}\right)\right] + \mathrm{div}\, q_{\mathrm{NS}}$$
$$= \left(\rho u \cdot F\right) + \mathrm{div}(\varPi_{\mathrm{NS}} \cdot u). \tag{1.41}$$

The first relation (1.39) is called the mass-balance equation or the continuity equation. Equations (1.40) and (1.41) express the momentum and total energy conservation laws, respectively.

The system becomes closed if we complement it by boundary and initial conditions and the state equations

$$p = p(\rho, T), \quad \varepsilon = \varepsilon(\rho, T), \tag{1.42}$$

and also by the expressions for the calculation of positive coefficients of the dynamical viscosity μ, the second (bulk) viscosity ζ, and the heat conductivity $\varkappa$.

In the case of an ideal polytropic gas consisting of elastic small balls, functions (1.42) are chosen in the form

$$p = \rho \mathcal{R} T, \quad \varepsilon = c_v T. \tag{1.43}$$

The first relation in Eq. (1.43) is called the Mendeleev–Clapeyron equation or the state equation of the ideal gas. The second relation characterizes the gas as a polytropic gas. In this case, the specific thermodynamic entropy is expressed by the formula

$$s = c_v \ln \frac{\mathcal{R}T}{\rho^{(\gamma-1)}} + \text{const,} \tag{1.44}$$

where

$$c_v = \frac{\mathcal{R}}{\gamma - 1}, \quad c_p = \frac{\gamma \mathcal{R}}{\gamma - 1}.$$

The functions $\mu = \mu(\rho, T)$ and $\varkappa = \varkappa(\rho, T)$ can be found either experimentally or by using the methods of kinetic gas theory. For an ideal polytropic gas, the viscosity and the heat conductivity coefficients depend only on the temperature and can be approximated by the functions

$$\mu = \mu_1 \left(\frac{T}{T_1} \right)^{\omega}, \quad \varkappa = \frac{\mu c_p}{\mathrm{Pr}}, \tag{1.45}$$

in which μ_1 is the known value of the dynamical viscosity coefficient for the temperature T_1, ω is a given exponent of the temperature dependence from the closed interval $[0.5, 1]$, and Pr is the Prandtl number. The coefficient of the second (bulk) viscosity can be approximated by the formula

$$\zeta = \mu \left(\frac{5}{3} - \gamma \right) \geq 0.$$

This coefficient is always positive and is related to the existence of internal degrees of freedom of the molecule. For a monoatomic gas, $\gamma = 5/3$ and $\zeta = 0$. Otherwise, $1 < \gamma < 5/3$ and $\zeta > 0$. The above formula was obtained by using kinetic theory in [205] for a gas with rotational degrees of freedom. The same formula is obtained on the basis of QGD equations for arbitrary γ (see p. 44). The influence of the second viscosity coefficient on the form of the density profile in shock waves is discussed in Appendix B.

For other media (for example, for the van der Waals gas), functions (1.43) and (1.45) can change their form.

Equations (1.39), (1.40), and (1.41) are invariant with respect to the Galileo transforms. This corresponds to the Galileo relativity principle on the same form of laws of motion in different inertial coordinate systems.

The system obtained satisfies the angular momentum conservation law and the entropy balance equation in the form

$$\frac{\partial(\rho s)}{\partial t} + \mathrm{div}(\rho u s) = \mathrm{div} \left(\varkappa \frac{\nabla T}{T} \right) + \varkappa \left(\frac{\nabla T}{T} \right)^2 + \frac{\Phi}{T}, \tag{1.46}$$

where the dissipative function has the form

$$\Phi = \frac{(\Pi_{\mathrm{NS}} : \Pi_{\mathrm{NS}})}{2\mu}. \tag{1.47}$$

The total entropy nondecrease law in a closed, adiabatically isolated volume, which follows from this, shows the noninvertible (dissipative) character of the Navier–Stokes system.

If, in Eqs. (1.39), (1.40), and (1.41), we neglect the viscosity and heat conductivity effects, then we arrive at the classical Euler system.

1.5 Quasi-gas-dynamic and Quasi-hydrodynamic Equations

Relation (1.32) holds for instant spatial averages. In the general case, this relation does not hold for spatial-time averages (see Sect. 1.1.4). The possible choice of quantities j_m, P, A, q, and X under the assumption that, in general, j_m is not equal to ρu is presented below.

By analogy with expressions (1.33) and (1.36), we assume that the stress tensor and the work of pressure and viscous friction forces are connected with the viscous stress tensor by the relations of the form

$$P = \Pi - pI, \quad A = (\Pi \cdot u) - p\frac{j_m}{\rho}.$$

For spatial-time averages, two variants of closing the general system (1.24), (1.25), (1.26), (1.27), and (1.28) were constructed. The resulting systems are called the quasi-gas-dynamic (QGD) and quasi-hydrodynamic (QHD) systems of equations. The additional terms in the QGD/QHD equations, which are proportional to a small parameter τ, are related to an additional averaging (smoothing) in time in determination of the gas-dynamic parameters.

The QGD system describes the behavior of the ideal polytropic gas. The first variant of this system was obtained on the basis of kinetic model in the 1980s (see, e.g., [40, 54, 55, 77]). Later on, this system was represented in the form of conservation laws [181, 184]. The QHD system was obtained later by Yu. V. Sheretov by the analysis of the conservation equations in the differential form [180, 181]. This system describes the gas flow with a more general state equation, and in the approximation where $\rho = $ const, it can be used for modelling the viscous incompressible fluid flows.

1.5.1 Quasi-gas-dynamic System

For a physically infinitely small time, let the instant values of the density, mean impulse, and energy of the volume unit change. For an ideal polytropic gas, i.e., for a gas with the state equation

$$p = \rho \mathcal{R} T, \quad \varepsilon = \frac{p}{\rho(\gamma - 1)},$$

a method for closing the general system (1.24), (1.25), (1.26), (1.27), and (1.28) was found, which leads to the QGD system. A variant of the construction of this system based on kinetic models is presented in Chap. 3. Here, we present the closing relations, which have the form

$$j_m = \rho(u - w), \tag{1.48}$$

$$P = -pI + \Pi = -pI + \Pi_{\mathrm{NS}} + \tau u \otimes [\rho(u \cdot \nabla)u + \nabla p - \rho F]$$
$$+ \tau I [(u \cdot \nabla)p + \gamma p \operatorname{div} u], \tag{1.49}$$

$$q = -\varkappa \nabla T - \tau \rho u \left[(u \cdot \nabla)\varepsilon + p(u \cdot \nabla)\left(\frac{1}{\rho}\right) \right] = -\varkappa \nabla T - \tau \rho T u(u \cdot \nabla)s, \tag{1.50}$$

where

$$w = w_{\mathrm{QGD}} = \frac{\tau}{\rho}[\operatorname{div}(\rho u \otimes u) + \nabla p - \rho F], \tag{1.51}$$

τ is a certain small coefficient having dimensions of time, which is called the relaxation parameter or the smoothing parameter in what follows. For $\tau = 0$, the expressions for j_m, P, and q presented above degenerate to the corresponding quantities for the Navier–Stokes equations. Methods for finding τ are discussed below.

The vector A and the nonnegative entropy production X are written in the form

$$A = (\Pi_{\mathrm{NS}} \cdot u) - p(u - w)$$
$$+ \tau u \left[\rho(u \cdot \nabla)\left(\frac{u^2}{2}\right) + (u \cdot \nabla)p \right] + \tau u[(u \cdot \nabla)p + \gamma p \operatorname{div} u] \tag{1.52}$$

and

$$X = \varkappa \left(\frac{\nabla T}{T}\right)^2 + \frac{(\Pi_{\mathrm{NS}} : \Pi_{\mathrm{NS}})}{2\mu T} + \frac{p\tau}{\rho^2 T}[\operatorname{div}(\rho u)]^2$$
$$+ \frac{\tau}{\rho T}[\rho(u \cdot \nabla)u + \nabla p - \rho F]^2 + \frac{\tau}{\rho \varepsilon T}[\rho(u \cdot \nabla)\varepsilon + p \operatorname{div} u]^2. \tag{1.53}$$

Note that the entropy production for the QGD system is the entropy production for the Navier–Stokes equations with additional terms that are the squares of the left-hand sides of the classical Euler equations in the stationary case with positive coefficients. For the QGD equations, the entropy production is nonnegative.

Substituting the values of the vectors and the viscous stress tensor written above in the general system (1.24), (1.25), (1.26), (1.27), and (1.28), we obtain the QGD system in the form

$$\frac{\partial \rho}{\partial t} + \operatorname{div} j_m = 0, \tag{1.54}$$

$$\frac{\partial(\rho u)}{\partial t} + \operatorname{div}\left(j_m \otimes u\right) + \nabla p = \rho_\star F + \operatorname{div} \Pi, \tag{1.55}$$

$$\frac{\partial}{\partial t}\left[\rho\left(\frac{u^2}{2} + \varepsilon\right)\right] + \operatorname{div}\left[j_m\left(\frac{u^2}{2} + \varepsilon + \frac{p}{\rho}\right)\right] + \operatorname{div} q$$
$$= \left(j_m \cdot F\right) + \operatorname{div}\left(\Pi \cdot u\right), \tag{1.56}$$

where $\rho_\star = \rho - \tau \operatorname{div}(\rho u)$ is the approximate value of the density at the point $(x, t + \tau)$. The quantity $\rho_\star$ is found by the choice of the value of the density in the first term of the right-hand side of Eq. (1.20) at the point shifted in time by the relation $\rho_\star = \rho + \tau \partial \rho / \partial t$, where $\dfrac{\partial \rho}{\partial t} + \operatorname{div}(\rho u) = 0$.

In [185], the entropy balance equation was obtained by using the nondivergence form of the QGD system in the form (1.28) with X defined by Eq. (1.53):

$$\frac{\partial(\rho s)}{\partial t} + \operatorname{div}(j_m s) = -\operatorname{div}\left(\frac{q}{T}\right) + X. \tag{1.57}$$

A variant of the construction of it for flows with exterior energy sources is presented in the last section of Chap. 3.

1.5.2 Quasi-hydrodynamic System

The second method of the closure of the problem (1.24), (1.25), (1.26), (1.27), and (1.28) was suggested by Yu. V. Sheretov in [181, 184].

For any physically infinitely small time, let only the instant value of the mean momentum of the volume unit change, and let the changes of instant values of the density and temperature be neglected. In this case, for the gas with the state equations (1.16) satisfying the Gibbs identity (1.37), the quantities j_m, P, q, A, and X are constructed in the form

$$j_m = \rho(u - w), \tag{1.58}$$
$$P = -pI + \Pi = -pI + \Pi_{\mathrm{NS}} + \rho u \otimes w, \tag{1.59}$$
$$q = -\varkappa \nabla T, \tag{1.60}$$
$$A = (\Pi_{\mathrm{NS}} \cdot u) + \rho u(w \cdot u) - p(u - w), \tag{1.61}$$
$$X = \varkappa\left(\frac{\nabla T}{T}\right)^2 + \frac{(\Pi_{\mathrm{NS}} : \Pi_{\mathrm{NS}})}{2\mu T} + \frac{\rho w^2}{\tau T}, \tag{1.62}$$

and, in addition,

$$w = w_{\mathrm{QHD}} = \frac{\tau}{\rho}[\rho(u \cdot \nabla)u + \nabla p - \rho F]. \tag{1.63}$$

Substituting expressions (1.58), (1.59), and (1.61) for the quantities j_m, P, and A, respectively, in Eqs. (1.24), (1.25), and (1.26), we obtain the QHD system (the Sheretov system)

$$\frac{\partial \rho}{\partial t} + \operatorname{div} j_m = 0, \tag{1.64}$$

$$\frac{\partial (\rho u)}{\partial t} + \operatorname{div}(j_m \otimes u) + \nabla p = \rho F + \operatorname{div} \Pi, \tag{1.65}$$

$$\frac{\partial}{\partial t}\left[\rho\left(\frac{u^2}{2} + \varepsilon\right)\right] + \operatorname{div}\left[j_m\left(\frac{u^2}{2} + \varepsilon + \frac{p}{\rho}\right)\right] + \operatorname{div} q = (j_m \cdot F) + \operatorname{div}(\Pi \cdot u). \tag{1.66}$$

The QHD system (1.64), (1.65), and (1.66) becomes closed if we complement it by the state equations (1.16) and represent the coefficients μ, $\varkappa$, and τ as functions of the macroscopic parameters of the medium.

The substitution of expressions (1.58), (1.60), and (1.62) in Eq. (1.28) yields the entropy balance equation

$$\frac{\partial (\rho s)}{\partial t} + \operatorname{div}(\rho u s) = \operatorname{div}(\rho w s) + \operatorname{div}\left(\varkappa \frac{\nabla T}{T}\right) + \varkappa\left(\frac{\nabla T}{T}\right)^2 + \frac{\Phi_{\mathrm{QHD}}}{T}, \tag{1.67}$$

in which

$$\Phi_{\mathrm{QHD}} = \frac{(\Pi_{\mathrm{NS}} : \Pi_{\mathrm{NS}})}{2\mu} + \frac{\rho w^2}{\tau}$$

is a nonnegative dissipative function.

1.5.3 Vector of the Mass Flux Density and the Parameter τ

Formally, in all equations of the QGD/QHD system, there are additional (as compared with the Navier–Stokes system) dissipative terms that are the spatial derivatives of the density, the velocity, and the pressure multiplied by the numerical coefficient τ. As $\tau \to 0$, the QGD/QHD equations pass to the Navier–Stokes equations.

The coefficient $\tau = \tau(\rho, T)$ called the relaxation (or smoothing) parameter is related to the introduction of averaging in time into the definition of the gas-dynamic quantities. This additional averaging allows us to take into account the influence of small fluctuations of the number of particles in the volume ΔV, which are neglected in the classical hydrodynamics. The value of the smoothing parameter can vary in wide limits depending on the type of the flow considered.

To find the value of the parameter τ, we consider the structure of the vector of the mass flux density. For definiteness, we dwell on the QGD system for the stationary flow of the ideal polytropic gas. In this case, the mass flux density is calculated according to Eqs. (1.48) and (1.51):

$$j_m = \rho(u - w) = \rho u - \tau\left[\operatorname{div}(\rho u \otimes u) + \nabla p - \rho F\right]. \tag{1.68}$$

Let us transform this expression to the form

$$j_m = \rho u + \tau \rho F - \tau \mathcal{R} T \nabla \rho - \tau \mathcal{R} \rho \nabla T - \tau \operatorname{div}(\rho u \otimes u). \qquad (1.69)$$

The first term in the right-hand side describes the mass flux density related to the convective motion of the gas. The second term is the flow determined by the motion of particles in the exterior field. The third term is the mass flux that accounts for the self-diffusion. The fourth term is the so-called thermodiffusion flow. The last term is the contribution to the mass flux that accounts for the gradient of velocity. In real flows, all these fluxes are closely related to each other and cannot be separated.

Let us dwell on the third term in more detail. The mass flux density that accounts for the self-diffusion has the form

$$j_\rho = -D \nabla \rho.$$

The self-diffusion coefficient D is well known for many media from experiments with isotopes. For example, according to [28, 160], in the polytropic gas, the self-diffusion coefficient is related to the viscosity coefficient as follows:

$$D = \frac{\mu}{\rho \operatorname{Sc}}, \qquad (1.70)$$

where Sc is the Schmidt number. According to [28], the Schmidt number in the gas is close to unity and can approximately be obtained as

$$\operatorname{Sc} = \frac{5}{7 - \omega}. \qquad (1.71)$$

Comparing the self-diffusion coefficient (1.70) and the expression of this coefficient in Eq. (1.69), we obtain the relaxation parameter for the gas with the state equation $p = \rho \mathcal{R} T$,

$$\tau = \frac{\mu}{p \operatorname{Sc}}. \qquad (1.72)$$

Therefore, with accuracy up to the Schmidt number, the quantity τ coincides with the so-called Maxwell relaxation time $\tau_m = \mu/p$, i.e., it is close to the mean collision time of particles in the gas.

Now let us consider the second term in Eq. (1.69). If we consider each molecule as a Brownian particle, then the mass flux density of these particles can be connected with the mass density of exterior forces by the relation

$$j_F = \rho b m_0 F,$$

where the coefficient b is called the mobility of the molecule. The mobility of the molecule is connected with the self-diffusion coefficient by the Einstein relation

$$D = bk_B T.$$

Substituting the mobility of the molecule through the Einstein relation in the expression for the mass flux density in the exterior field, we again arrive at formula (1.72) for the relaxation coefficient τ.

According to [134, 142], the thermodiffusion flow described by the fourth term is represented in the form

$$j_T = -\rho D \frac{k_T}{T} \nabla T,$$

where k_T is a dimensionless quantity called the thermodiffusion quotient, which determines the relation between the thermodiffusion coefficients D_T and self-diffusion coefficients D in the form $D_T = Dk_T$. Comparing the fourth term in Eq. (1.69) and the expression for j_T, we again arrive at the expression for the relaxation parameter (1.72) obtained earlier in the form $\tau = k_T \mu/(p\,\mathrm{Sc})$ with accuracy up to the coefficient k_T.

In [184], the following more general formula was proposed for calculating the relaxation parameter:

$$\tau = \frac{\gamma}{\mathrm{Sc}} \frac{\mu}{\rho c^2}. \tag{1.73}$$

When taking into account the Laplace formula $c^2 = \gamma p/\rho$ for the sound speed in the gas, expression (1.73) transforms into the form (1.72).

For dense gases and fluids, the value of the smoothing parameter chosen in accordance with Eq. (1.73) turns out to be very small, and we can neglect the influence of terms in the QGD/QHD equations containing τ. For example, for air under the temperature $T = 20°\mathrm{C}$, we have $\gamma = 1.4$, $\mathrm{Sc} = 0.74$, $c = 3.4 \times 10^4$ cm/s, $v = \mu/\rho = 0.15$ cm^2/s, and $\tau = 2.45 \times 10^{-10}$ s. For water, under similar conditions, $\gamma = 1$, $\mathrm{Sc} = 1$, $c = 1.45 \times 10^5$ cm/s, $v = 0.01$ cm^2/s, and $\tau = 4.75 \times 10^{-13}$ s.

For flows of a rarefied gas, the smoothing parameter can be sufficiently large. A variant of the calculation of it for rarefied flows in channels is considered in Appendix C. When describing fastly varying or turbulent flows, the contribution of additional viscous terms can also be substantial. The numerical simulation of turbulent flows behind a ledge in a plane channel and the method for choosing the smoothing parameter in this problem is considered in Appendix D.

In performing simulations, the terms with coefficient τ can be used as convenient and effective regularizers for finding a numerical solution. In this case, the value of the parameter τ is no longer related to the molecular properties of the gas but is determined by the step of a spatial grid and is chosen from the convergence conditions and the accuracy of a numerical solution of the problem.

The expression for the mass flux density (1.68) involves the spatial derivative of the pressure, which makes the continuity equation a second-order equation for both QGD and QHD systems. Therefore, when stating the initial-boundary-value problem for the QGD/QHD systems, an additional (as compared with the Navier–Stokes system) boundary condition is required. This additional condition can be obtained from the conditions for the vector of mass flux density j_m ((1.48) and (1.58)) on the boundary.

Assume that the boundary is an impenetrable rigid wall and that the exterior force vanishes. Then the non-leakage condition $(j_m \cdot n) = 0$ for the mass flux and the non-leakage condition $(u \cdot n) = 0$ for the velocity lead to the condition for the pressure on the boundary in the form $\partial p/\partial n = 0$.

It follows from the formulas for the calculation of the viscous stress tensor and the heat flow for both QGD and QHD systems ((1.49), (1.50), (1.51) and (1.59), (1.60), (1.63)) that the non-leakage condition $(u \cdot n) = 0$ for the velocity leads to the fact that additional terms in the heat flow and the viscous stress tensor proportional to τ vanish and the following condition holds for the QGD/QHD systems on the wall:

$$\Pi = \Pi_{\mathrm{NS}}, \quad q = q_{\mathrm{NS}}.$$

Thus, the formulas for the calculation of the heat flow and the friction force on the rigid wall for the QGD/QHD equations coincide with the traditional expressions obtained in the framework of the Navier–Stokes equations.

1.5.4 Comparison of the Models and the Barometric Formula

The QGD/QHD systems differ from other systems with nonclassical continuity equation that have been suggested in [4, 5, 30, 31, 119–121, 154, 194, 204]. The QGD/QHD equations principally differ from the Barnett equations [34, 141] in which additional terms have the form of the third spatial derivatives and have not dissipative but the dispersive character.

For a number of problems, the comparison of numerical results obtained for the Navier–Stokes and QGD/QHD equations and experiments is presented in the subsequent chapters. For the simulation of rarefied flows, such a comparison with calculations for kinetic models have been performed.

We carry out a detailed comparison of the result obtained in calculations with QGD equations with the data obtained for the Navier–Stokes equations and kinetic approaches for stationary flows in neighborhoods of planar plates [61], disks [43], and hollow cylinders [110]. In these calculations, all three models yield very close results for flows of sufficiently dense gases. Moreover, the numerical algorithm based on the QGD equations turns out to be substantially simpler in the numerical implementation and has a large stability store. If the rarefaction increases, the results of calculation become different. In this case, the data of the QGD model are, as a rule, located between the data obtained by using the kinetic approach and the results of simulation for the Navier–Stokes equations.

As a clear example of the comparison of the QGD/QHD equation with the Navier–Stokes model, we consider the classical hydrostatics problem for the distribution of the pressure in an ideal polytropic gas in a homogeneous gravitational field [134, 144]. In the equilibrium state, the macroscopic velocity u is equal to zero, and the distribution of the gas parameters is independent of time. In the gravity field of the Earth, $F = g$, where $g = 9.8 \times 10^2$ cm/s^2.

In this case, both QGD and QHD systems are substantially simplified and take the same form:

$$\operatorname{div}\{\tau(\nabla p - \rho g)\} = 0, \quad \nabla p = \rho g,$$

$$\operatorname{div}\left\{\tau\left(\nabla p - \rho g\right)\left(\varepsilon + \frac{p}{\rho}\right)\right\} + \operatorname{div}(\varkappa \nabla T) = \tau g \cdot \left(\nabla p - \rho g\right).$$

This directly implies that the mechanical equilibrium conditions follow from the Navier–Stokes system:

$$\nabla p = \rho g, \quad \operatorname{div}(\varkappa \nabla T) = 0.$$

Let the temperature of the gas be constant. Then we arrive at the classical formula determining the distribution of the pressure in the gas:

$$p = p_0 \exp\left(\frac{(g \cdot x)}{\mathcal{R}T}\right), \tag{1.74}$$

where p_0 is a given value of the pressure at the point $x = 0$. Formula (1.74) is called the Laplace barometric formula. Therefore, the Laplace formula is an exact solution of the Navier–Stokes equations and both QGD and QHD systems.

Another exact solution common for the QGD and QHD systems and the Navier–Stokes equation is a solution of the problem for the Couette flow presented in [184]. In [184], a number of other exact solutions of hydrodynamic problems are considered and the connections of these solutions constructed in the framework of the classical model and QGD/QHD systems are traced.

1.5.4.1 Historical Facts

Separate experimental facts of the fluid and gas flows were found by Archimedes (287–212 BC), Blaise Pascal (1588–1651), Evangelista Torricelli (1608–1647), and Isaac Newton (1643–1727).

Leonard Euler and Daniel Bernoulli can be considered as founders of theoretical hydrodynamics.

The term "hydrodynamics" was introduced by Daniel Bernoulli (1700–1783). Jean le Rond D'Alembert (1717–1783) introduced the mass conservation law for a fluid in the form of the equation of continuity.

In 1755, Leonard Euler (1707–1783) obtained the equations of motion of an ideal fluid and developed a mathematical theory of these equations. He deduced the continuity equation expressing the preservation property of a mass moving together with the fluid in a material volume. Also, he obtained the momentum balance equation in the local form without accounting for the influence of the viscosity. That time, fluids and gases were considered as continuous media in a literal sense. The molecular consistence of the substance was not taken into account. The density was defined as a formal mathematical limit of the quotient of the fluid mass at an instant of time t in the volume and the value of this volume as it tends to zero.

The works of Euler were continued by Joseph Louis Lagrange (1736–1813).

Claude Louis Navier (1785–1836) deduced the equation of motion of a viscous fluid using the conjecture on the interaction of molecules. The history of the equations of viscous fluid began from Navier's report (1822) about their simplest variant in the incompressible case. The corresponding paper was published 5 years later.

George Gabriel Stokes (1819–1903) obtained the equation of motion of a viscous fluid based on axiomatics. From the modern viewpoint, as postulates, he used the integral conservation laws of mass, momentum, and total energy in a material volume moving along the integral curves of the velocity field. He can be considered as a founder of the modern hydrodynamics.

Studying the motion of a viscous fluid, Osborne Reynolds (1842–1912) introduced the concepts of the laminar and turbulent flows and pointed to a possibility of a sharp transition from one form of flow to the other.

The kinetic justification of the hydrodynamic equations was constructed on the basis of the Boltzmann equation. This equation for the description of the behavior of the distribution function of particles of a monoatomic gas with binary collision was written by Ludwig Boltzmann (1844–1906) in 1872 (see [179, 197]).

Chapter 2
Elements of Kinetic Gas Theory

In this chapter, we present some aspects of kinetic theory. They will be used in deducing the quasi-gas-dynamic equations in Chap. 3, in constructing their generalizations (Chaps. 8 and 9), and in considering problems on the structure of a shock wave and the flow in microchannels (Appendices B and C). We present a schematic description of the kinetic DSMC algorithm,[1] which is widely used in numerical modelling of rarefied gas flows. Simulations within the framework of this approach were used for the verification of the QGD algorithm in modelling moderately rarefied flows. In the last section, we present a method for constructing kinetically consistent difference schemes whose differential analogs served as a basis for first variants of the QGD equations. The presentation in this chapter is based on [28, 51, 52, 54–56, 122, 127, 160, 184, 190].

2.1 Boltzmann Equation

In 1872, L. Boltzmann suggested the integro-differential kinetic equation, which became the classical model in rarefied monoatomic gas theory [27, 28, 118, 122, 141]. This equation has the form

$$f_t + (\xi \cdot \nabla_x)f + (F \cdot \nabla_\xi)f = \mathcal{I}(f, f) \tag{2.1}$$

and describes the evolution of the one-particle distribution function $f = f(x, \xi, t)$. Here, ξ is the velocity of a separate particle, which is considered as a spherical atom of mass m_0, F is the exterior force acting on the particles relative to the mass unit, and ∇ is the Hamilton operator. The function f is normalized in such a way that the relation

$$m_0 dN = f(x, \xi, t)\, dx\, d\xi$$

defines a probable (or expected) number dN of particles in the volume element $dx\, d\xi$ near a point (x, ξ) of the phase space of coordinates and velocities at a fixed instant of time t.

---

[1] Direct simulation Monte Carlo algorithm.

T.G. Elizarova, *Quasi-Gas Dynamic Equations*, Computational Fluid and Solid Mechanics, DOI 10.1007/978-3-642-00292-2_2,
© Springer-Verlag Berlin Heidelberg 2009

The collision integral $\mathcal{I}(f, f)$ is a nonlinear functional determining the variation of the distribution function resulting from pair collisions. A concrete form of this integral can be found in [27, 28, 122, 141].

An important property of the collision integral, which we need in what follows, is its orthogonality to any of the so-called collision (or summator) invariants

$$h(\xi) = 1, \ \xi, \ \frac{\xi^2}{2}.$$

In other words, we can write

$$\int h(\xi)\mathcal{I}(f, f)\, d\xi = 0. \tag{2.2}$$

This relation expresses the conservation laws of mass, momentum, and energy of particles under their pair collision. Here and in what follows, the integration is performed over the whole three-dimensional velocity space of the particle.

If the distribution function f is known, we can define the gas-dynamic quantities—density ρ, velocity u, pressure p, temperature T, specific internal energy ε, viscosity stress tensor $\varPi$, and heat flux q—using the expressions

$$\rho = \int f\, d\xi, \quad \rho u = \int \xi f\, d\xi, \quad p = \int \frac{c^2}{3} f\, d\xi,$$

$$\rho c_v T = \rho \varepsilon = \int \frac{c^2}{2} f\, d\xi, \quad q = \int \frac{c^2}{2} cf\, d\xi, \tag{2.3}$$

$$\varPi = \int \left[I\frac{c^2}{3} - c \otimes c \right] f\, d\xi.$$

Here, $c = \xi - u$ is the velocity of the chaotic motion of a gas particle or thermal velocity, and $c_v = 3\mathcal{R}/2$ is the specific heat capacity at constant volume for monoatomic gases.

Integrating Eq. (2.1) with weights 1, ξ, and $\xi^2/2$ and using property (2.2), we obtain the following system of equations for the macroscopic parameters:

$$\frac{\partial \rho}{\partial t} + \operatorname{div} \rho u = 0, \tag{2.4}$$

$$\frac{\partial(\rho u)}{\partial t} + \operatorname{div}(\rho u \otimes u) + \nabla p = \rho F + \operatorname{div} \varPi, \tag{2.5}$$

$$\frac{\partial}{\partial t}\left[\rho\left(\frac{u^2}{2} + \varepsilon\right)\right] + \operatorname{div}\left[\rho u\left(\frac{u^2}{2} + \varepsilon + \frac{p}{\rho}\right)\right] + \operatorname{div} q = (\rho u \cdot F) + \operatorname{div}(\varPi \cdot u), \tag{2.6}$$

which, however, is not closed.

For positive solutions $f = f(x, \xi, t)$ of Eq. (2.1), under the assumptions that they exist, have a necessary smoothness, and tend to zero as $|\xi| \rightarrow \infty$, Boltzmann proved his famous H-theorem.

Assume that a monoatomic gas is in a bounded volume V_0 with interior wall that is mirror reflecting. Let the corresponding initial and boundary conditions for the particle distribution function be given in this volume. Then for the Boltzmann function

$$H(t) = \int_{V_0} dx \int f \ln f \, d\xi,$$

the following inequality holds for all $t \geq 0$:

$$\frac{dH(t)}{dt} \leq 0. \tag{2.7}$$

Therefore, the motion of the gas in the vessel considered is accompanied with the nonincrease of the quantity $H(t)$ in time, which indicates its nonreversible character.

2.2 Equilibrium Distribution Function and the Euler System

An exact solution of the Boltzmann equation is the distribution function called the locally Maxwell distribution function, which in dimensional quantities has the form

$$f^{(0)}(x, \xi, t) = \frac{\rho}{(2\pi \mathcal{R}T)^{3/2}} \exp\left(-\frac{(u - \xi)^2}{2\mathcal{R}T} \right). \tag{2.8}$$

For the function $f^{(0)}$, the relation $\mathcal{I}(f^{(0)}, f^{(0)}) = 0$ holds, and it is connected with f by the relations

$$\rho = \int f \, d\xi = \int f^{(0)} \, d\xi,$$

$$\rho u = \int \xi f \, d\xi = \int \xi f^{(0)} \, d\xi, \tag{2.9}$$

$$\rho c_v T = \rho \varepsilon = \int \frac{c^2}{2} f \, d\xi = \int \frac{c^2}{2} f^{(0)} \, d\xi.$$

By a direct substitution, we can verify that for the locally Maxwell distribution function, one has

$$q = \int \frac{c^2}{2} c f^{(0)} \, d\xi = 0, \quad \Pi = \int \left[I \frac{c^2}{3} - c \otimes c \right] f^{(0)} \, d\xi = 0.$$

The function $f^{(0)}$ is also called the locally equilibrium distribution function.

The integration of the Boltzmann equation with weights 1, ξ, and $\xi^2/2$ in the zero approximation, i.e., when f is assumed to be equal to $f^{(0)}$, allows us to close system (2.4), (2.5), and (2.6) and obtain the Euler system.

2.3 Navier–Stokes Equations

In 1916–1917, S. Chapman and D. Enskog suggested an asymptotic method for solving the Boltzmann equation, which allows one to close system (2.4), (2.5), and (2.6) and obtain a system of equations in the first approximation for describing flows of a viscous heat-conducting gas, the Navier–Stokes system [28, 122, 141].

The essence of the method is that a solution of Eq. (2.1) reduced to a dimensionless form is sought for in the form of a formal asymptotic series in powers of a small positive parameter, the Knudsen number Kn, in the form

$$f = f^{(0)}\left(1 + \text{Kn}\, f^{(1)} + \text{Kn}^2\, f^{(2)} + \cdots\right),$$

where

$$\text{Kn} = \frac{\lambda}{L}, \tag{2.10}$$

λ is the mean free path of particles in the unperturbed flow and L is the characteristic size of the flow domain. As the zero approximation, the locally Maxwell function (2.8) is used.

In the first approximation by the number Kn, the calculations via the Chapman–Enskog method lead to the so-called Navier–Stokes distribution function

$$f_{\text{NS}} = f^{(0)}\left[1 - \frac{1}{p\mathcal{R}T}\left(1 - \frac{c^2}{5\mathcal{R}T}\right)(c \cdot q_{\text{NS}}) - \frac{1}{2p\mathcal{R}T}\Pi_{\text{NS}} : (c \otimes c)\right]. \tag{2.11}$$

The quantities Π_{NS} and q_{NS} were written earlier (see Eqs. (1.34) and (1.35)).

Sequentially integrating the Boltzmann equation with the summator invariants 1, ξ, and $\xi^2/2$, under the assumption that f coincides with f_{NS}, we obtain the Navier–Stokes system written above in Sect. 1.4. The Chapman–Enskog procedure allows us to perform an approximate calculation of the viscosity and heat conduction coefficients. For the hard-sphere gas, the approximate calculation of these coefficients leads to the expressions

$$\mu = \frac{5}{64}\frac{m_0}{r_0^2}\sqrt{\frac{\mathcal{R}T}{\pi}}, \quad \varkappa = \frac{c_p\mu}{\text{Pr}}, \tag{2.12}$$

where r_0 is the radius of the hard-sphere particle. The number Pr in Eq. (2.12) turns out to be equal to 2/3, and the coefficients themselves depend only on the temperature, which is in concordance with the known experimental data.

Using the next approximation in the expansion of the distribution function with respect to the Knudsen number, we can obtain the Burnett equations. These equations involve the third-order partial derivatives, which lead to essential difficulties in their numerical solution and in the substitution of the boundary conditions.

2.4 Bhatnagar–Gross–Krook Equation

In 1954, P. Bhatnagar, E. Gross, and M. Krook [25] suggested the approximate kinetic equation of the form

$$f_t + \left(\xi \cdot \nabla_x\right)f + \left(F \cdot \nabla_\xi\right)f = \frac{f^{(0)} - f}{\tau}, \tag{2.13}$$

i.e., Eq. (2.1) in which the collision integral $\mathcal{I}(f, f)$ was approximated by the expression

$$\mathcal{I}(f, f) = \frac{f^{(0)} - f}{\tau}. \tag{2.14}$$

At present, Eq. (2.13) is called the Bhatnagar–Gross–Krook model kinetic equation (BGK). Approximately at that time, Eqs. (2.13) and (2.14) were independently published by P. Velander [141]. The positive parameter τ in the right-hand side of relation (2.14) is interpreted as the characteristic time of relaxation of the function f to the locally Maxwell distribution $f^{(0)}$ function defined by formula (2.8) and is assumed to be a given function of the density and the temperature. The quantity τ has the order of the mean collision time of molecules in the gas and is called the Maxwell relaxation time. The macroscopic quantities entering the formula for calculating τ are also quadratures of f. An analog of the Boltzmann H-theorem holds for the BGK model.

The application of the Chapman–Enskog method to the BGK equation also leads to the Navier–Stokes system [28, 122, 141]. In this case, the dynamical viscosity coefficient μ and the heat conduction coefficient $\varkappa$ are calculated by the formulas

$$\mu = p\tau, \quad \varkappa = c_p p\tau. \tag{2.15}$$

It follows from the presented formulas that in the BGK approximation, the Prandtl number is equal to unity.

At present, improved models of the BGK approximation type are elaborated. In particular, the Shakhov S-model has been suggested; it allows one to consider the real value of the Prandtl number. Moreover, instead of the equilibrium distribution function, in the collision integral (2.14), one chooses the distribution function of the form

$$f_S = f_0\big(1 + (1 - \Pr)\psi(c, \rho, T, u)\big),$$

where ψ is a certain function [176].

There are generalizations of the BGK approximation to the case where the characteristic time of relaxations depends on the velocity of the particles $\tau = \tau(\xi)$. A variant of the relaxation equation that takes into account the nonequilibrium character with respect to internal degrees of freedom of the particles [28] is used in Chap. 8 for the construction of the gas-dynamic equations. A variant of the BGK approximation for a gas mixture and its use for constructing the moment equations is discussed in Chap. 9.

2.5 Mean Collision Quantities of the Particle Motion

Let us present the definitions of the basic quantities characterizing the chaotic motion of particles in a gas with distribution function f. The expressions obtained are used in the sequel.

The *mean thermal velocity* of particles is calculated as follows:

$$\langle \bar{c} \rangle = \frac{1}{\rho} \int \bar{c} f \, dc, \quad \text{where} \quad \bar{c} = \sqrt{c_x^2 + c_y^2 + c_z^2}. \tag{2.16}$$

The *mean relative velocity* of particles is defined in the form

$$\langle \bar{c}_r \rangle = \frac{1}{\rho^2} \iint \bar{c}_r f_1 f_2 \, dc_1 dc_2, \tag{2.17}$$

where $c_r = \xi_1 - \xi_2$ is the relative velocity of two colliding molecules,

$$\bar{c}_r = [(c_{x1} - c_{x2})^2 + (c_{y1} - c_{y2})^2 + (c_{z1} - c_{z2})^2]^{1/2},$$

and f_1 and f_2 are the corresponding one-particle distribution functions.

In the case of the equilibrium distribution function $f = f^{(0)}$ defined according to Eq. (2.8), integrals (2.16) and (2.17) can be calculated analytically, and the values of the corresponding means are defined as follows:

$$\langle \bar{c} \rangle = \sqrt{\frac{8}{\pi} \mathcal{R} T}, \quad \langle \bar{c}_r \rangle = 4 \sqrt{\frac{\mathcal{R} T}{\pi}}. \tag{2.18}$$

The *most probable velocity* of particles is determined by the "width" of the distribution function, and in the equilibrium case (2.8), it is

$$c_m = \sqrt{2 \mathcal{R} T}.$$

The mean frequency of collisions (the mean collision rate) ν is given by

$$\nu = \frac{\rho}{m_0} \langle \sigma c_r \rangle, \tag{2.19}$$

where σ is the collision cross-section and

$$\langle \sigma c_r \rangle = \frac{1}{\rho^2} \iint \sigma \bar{c}_r f_1 f_2 \, dc_1 dc_2$$

is the scattering cross-section.

For gases with the Maxwell distribution function $f_1 = f_1^{(0)}$, $f_2 = f_2^{(0)}$ in the VHS approximation [28], we have

$$\langle \sigma c_r \rangle = 4\sigma_{\text{ref}} \sqrt{\frac{\mathcal{R}T_{\text{ref}}}{\pi}} \left(\frac{T}{T_{\text{ref}}} \right)^{1-\omega}, \tag{2.20}$$

where σ_{ref} is the value of the collision cross-section at the temperature T_{ref}. For the hard-sphere gas ($\sigma = \sigma_0$, $\omega = 0.5$), expression (2.20) takes the form

$$\langle \sigma c_r \rangle = \sigma_0 \langle c_r \rangle = 4\sigma_0 \sqrt{\frac{\mathcal{R}T}{\pi}}. \tag{2.21}$$

The mean collision time is the inverse of the frequency of collisions:

$$\tau_c = \frac{1}{\nu}. \tag{2.22}$$

For the hard-sphere gas,

$$\tau_c = \frac{\sqrt{\pi}\, m_0}{4\rho\sigma_0 \sqrt{\mathcal{R}T}}. \tag{2.23}$$

In the VHS approximation the Maxwell relaxation time is connected with the mean collision time (see [28]):

$$\tau = \frac{\mu}{p} = \Omega(\omega)\tau_c, \quad \text{where} \quad \Omega(\omega) = \frac{30}{(7-\omega)(5-\omega)}. \tag{2.24}$$

For the hard-sphere gas, $\Omega(\omega = 1/2) = 5/4$.

The mean free path λ of particles is determined by the mean collision time τ_c and the mean thermal velocity (2.16):

$$\lambda = \tau_c \langle \bar{c} \rangle. \tag{2.25}$$

For the hard-sphere gas with Maxwell distribution function, the substitution of relations (2.18) and (2.23) leads to the expression

$$\lambda = \frac{m_0}{\sqrt{2}\rho\sigma_0}. \tag{2.26}$$

Let us present certain estimates for the characteristic parameters of air. At atmospheric pressure at sea level, the number of particles is $n = 2.4 \times 10^{25}$ m^{-3}, the mean distance between particles is $r_\star = n^{-1/3} = 3 \times 10^{-7}$ m, the mean free path is $\lambda = 10^{-7}$ m, the mean collision time $\tau_c = 2.5 \times 10^{-10}$ s, the mean thermal velocity is $\langle \bar{c} \rangle = 300$ m/s, and the collision cross-section is $\sigma = 10^{-18}$ m^2.

At a height of 300 km from the Earth's surface the concentration of particles is $n = \rho/m_0 \approx 10^{15}$ m^{-3}, the mean distance between molecules is $r_\star \approx 3 \times 10^{-5}$ m, the mean free path is $\lambda \approx 10^3$ m, and the mean collision time is $\tau_c \approx 1$ s.

The estimates presented show clearly that the mean values in gases vary strongly when the density of the particles varies. In particular, this refers to the relation between the mean free path and the mean distance between molecules. This explains the fact that the scales of the spatial average V and the time average Δt introduced in Sect. 1.1 can substantially change depending on the concrete problem considered.

2.6 Transport Coefficients in Equilibrium Gases

The chaotic motion of particles considered on the microscopic level is accompanied by the redistribution of their number and also by the transport of the momentum and energy by each particle. Thus, on the macroscopic level, the description of the motion of molecules generates three interconnected transport processes: diffusion or self-diffusion, viscosity, and heat conduction.

These three transport processes are closely related to each other. In all three cases, the fluxes are proportional to the gradients of the corresponding quantities. According to [160], these processes can be uniformly described by using basic concepts of kinetic theory, namely, in terms of the mean thermal velocity of particles and the mean free path.

Indeed, let the mean free path λ be much less than the characteristic size of the problem, which is related to the gradients of the macroscopic quantities, the density, the velocity, and the pressure. Consider the transport of a certain scalar quantity A through a unit area, which is perpendicular to the axis z, for a unit time. Then the normal component of the flow Γ_A through this small area due to the chaotic motion of particles at a unit time is defined as follows:

$$\Gamma_A = -\frac{1}{3}(n\bar{v}\lambda)\frac{dA}{dz}, \tag{2.27}$$

where n is the density of particles, $\bar{v}$ is the mean velocity of chaotic motion, and λ is the mean free path. The coefficient $1/3$ is chosen from the assumption that all three coordinate directions are equally probable in the chaotic motion.

If the quantity A is the concentration, $A = n_1/n$, then we find the diffusion flux through the unit area:

$$\Gamma_n = j_\rho = -\frac{1}{3}(n\bar{v}\lambda)\frac{dn_1/n}{dz} = -\frac{1}{3}(\bar{v}\lambda)\frac{dn_1}{dz}. \tag{2.28}$$

This yields the expression for the diffusion coefficient in the form

$$D = \frac{1}{3}\bar{v}\lambda. \tag{2.29}$$

To calculate the momentum transport through the unit small area, we take $A = m_0 u$, where u is the macroscopic velocity of the gas flow along the small area. Then the flow of the quantity A transported in the random walk corresponds to the component of the viscous stress tensor:

$$\Gamma_A = \Pi = -\frac{1}{3}(n\bar{v}\lambda)\frac{m_0 du}{dz} = -\frac{1}{3}(\rho\bar{v}\lambda)\frac{du}{dz}. \tag{2.30}$$

Therefore, we obtain the kinetic estimate for the coefficient of viscosity, the Maxwell formula

$$\mu = \frac{1}{3}\rho\bar{v}\lambda. \tag{2.31}$$

If A is the heat energy of a particle, $A = m_0\bar{v}^2/2$. Replacing $\bar{v}^2$ by the most probable velocity $\bar{v}^2 = c_m^2 = 2\mathcal{R}T$, we obtain the following expression for the heat flux through the unit area:

$$\Gamma_A = q_z = -\frac{1}{3\cdot 2}(n\bar{v}\lambda)\frac{d}{dz}m_0 2\mathcal{R}T = -\frac{1}{3}(n\bar{v}\lambda)k_B\frac{dT}{dz}. \tag{2.32}$$

In other words, the heat conduction coefficient has the form

$$\varkappa = \frac{1}{3}n\bar{v}\lambda k_B = \frac{1}{9}\rho\bar{v}\lambda c_v, \tag{2.33}$$

where $c_v = 3k_B/m_0$ is the specific heat capacity at constant volume for the monoatomic gas.

The values of coefficients of diffusion, viscosity, and heat conduction obtained on the basis of simple kinetic estimates turn out to be connected with each other and are proportional to the mean free path λ and the mean velocity $\bar{v}$ of chaotic motion of particles. Defining again $\bar{v}$ as the mean thermal velocity (2.18) or the most probable velocity, we obtain the well-known expressions for these three transport coefficients for Sc $= 1$ and Pr $= 1$ and also the connection of the mean free path with the coefficient of viscosity, which differs from the Chapman formula (see Sect. 3.4) by a numerical coefficient of unit order.

Therefore, the simplified consideration of the transport processes in an equilibrium gas allows us to obtain qualitatively correct expressions for the coefficients of diffusion, viscosity, and heat conduction. These transport processes have equal rights and explicitly participate in the QGD/QHD equations. The Navier–Stokes system contains only two of them, the momentum and heat energy transports related to the coefficients of viscosity and heat conduction.

2.7 Numerical Simulation of Flows of Rarefied Gases

2.7.1 General Remarks

A simple characteristic of the rarefaction degree of a gas-dynamic flow is the Knudsen number Kn $= \lambda/L$, which is the quotient of the mean free path λ of molecules and the characteristic size L of the problem considered. As usual, a gas is assumed to be dense if Kn $\to 0$ (in practice, Kn < 0.01). The conditions under which

$Kn \to \infty$ (in practice, $Kn > 10$) are characteristic for free molecule flows when collisions between particles are practically absent. For intermediate values of Kn, the gas is considered to be rarefied.

The methods of simulation of free molecule regimes are sufficiently well elaborated at present. For these problems, we can neglect collisions of particles between each other and take into account only the interaction of particles with walls. We can assume that the velocity distribution of particles is known with large accuracy, for example, we can assume that it is in equilibrium with the distribution function $f^{(0)}$. In this case, the main problem is the description of the interaction of particles with walls. This process can be approximately described by using the accommodation coefficient σ. The simplest models are the complete accommodation of particles on the wall, or the so-called diffusion reflection, which corresponds to $\sigma = 1$, and the model of mirror reflection, in which $\sigma = 0$.

By a moderately rarefied flow of a gas one means flows for which the Knudsen number lies in the interval between 0.01 and 0.1, depending on the problem considered. The flows of moderately rarefied gases present a domain lying on the limits of applicability of the kinetic approach and the approach related to the solution of the moment equations. The simulation of such flows by methods of kinetic theory requires unwarranted large computational resources, which is stipulated by a high density of the gas. At the same time, the Navier–Stokes equations obtained in the approximation $Kn \to 0$ lose their accuracy in analyzing the regimes discussed.

To calculate flows of moderately rarefied gases in the framework of the moment equations, there arises a necessity to take into account the deviation from the continuous flow regimes, first of all, near the flowing around a surface. For this purpose, one uses special boundary conditions.

For all Knudsen numbers, whatever small, in the boundary near the wall, there exists a gas layer whose thickness is of order of the mean free path of molecules, the so-called Knudsen layer. To take into account the influence of this layer on the flow field in the framework of the macroscopic equations, one introduces the boundary conditions being slip conditions for the velocity and jump conditions for the temperature.

The first variant of such conditions was written by Maxwell under the assumption of the diffusion reflection of molecules from the wall [148]. At present, in the literature, there are different variants of such conditions (see, e.g., [28, 122, 142]). They have a similar structure and differ from each other only by numerical coefficients of order 1. Here, we present the condition in the Smoluchowski form in which, as compared with the classical Maxwell conditions, the value of accommodation coefficients for the velocity σ_u and the energy σ_e, which can be different, are taken into account, and the influence of the gradient of temperature along the wall (temperature crip) is also taken into account:

$$u_s = \frac{2 - \sigma_u}{\sigma_u} \lambda \left(\frac{\partial u_n}{\partial n} \right)_s + \frac{4}{3} \left(\frac{\mu}{\rho T} \frac{\partial T}{\partial s} \right)_s,$$

$$T_s - T_w = \frac{2 - \sigma_e}{2\sigma_e} \frac{2\gamma}{\gamma + 1} \frac{\lambda}{\mathrm{Pr}} \left(\frac{\partial T}{\partial n} \right)_s,$$

where u_n and u_s are the normal component of the gas velocity near the wall and the slip velocity along the wall, respectively, n and s, respectively, are the coordinates along the exterior normal to the wall and along the wall itself, T_s is the temperature of the gas near the wall, and T_w is the temperature of the wall.

For most materials, under the supersonic flowing around, the velocity and energy accommodation coefficients can be assumed to be equal and close to the unit. In the formula for the slip velocity, the second term plays an appreciable role only for Knudsen numbers approaching the unit (see [99]).

Flows of gases in the range 0.1–10 of the Knudsen numbers are substantially difficult for the analytical study and the numerical simulation since in this range, it is impossible to introduce a small parameter with respect to Knudsen number of the type Kn or $1/$ Kn. In this range, methods of kinetic theory are applied. The numerical analysis of flows is performed either by the direct solution of the Boltzmann equation or its simplified variants, or by using methods of direct numerical simulation, the Monte Carlo methods or DSMC methods (see [27, 28]).

The difficulty in using these approaches is related to large expenditures of computer time in modelling the particle collisions and a large dimension of the problem as a whole, which is considered in the space of seven dimensions (x, ξ, t). An additional difficulty is stipulated by the necessity of the computation of averaged characteristics in order to obtain the gas-dynamic quantities being measured: the velocity, the density, and the pressure. In the framework of DSMC approaches, additional difficulties are the computations of nonstationary flows and flows with small (i.e., subsonic) velocities since such computations are accompanied with considerable fluctuations of average quantities.

2.7.2 Monte Carlo Method

In a wide sense, the Monte Carlo methods are methods based on random variables. Monte Carlo methods are successfully used for the solution of various problems of statistical physics, computational mathematics, game theory, mathematical economics, etc. (see [93]).

The direct-simulation method for computation of flows of a rarefied gas (the direct simulation Monte Carlo, DSMC) was first elaborated by G. A. Bird in the 1960s and was then improved and developed (see [27, 28]).

In gas dynamics, a version of the Monte Carlo method is applied; it is based on modelling a real gas flow by using a relatively small number of molecules. In other words, a numerical experiment is performed in which the history of a bounded number of particles is traced; each of them is representative of a large number W of real molecules. The quantity W is called the weighting factor.

For each molecule, its coordinates, velocity, and energy are stored. According to these quantities, by averaging, the gas-dynamic parameters of the flow are found.

For stationary problems, the computation starts from prescribing an arbitrary distribution in the computational domain, which develops to its equilibrium state with time. Now we list the main steps of the DSMC method.

2.7.2.1 Discretization and Modelling of the Particle Motion

The flow domain is partitioned into spatial cells such that the change of gas-dynamic parameters in each of the cells is small. For the efficiency of simulation, the number of particles in each cell must not be very different and is of order equal to several tens.

Modelling the physical motion of molecules is performed by discrete steps in time Δt, which are small as compared with the mean free path time, $\Delta t < \tau_c$. The motion of molecules and molecule collisions on the interval of time are sequentially modelled. At each time step Δt, the motion of particles is partitioned into two steps and is described in the framework of the kinetic model, which is a cyclically repeating process of free collision scattering and subsequent collisions that are considered as instantaneous. This model corresponds to two steps of computation.

2.7.2.2 Movement

At the first step, all molecules move by a distance determined by their velocities $\xi \Delta t$. The intersection of molecules and rigid bodies, lines, symmetry planes, and boundaries of the flow is taken into account. If there is a flow inside the domain, new molecules are generated. If a molecule leaves the computational domain, then it disappears.

2.7.2.3 Collisions

At the second step, the number of collisions between molecules are computed resulting in a change in their velocities. The choice of the colliding pair of particles is performed in the same cell randomly, taking into account that the probability of a pair of molecules to collide is proportional to σc_r.

An important part of the direct modelling method is the computation of a correct number of collisions (collision rate). The frequency of collisions ν is determined by properties of the real gas, for which the problem is solved, and precisely this quantity determines the dissipative characteristics of the flow, the coefficients of viscosity μ and heat conduction $\varkappa$ of the gas modelled.

2.7.2.4 Computation of Macroscopic Characteristics

When the computation is completed, a number of quantities (Σc, Σc^2, etc.) are accumulated to obtain the macroscopic parameters of a gas (ρ, u, p, and T). The longer the cumulation process, the smaller the statistical fluctuations of macroscopic parameters.

2.8 Difference Approximation of the Boltzmann Equation and Kinetically Consistent Difference Schemes

The numerical solution of the Boltzmann equation (2.1) is severely limited by computational requirements, due to the large number of independent variables and problems of approximation and calculation of the collision integral. Corresponding numerical algorithms are presented, e.g., in [11, 12].

The left-hand side of Eq. (2.1) has the form of the transport equation for the function f with velocity ξ, and it can be approximated by a first-order finite-difference scheme with up-wind differences, the Courant–Isaacson–Rees (CIR) scheme. For a consistent choice of steps in time and space, $h = |\xi|\Delta t$, this finite-difference scheme describes the transport of a given perturbation of the function f without distortion (see [163, 167]).

Let us present the form of the corresponding finite-difference scheme for a one-dimensional spatial flow without exterior forces on a uniform spatial grid with space step $h = x_{i+1} - x_i$ and time step $\Delta t = t^{j+1} - t^j$:

$$\frac{f_i^{j+1} - f_i^j}{\Delta t} + \xi_x \frac{f_i - f_{i-1}}{h} = \mathcal{I}(f, f)_{h,i}, \qquad \xi_k > 0,$$

$$\frac{f_i^{j+1} - f_i^j}{\Delta t} + \xi_x \frac{f_{i+1} - f_i}{h} = \mathcal{I}(f, f)_{h,i}, \qquad \xi_k < 0. \tag{2.34}$$

Here ξ_k is a component of the particle velocity, $f = f^j = f(x_i, \xi, t^j)$, and $\mathcal{I}(f, f)_{h,i}$ is the difference analog of the collision integral.

The finite-difference scheme (2.34) can be identically transformed to the equivalent form

$$\frac{f_i^{j+1} - f_i^j}{\Delta t} + \xi_k \frac{f_{i+1} - f_{i-1}}{2h} - \frac{h}{2}|\xi_k|\frac{f_{i+1} - 2f_i + f_{i-1}}{h^2} = \mathcal{I}(f, f)_{h,i}. \tag{2.35}$$

For the finite-difference analog of the collision integral $\mathcal{I}(f, f)_{h,i}$, let the orthogonality conditions with summator invariants $h(\xi)$ hold. The finite-difference scheme (2.35) can be averaged in the velocity space assuming that $f = f^{(0)}$. In this case, all the integrals can be analytically calculated, and it is possible to construct immediately the finite-difference scheme for the gas-dynamic quantities—density ρ, velocity u, and energy E (see [40, 41, 51, 55, 184]).

The scheme obtained is rather cumbersome since it involves error integrals arising in averaging the modules of the velocities ξ_k. However, this scheme turned out to be very efficient for solving the Euler equations, and it was used in performing simulations of one-dimensional, as well as some two-dimensional, supersonic gas-dynamic flows. Later on, finite-difference equations having a close structure were obtained in [48].

In the finite-difference scheme (2.35), let us replace the coefficient of the last term by setting $|\xi| \sim c$, where c is the sound speed. Taking into consideration a time $\tau = h/2c$ characterizing the time to cross a computational cell by a particle, we obtain

$$\frac{h}{2}|\xi_k| = \frac{h\xi_k^2}{2|\xi_k|} \sim \frac{h}{2c}\xi_k^2 = \tau\xi_k^2. \tag{2.36}$$

Then scheme (2.35) takes the form

$$\frac{f_i^{j+1} - f_i^{j}}{\Delta t} + \xi_k \frac{f_{i+1} - f_{i-1}}{2h} - \tau\xi_k^2 \frac{f_{i+1} - 2f_i + f_{i-1}}{h^2} = \mathcal{I}(f, f)_{h,i}. \tag{2.37}$$

Expression (2.37) in the case $f = f^{(0)}$ also admits an analytical averaging with the summator invariants $h(\xi)$ and leads to constructing a more elegant, as compared with the previous case, system of finite-difference equations for describing the macroscopic quantities in the gas. The schemes thus obtained are called kinetically consistent difference schemes.

Using the notation for central difference derivatives of the first and second order in space accepted in [170], we rewrite the difference scheme (2.37) in the form

$$\frac{f_i^{j+1} - f_i^{j}}{\Delta t} + (f\xi_k)_{\overset{\circ}{x}} - \tau(f\xi_k\xi_k)_{\bar{x}x} = \mathcal{I}(f, f)_{h_i}. \tag{2.38}$$

Here, $f_{\overset{\circ}{x}} = (f_{i+1} - f_{i-1})/h$ is the first-order central difference derivative, $f_x = (f_{i+1} - f_i)/h$ and $f_{\bar{x}} = (f_i - f_{i-1})/h$ are the left and right difference derivatives, respectively, and $f_{\bar{x}x} = (f_x - f_{\bar{x}})/h$ is the second-order central difference derivative.

Using the notation introduced, we write the kinetically consistent difference schemes for the case of a plane one-dimensional flow in the form

$$\frac{\rho_i^{j+1} - \rho_i^{j}}{\Delta t} + (\rho u)_{\overset{\circ}{x}} = \tau(\rho u^2 + p)_{\bar{x}x},$$

$$\frac{(\rho u)_i^{j+1} - (\rho u)_i^{j}}{\Delta t} + (\rho u^2 + p)_{\overset{\circ}{x}} = \tau(\rho u^3 + 3pu)_{\bar{x}x}, \tag{2.39}$$

$$\frac{E_i^{j+1} - E_i^{j}}{\Delta t} + (u(E + p))_{\overset{\circ}{x}} = \tau(u^2(E + 2p))_{\bar{x}x} + \tau\left(\frac{p}{\rho}(E + p)\right)_{\bar{x}x}.$$

The schemes of type (2.39) modified later turned out to be very efficient in modelling a wide range of gas-dynamic flows [8–10, 40, 41, 51, 52, 54–56, 104, 105].

Differential analogs of these schemes served as a base of the first variant of the QGD equations.

Chapter 3
Quasi-gas-dynamic Equations

In this chapter, we present two variants of constructing the quasi-gas-dynamic system, which allow us to obtain a concrete form of expressions for the vectors of mass flux density j_m, the viscous stress tensor Π, and the vector of heat flux q written early without deduction.

The first variant is based on the use of a rather simple kinetic model of particle motion. This model is a cyclically repeating process of collision-free scattering and subsequent collisions of particles with attaining the Maxwell equilibrium that leads to the regularized kinetic equation. The deduction of the QGD equations presented states a connection of the smoothing parameter τ with the Maxwell relaxation time. The QGD equations thus constructed are then represented in the form of conservation laws. Moreover, the expressions for j_m, Π, and q are written as the corresponding expressions for the Navier–Stokes equations with additional terms proportional to τ. It is shown that for stationary flows, these additional terms are second-order terms in τ. Based on this, we obtain a form of dissipative coefficients and their generalizations.

The second variant is based on the consideration of conservation laws of mass, momentum, and energy for a small but finite immovable volume, where in averaging, we take into account the nonstationarity of the gas-dynamic quantities in time. This approach allows us to construct the QGD system for flows with the presence of exterior forces and heat sources. Also, we present here the deduction of the entropy-balance equation. This equation demonstrates the dissipative character of arising additional τ-terms.

The presentation of these results is in accordance with [54–56, 61, 69, 72, 79, 84, 108, 184, 185].

3.1 Regularized Kinetic Equation

The solution of the integro-differential Boltzmann equation is a complicated problem, and, therefore, at present, some simplified kinetic models are developed; they allow one to find approximate solutions of some problems. Many models are based on the idea of splitting the problem with respect to physical processes; this idea is

T.G. Elizarova, *Quasi-Gas Dynamic Equations*, Computational Fluid and Solid Mechanics, DOI 10.1007/978-3-642-00292-2_3,

also called the summary approximation principle. Precisely this approach serves as a base in the DSMC method described in Sect. 2.7.2.

Consider the kinetic model, which was used in [12] for numerical modelling of rarefied flows in the framework of the Boltzmann equation; on the basis of this model, for the first time, the system of quasi-gas-dynamic equations was obtained.

This model represents the gas motion as a cyclically repeating process consisting of the following two stages: the collision-free scattering of gas molecules and the subsequent instantaneous attaining of a thermodynamic equilibrium that accounts for the particle collision, the stage of instantaneous maxwellization. A schematic picture of this model is presented in Fig. 3.1. A detailed analysis of this model and variants of kinetic equations obtained on its basis are presented in [40, 41, 55, 56] and are not discussed here.

At a certain instant of time $t = t_1$, let the distribution function have the locally Maxwell form

$$f^{(0)}(x, \xi, t) = \frac{\rho}{(2\pi \mathcal{R} T)^{3/2}} \exp\left(-\frac{(\xi - u)^2}{2\mathcal{R} T}\right). \tag{3.1}$$

Then, during time Δt, there is a collision-free scattering of molecules, which is described by the Boltzmann equation for the free molecule flow:

$$\frac{\partial f}{\partial t} + (\xi \cdot \nabla)f = 0.$$

This equation is a linear transport equation, and it has an exact solution

$$f(x, \xi, t) = f_0(x - \xi t, \xi), \tag{3.2}$$

where $f_0 = f_0(x, \xi)$ is the known distribution function at the instant of time $t = 0$.

Furthermore, at the instant of time $t = t_2 = t_1 + \Delta t$, the distribution function f again momentarily becomes locally Maxwell (3.1) but with new values of the macroscopic parameters ρ, u, and T. The instantaneous maxwellization imitates the stage of molecule collision, which is described by the collision integral $\mathcal{I}(f, f)$ in the Boltzmann equation. Then both stages are cyclically repeated.

Assuming that the collision-free scattering time Δt is sufficiently small, we take the Taylor series expansion of the distribution function (3.2) assuming that f_0 is a

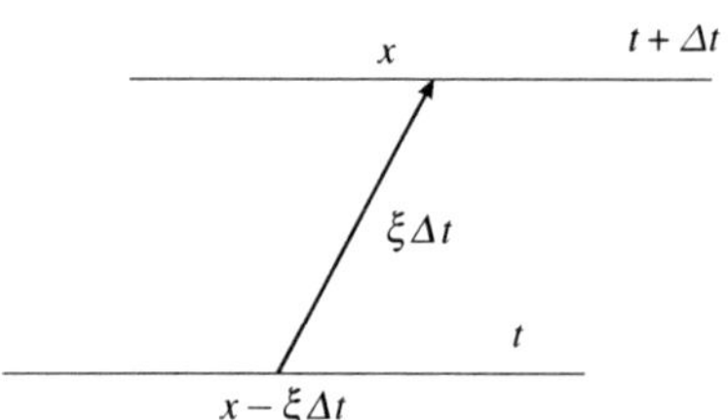

Fig. 3.1 Scheme explaining the kinetic model

Maxwell function, $f_0 = f^{(0)}$. Restricting ourselves to the second-order terms in Δt, we obtain

$$f(x, \xi, t) = f^{(0)} - \Delta t(\xi \cdot \nabla)f^{(0)} + \frac{\Delta t^2}{2}(\xi \cdot \nabla)(\xi \cdot \nabla)f^{(0)}. \tag{3.3}$$

The expansion parameter $\Delta t \cdot \xi$ for large values of velocities ξ cannot be considered as small, and the terms rejected in such an expansion can be important. However, in this case, we can neglect their contribution for $|\xi| \gg \sqrt{2\mathcal{R}T}$, since the distribution function $f^{(0)}$ itself and all its derivatives exponentially decay when ξ grows.

In relation (3.3), let us transport all terms to the left-hand side and divide both sides of the equation by Δt replacing the finite-difference derivative in time in the first term by the differential derivative. Identifying the interval of time $\Delta t/2$ with the Maxwell relaxation time τ, we obtain the following regularized kinetic equation for the distribution function:

$$\frac{\partial f}{\partial t} + (\xi \cdot \nabla)f^{(0)} - (\xi \cdot \nabla)\tau(\xi \cdot \nabla)f^{(0)} = \mathcal{I}(f, f).$$

Here, we add the collision integral to the right-hand side of the obtained equation, which ensures the relaxation of the distribution function to the Maxwell distribution on the new time layer. Writing down the collision integral in the Bhatnagar–Gross– Krook (BGK) approximation, we obtain the final form of the regularized kinetic equation as

$$\frac{\partial f}{\partial t} + (\xi \cdot \nabla)f^{(0)} - (\xi \cdot \nabla)\tau(\xi \cdot \nabla)f^{(0)} = \frac{f^{(0)} - f}{\tau}. \tag{3.4}$$

Note that this equation is a differential analog of the Courant–Isaacson–Rees finite-difference scheme for the BGK equation written in the form (2.38) for the case $\tau = \text{const}$.

As shown in what follows, in averaging Eq. (3.4) with summator invariants, we obtain the system of QGD equations for $\gamma = 5/3$, $\text{Pr} = 1$, and $\text{Sc} = 1$.

We can formally write Eq. (3.4) starting from the Boltzmann equation in the BGK approximation:

$$\frac{\partial f}{\partial t} + (\xi \cdot \nabla)f = \frac{f^{(0)} - f}{\tau}, \tag{3.5}$$

replacing the distribution function in the convective term of Eq. (3.5) by its approximate value of the form

$$f = f^{(0)} - \tau(\xi \cdot \nabla)f^{(0)}. \tag{3.6}$$

Note that the formal replacement of the distribution function f in the convective term of Eq. (3.5) by the Navier–Stokes distribution function (2.11) allows us to obtain the Navier–Stokes system after the averaging with summator invariants.

In [108], it was shown that relations (2.11) and (3.6) can be represented in the identical form as follows:

$$f = f^{(0)}(1 + \tau \mathcal{P}^3(\xi_i)),$$

where $\mathcal{P}^3(\xi_i)$ is a polynomial of the third degree. Moreover, the coefficients of polynomials for both distribution functions turn out to be close.

The regularized kinetic equation (3.4) is constructed here by using a large number of assumptions, but it has a number of properties similar to those of the BGK equation. In particular, for stationary flows, it is shown that if the distribution function f satisfies the BGK equation, then it satisfies Eq. (3.4) with accuracy $O(\tau^2)$. Vice versa, if the distribution function f satisfies the model kinetic equation, then it satisfies the stationary BGK equation with accuracy $O(\tau^2)$. Therefore, the solutions of these equations must be close. For Eq. (3.4), an analog of the Boltzmann H-theorem is proved (see [184]).

On the basis of the regularized kinetic equation (3.4), we succeed in constructing the QGD system and the moment equations contiguous to it for describing weakly nonequilibrium gas-dynamic flows (see Chaps. 8 and 9).

3.2 Kinetic Deduction of the QGD Equations

Let us present a method for constructing the QGD system on the basis of the regularized kinetic equation written in the previous section.

Sequentially integrating the model kinetic equation (3.4) with weights 1, ξ, and $\xi^2/2$ and using the expressions

$$\rho = \int f \, d\xi = \int f^{(0)} \, d\xi,$$

$$\rho \left(\frac{u^2}{2} + \varepsilon \right) = \int \frac{\xi^2}{2} f \, d\xi = \int \frac{\xi^2}{2} f^{(0)} \, d\xi, \quad c = \xi - u,$$

$$\frac{3}{2}\rho RT = \int \frac{c^2}{2} f \, d\xi = \int \frac{c^2}{2} f^{(0)} \, d\xi, \quad Ip = \int (c \otimes c) f^{(0)} \, d\xi, \qquad (3.7)$$

$$\int c f^{(0)} \, d\xi = 0, \quad \int (c \otimes c \otimes c) f^{(0)} \, d\xi = 0,$$

$$\int cc^2 f^{(0)} \, d\xi = 0, \quad \int (c \otimes c)\frac{c^2}{2} f^{(0)} \, d\xi = \frac{5}{2} I \frac{p^2}{\rho},$$

we obtain the system of the QGD equations in the following form:

$$\frac{\partial \rho}{\partial t} + \mathrm{div}(\rho u) = \mathrm{div}\left\{\tau\left[\mathrm{div}(\rho u \otimes u) + \nabla p\right]\right\}, \tag{3.8}$$

$$\frac{\partial(\rho u)}{\partial t} + \mathrm{div}(\rho u \otimes u) + \nabla p$$
$$= \mathrm{div}\left\{\tau\left[\mathrm{div}(\rho u \otimes u \otimes u) + (\nabla \otimes pu) + (\nabla \otimes pu)^T\right]\right\} + \nabla\left\{\tau[\mathrm{div}(pu)]\right\}, \tag{3.9}$$

$$\frac{\partial}{\partial t}\left[\rho\left(\frac{u^2}{2} + \varepsilon\right)\right] + \mathrm{div}\left[\rho u\left(\frac{u^2}{2} + \varepsilon\right) + pu\right]$$
$$= \mathrm{div}\left\{\tau\left\{\mathrm{div}\left[\rho\left(\frac{u^2}{2} + \varepsilon + 2\frac{p}{\rho}\right)u \otimes u\right] + \nabla\left[p\left(\frac{u^2}{2} + \varepsilon + \frac{p}{\rho}\right)\right]\right\}\right\}. \tag{3.10}$$

The closing of the form (3.7) based on the locally Maxwell distribution function automatically leads to constructing the system of moment equations for the hard-sphere gas, i.e., for $\gamma = 5/3$. The generalization to the case of arbitrary molecules is performed by distinguishing the terms with specific internal energy under the assumption that $\gamma = 5/3$, and then by generalizing this expression to the case of an arbitrary value of γ.

System (3.8), (3.9), and (3.10) deduced for the case of arbitrary curvilinear Euler coordinates is closed by the state equations of the ideal polytropic (dilute) gas:

$$p = \rho \mathcal{R}T, \quad \varepsilon = \frac{p}{\rho(\gamma - 1)}.$$

In writing the energy equation, in the last term, terms containing no divergence operations for the velocity are present. From this term, we can immediately distinguish the terms in the form of the Navier–Stokes heat flux q_{NS} and introduce the Prandtl number $\mathrm{Pr} \neq 1$

$$\frac{\partial}{\partial t}E + \nabla_i u^i(E + p) = \nabla_i \tau\left(\nabla_j(E + 2p)u^i u^j + \frac{1}{2}\nabla^i u_k u^k p\right)$$
$$+ \frac{\gamma}{\gamma - 1}\nabla\left(\tau\frac{p}{\rho}\nabla p\right) + \mathrm{Pr}^{-1}\frac{\gamma}{\gamma - 1}\nabla\left(\tau p\nabla\frac{p}{\rho}\right). \tag{3.11}$$

An example of constructing the QGD equations for the planar one-dimensional gas flow is written in Appendix A in detail, where all the used transformations are clearly traced.

On the basis of the averaging of the regularized kinetic equation with account for the external force F,

$$\frac{\partial f}{\partial t} + \left[(\xi \cdot \nabla_x) + (F \cdot \nabla_\xi)\right]f^{(0)}$$
$$- \left[(\xi \cdot \nabla) + (F \cdot \nabla_\xi)\right]\tau\left[(\xi \cdot \nabla) + (F \cdot \nabla_\xi)\right]f^{(0)} = \frac{f^{(0)} - f}{\tau}. \tag{3.12}$$

Yu. V. Sheretov constructed the QGD equations for describing gas flows in the exterior force field (see [183, 184]).

The left-hand sides of system (3.8), (3.9), and (3.10) are the left-hand sides of the Euler equations. The right-hand sides have the divergence form and are proportional to the small parameter τ. In this case, Eqs. (3.8), (3.9), and (3.10) do not have the form of conservation laws, i.e., in these equations the viscous stress tensor and the vectors of mass flux density and heat flux are not distinguished in the explicit form. The dissipative terms entering the Navier–Stokes equations are also not explicitly distinguished. The transformation of the QGD system constructed above into the form of conservation laws is performed in the next section.

3.3 QGD Equations in the Form of Conservation Laws

Let us represent the QGD system in the form of the conservation laws (1.54), (1.55), and (1.56), which, in the case where $F = 0$, have the form

$$\frac{\partial \rho}{\partial t} + \operatorname{div} j_m = 0, \tag{3.13}$$

$$\frac{\partial(\rho u)}{\partial t} + \operatorname{div}(j_m \otimes u) + \nabla p = \operatorname{div} \Pi, \tag{3.14}$$

$$\frac{\partial}{\partial t}\left[\rho\left(\frac{u^2}{2} + \varepsilon\right)\right] + \operatorname{div}\left[j_m\left(\frac{u^2}{2} + \varepsilon\right)\right] = \operatorname{div} A - \operatorname{div} q. \tag{3.15}$$

3.3.1 Equation of Continuity and the Vector of the Mass Flux Density

Comparing the mass transport equations (3.8) and (3.13), we find the vector of the mass flux density

$$j_m = \rho u - \tau \left(\operatorname{div}(\rho u \otimes u) + \nabla p\right).$$

Denoting the addition to the velocity by

$$w = \frac{\tau}{\rho}\left(\operatorname{div}(\rho u \otimes u) + \nabla p\right), \tag{3.16}$$

we represent j_m in the form suggested earlier in Chap. 1 and convenient for the further analysis of the equations:

$$j_m = \rho(u - w).$$

3.3.2 Momentum Equation and the Viscous Shear-Stress Tensor

Comparing the momentum equations (3.9) and (3.14), we find the form of the viscous shear-stress tensor Π. For this purpose, we represent

$$\operatorname{div}(j_m \otimes u) = \operatorname{div}(\rho u \otimes u) - \operatorname{div}(\tau \operatorname{div}(\rho u \otimes u) \otimes u) - \operatorname{div}(\tau \nabla p \otimes u).$$

Then Eq. (3.9) can be rewritten in the form

$$\begin{aligned}
\frac{\partial(\rho u)}{\partial t} + \operatorname{div}(j_m \otimes u) + \nabla p = & - \operatorname{div}(\tau \operatorname{div}(\rho u \otimes u) \otimes u) - \operatorname{div}(\tau \nabla p \otimes u) \\
& + \operatorname{div}\left\{\tau\left[\operatorname{div}(\rho u \otimes u \otimes u) + (\nabla \otimes pu)\right.\right. \\
& \left.\left. + (\nabla \otimes pu)^T\right]\right\} + \operatorname{div}\left\{I\tau \operatorname{div}(pu)\right\}, \qquad (3.17)
\end{aligned}$$

where we have used the identity

$$\operatorname{div}\left\{I\tau \operatorname{div}(pu)\right\} = \nabla\left\{\tau \operatorname{div}(pu)\right\}. \qquad (3.18)$$

It follows from Eqs. (3.17) and (3.14) that the viscous shear-stress tensor Π has the form

$$\begin{aligned}
\Pi = & - \tau \operatorname{div}(\rho u \otimes u) \otimes u - \tau(\nabla p \otimes u) + \tau\left[\operatorname{div}(\rho u \otimes u \otimes u)\right. \\
& \left. + (\nabla \otimes pu) + (\nabla \otimes pu)^T\right] + I\tau \operatorname{div}(pu).
\end{aligned}$$

Let us represent this tensor in the form of the sum of the Navier–Stokes viscous shear-stress tensor

$$\Pi_{NS} = \mu[(\nabla \otimes u) + (\nabla \otimes u)^T - \tfrac{2}{3}I \operatorname{div} u] + \zeta\, I \operatorname{div} u \qquad (3.19)$$

and a certain additional term. To find the form of this term, we use the identities

$$\begin{aligned}
(\nabla \otimes pu) &= p(\nabla \otimes u) + (u \otimes \nabla p), \\
(\nabla \otimes pu)^T &= p(\nabla \otimes u)^T + (\nabla p \otimes u), \\
\operatorname{div}(pu) &= (u \cdot \nabla)p + p \operatorname{div} u, \\
\operatorname{div}(\rho u \otimes u \otimes u) &= \operatorname{div}(\rho u \otimes u) \otimes u + \rho u \otimes [(u \cdot \nabla)u].
\end{aligned}$$

Then Π takes the form

$$\begin{aligned}
\Pi = & \tau p\left[(\nabla \otimes u) + (\nabla \otimes u)^T\right] + \tau u \otimes [\rho(u \cdot \nabla)u + \nabla p] \\
& + \tau I\left[(u \cdot \nabla)p + p \operatorname{div} u\right].
\end{aligned}$$

Let us transform the obtained expression adding and subtracting the quantities $\tau I \gamma p \operatorname{div} u$ and $\frac{2}{3}\tau p I \operatorname{div} u$:

$$\Pi = \tau p \left[(\nabla \otimes u) + (\nabla \otimes u)^T - (2/3)I \operatorname{div} u \right]$$
$$+ \tau u \otimes [\rho(u \cdot \nabla)u + \nabla p] + \tau I \left[(u \cdot \nabla)p + \gamma p \operatorname{div} u \right]$$
$$+ \tau I p \operatorname{div} u - \tau I \gamma p \operatorname{div} u + \tfrac{2}{3}\tau p I \operatorname{div} u. \tag{3.20}$$

Collecting the summands, we obtain

$$\Pi = \tau p \left[(\nabla \otimes u) + (\nabla \otimes u)^T - \tfrac{2}{3}I \operatorname{div} u \right]$$
$$+ \tau p I (\tfrac{5}{3} - \gamma) \operatorname{div} u + \tau u \otimes [\rho(u \cdot \nabla)u + \nabla p]$$
$$+ \tau I \left[(u \cdot \nabla)p + \gamma p \operatorname{div} u \right]. \tag{3.21}$$

Comparing Eq. (3.21) with the form of the Navier–Stokes tensor (3.20), we see that

$$\mu = \tau p, \quad \zeta = \tau p \left(\frac{5}{3} - \gamma \right). \tag{3.22}$$

The first relation immediately implies that τ has the meaning of the Maxwell relaxation time (see [28]).

It follows from the obtained formula for the coefficients of the volume (or second, or bulk) viscosity ζ that the coefficient ζ is nonnegative and is related to the existence of molecular internal degrees of freedom, which corresponds to the theoretical ideas presented in [113, 134, 142, 212]. Indeed, for a monoatomic gas, $\gamma = 5/3$ and $\zeta = 0$, otherwise, under the existence of molecular oscillatory or rotational degrees of freedom, $\gamma < 5/3$ and $\zeta > 0$.

As a result of these transformations, we obtain that the viscous shear-stress tensor in the QGD system of equations is represented as the Navier–Stokes viscous shear-stress tensor Π_{NS}, (3.19), and an additional term proportional to τ; it vanishes for $u = 0$:

$$\Pi = \Pi_{\mathrm{NS}} + \tau u \otimes [\rho(u \cdot \nabla)u + \nabla p] + \tau I \left[(u \cdot \nabla)p + \gamma p \operatorname{div} u \right]. \tag{3.23}$$

3.3.3 Total Energy Equation and Heat Flux Vector

In conclusion, let us consider the third pair of Eqs. (3.10) and (3.15), from which we find the form of the vectors A and q. For this purpose, we write Eq. (3.10) in the form

$$\frac{\partial}{\partial t} \left[\rho \left(\frac{u^2}{2} + \varepsilon \right) \right] + \operatorname{div} \left[j_m \left(\frac{u^2}{2} + \varepsilon + \frac{p}{\rho} \right) \right]$$
$$= -\operatorname{div} \left[\rho w \left(\frac{u^2}{2} + \varepsilon + \frac{p}{\rho} \right) \right] + \operatorname{div} \tau \left\{ \operatorname{div} \left[\rho \left(\frac{u^2}{2} + \varepsilon + 2\frac{p}{\rho} \right) u \otimes u \right] \right.$$
$$\left. + \nabla \left[p \left(\frac{u^2}{2} + \varepsilon + \frac{p}{\rho} \right) \right] \right\},$$

and rewrite Eq. (3.15) in the form

$$\frac{\partial}{\partial t}\left[\rho\left(\frac{u^2}{2}+\varepsilon\right)\right]+\operatorname{div}\left[j_m\left(\frac{u^2}{2}+\varepsilon+\frac{p}{\rho}\right)\right]=\operatorname{div}\left(j_m\frac{p}{\rho}\right)+\operatorname{div}A-\operatorname{div}q.$$

Comparing the last two expressions, we find that

$$\begin{aligned}
A-q={}&-j_m\frac{p}{\rho}-[\tau\operatorname{div}(\rho u\otimes u)+\nabla p]\left(\frac{u^2}{2}+\varepsilon+\frac{p}{\rho}\right)\\
&+\tau\operatorname{div}\left[\rho\left(\frac{u^2}{2}+\varepsilon+2\frac{p}{\rho}\right)u\otimes u\right]+\tau p\nabla\left(\frac{u^2}{2}+\varepsilon+\frac{p}{\rho}\right)\\
&+\tau\left(\frac{u^2}{2}+\varepsilon+\frac{p}{\rho}\right)\nabla p;
\end{aligned}$$

here we have used representation (3.16) for w. Collecting similar terms, we obtain

$$\begin{aligned}
A-q={}&-j_m\frac{p}{\rho}-\tau\left[\frac{u^2}{2}+\varepsilon+\frac{p}{\rho}\right]\operatorname{div}(\rho u\otimes u)\\
&+\tau\left(\frac{u^2}{2}+\varepsilon+2\frac{p}{\rho}\right)\operatorname{div}(\rho u\otimes u)\\
&+\tau u(\rho u\cdot\nabla)\left(\frac{u^2}{2}+\varepsilon+2\frac{p}{\rho}\right)+\tau p\nabla\left(\frac{u^2}{2}+\varepsilon+\frac{p}{\rho}\right),
\end{aligned}$$

where we have used the identity

$$\begin{aligned}
\operatorname{div}&\left[\rho\left(\frac{u^2}{2}+\varepsilon+2\frac{p}{\rho}\right)u\otimes u\right]\\
&=\left(\frac{u^2}{2}+\varepsilon+2\frac{p}{\rho}\right)\operatorname{div}(\rho u\otimes u)+u(\rho u\cdot\nabla)\left(\frac{u^2}{2}+\varepsilon+2\frac{p}{\rho}\right).
\end{aligned}$$

Let us collect similar terms:

$$\begin{aligned}
A-q={}&-j_m\frac{p}{\rho}+\tau\left[\frac{u^2}{2}+\varepsilon+2\frac{p}{\rho}-\frac{u^2}{2}-\varepsilon-\frac{p}{\rho}\right]\operatorname{div}(\rho u\otimes u)\\
&+\tau u(\rho u\cdot\nabla)\left(\frac{u^2}{2}+\varepsilon+2\frac{p}{\rho}\right)+\tau p\nabla\left(\frac{u^2}{2}+\varepsilon+\frac{p}{\rho}\right).
\end{aligned}$$

Applying the identity

$$\begin{aligned}
\operatorname{div}(\rho u\otimes u)&=u\operatorname{div}(\rho u)+\rho(u\cdot\nabla)u\\
&=\rho u\operatorname{div}u+u(u\cdot\nabla)\rho+\rho(u\cdot\nabla)u,\tag{3.24}
\end{aligned}$$

we obtain

$$A - q = -\,j_m\frac{p}{\rho} + \tau pu\,\mathrm{div}\,u + \tau\frac{p}{\rho}u(u\cdot\nabla)\rho + \tau p(u\cdot\nabla)u$$
$$+ \tau\rho u(u\cdot\nabla)\left(\frac{u^2}{2}\right) + \tau\rho u(u\cdot\nabla)\varepsilon + 2\tau p\rho u(u\cdot\nabla)\left(\frac{1}{\rho}\right)$$
$$+ 2\tau u(u\cdot\nabla)p + \tau p\nabla\left(\frac{u^2}{2}\right) + \tau p\nabla\varepsilon + \tau p\nabla\left(\frac{p}{\rho}\right).$$

Let us collect terms as follows:

$$A - q = -\,j_m\frac{p}{\rho} + \tau p\left[\nabla\left(\frac{u^2}{2}\right) + (u\cdot\nabla)u\right]$$
$$+ \tau u\left[\rho(u\cdot\nabla)\left(\frac{u^2}{2}\right) + (u\cdot\nabla)p\right] + \tau u(u\cdot\nabla)p$$
$$+ \tau pu\,\mathrm{div}\,u + \tau\frac{p}{\rho}u(u\cdot\nabla)\rho + \tau p\rho u(u\cdot\nabla)\left(\frac{1}{\rho}\right)$$
$$+ \tau\rho u\left[(u\cdot\nabla)\varepsilon + p(u\cdot\nabla)\left(\frac{1}{\rho}\right)\right] + \tau p\nabla\left[\frac{p}{\rho} + \varepsilon\right].$$

Applying the identities

$$(\nabla\otimes u)u = \nabla\left(\frac{u^2}{2}\right), \quad (\nabla\otimes u)u = (u\cdot\nabla)u,$$

we obtain

$$A - q = -\,j_m\frac{p}{\rho} + \tau p\left[(\nabla\otimes u) + (\nabla\otimes u) - \frac{2}{3}I\,\mathrm{div}\,u\right]u$$
$$+ \tau u\,(u\cdot[\rho(u\cdot\nabla)u + \nabla p]) + \tau u\,[(u\cdot\nabla)p + \gamma p\,\mathrm{div}\,u]$$
$$+ \tau pu\left(\frac{5}{3} - \gamma\right)\mathrm{div}\,u + \tau p\nabla\left[\frac{p}{\rho} + \varepsilon\right]$$
$$+ \tau\rho u\left[(u\cdot\nabla)\varepsilon + p(u\cdot\nabla)\left(\frac{1}{\rho}\right)\right]$$
$$+ \tau\frac{p}{\rho}u(u\cdot\nabla)\rho - \tau\rho up\frac{1}{\rho^2}(u\cdot\nabla)\rho.$$

From the expression obtained, we distinguish the tensor Π in the form (3.21):

$$A - q = \Pi u - j_m\frac{p}{\rho} + \tau p\nabla\left[\frac{p}{\rho} + \varepsilon\right] + \tau\rho u\left[(u\cdot\nabla)\varepsilon + p(u\cdot\nabla)\left(\frac{1}{\rho}\right)\right].$$

$$(3.25)$$

In the Navier–Stokes theory (see [134])

$$A_{\mathrm{NS}} = \Pi_{\mathrm{NS}} u - p u$$

is the work of the surface pressure force and the internal viscous friction at unit time. By analogy, we set

$$A = \Pi u - p \frac{j_m}{\rho}.$$

Represent the other terms in Eq. (3.25) as the heat flux and denote it by

$$q = -\tau p \nabla \left[\frac{p}{\rho} + \varepsilon \right] - \tau \rho u \left[(u \cdot \nabla)\varepsilon + p(u \cdot \nabla) \left(\frac{1}{\rho} \right) \right].$$

In the heat flux q, we distinguish the part connected with the Navier–Stokes flux:

$$q_{\mathrm{NS}} = -\varkappa \nabla T.$$

As the state equations, we take the state equations of the ideal polytropic gas,

$$p = \rho \mathcal{R} T, \quad \varepsilon = \frac{\mathcal{R} T}{\gamma - 1}.$$

Represent the formula for the heat flux in the form

$$q = -\tau p \frac{\gamma \mathcal{R}}{\gamma - 1} \nabla T - \tau \rho u \left[(u \cdot \nabla)\varepsilon + p(u \cdot \nabla) \left(\frac{1}{\rho} \right) \right].$$

Comparing the formulas for q and q_{NS}, we define the heat conduction coefficient in the QGD model as follows:

$$\varkappa = \tau p \frac{\gamma \mathcal{R}}{\gamma - 1},$$

and the expression for the heat flux q in the QGD equations has the form

$$q = q_{\mathrm{NS}} - \tau \rho u \left[(u \cdot \nabla)\varepsilon + p(u \cdot \nabla) \left(\frac{1}{\rho} \right) \right]. \tag{3.26}$$

Therefore, system (3.8), (3.9), and (3.10) obtained on the basis of the regularized kinetic equation is represented in the form of conservation laws (3.13), (3.14), and (3.15). Moreover, the vector of the mass flux density, the viscous stress tensor, and the vector of the heat flux are represented in the form of the sum of the corresponding quantities in the Navier–Stokes form and additional terms of substantially nonlinear form proportional to the coefficient τ (3.16), (3.23), and (3.26).

3.4 Dissipation Coefficients

3.4.1 Formulas for Dissipative Coefficients and Their Generalizations

On the basis of the kinetic deduction of the QGD equations and their representation in the form of conservation laws, we obtain the form of the dissipative coefficients μ, ζ, and $\varkappa$. These coefficients are related with each other through the parameter τ and are expressed as follows:

$$\mu = \tau p, \quad \zeta = \tau p \left(\frac{5}{3} - \gamma \right), \quad \varkappa = \tau p \frac{\gamma \mathcal{R}}{\gamma - 1}. \tag{3.27}$$

Precisely in this form, the viscosity coefficient μ and the heat conduction coefficient $\varkappa$ are obtained in deducing the Navier–Stokes system from the BGK equation by using the Chapman–Enskog methods. The volume viscosity coefficient ζ coinciding with formula (3.27) was obtained in [205] on the basis of BGK approximation for a non-monoatom gas having rotational degrees of freedom. Indeed, the formula for this coefficient presented [205] has the form

$$\zeta = \mu \frac{2i}{3(i + 3)},$$

where i is the number of rotational degrees of freedom. Expressing the total number $n = i + 3$ of degrees of freedom through γ as $n = i + 3 = 2/(\gamma - 1)$, we obtain the volume viscosity coefficient of the form (3.27).

Introducing the Prandtl number $\Pr \neq 1$, we rewrite the heat conduction coefficient in the form

$$\varkappa = \mu \frac{\gamma \mathcal{R}}{\gamma - 1} \frac{1}{\Pr}.$$

In Sect. 1.5.3, with accuracy up to a numerical coefficient, we have found the connection of the relaxation parameter τ and the dynamical viscosity coefficient μ in the form

$$\tau = \frac{\mu}{p \, \mathrm{Sc}},$$

where Sc is the Schmidt number; for gases, its value is close to 1.

The improvement of the volume viscosity coefficient for gases having rotational degrees of freedom is presented in the next section.

3.4.2 Volume Viscosity Coefficient

Dissipative effects related to the relaxation of internal degrees of freedom in gases can substantially influence the flows in shock waves and the processes fast-variable in time. In gas-dynamic models, in the simplest way, these dissipative effects are described in terms of the so-called volume (or second, or bulk) viscosity. This description is sufficiently approximative, and its use is restricted by flows in which the characteristic relaxation times of the inner degrees of freedom are small as compared with the characteristic hydrodynamic times of the problem (see [134, 212]).

As was mentioned above, the volume viscosity coefficient ζ is substantially related to the existence of internal degrees of freedom of molecules and vanishes for one-atom gases. Meanwhile, in a number of cases, the value of this coefficient is close to the value of the dynamical viscosity coefficient μ.

The methods for calculating the second viscosity coefficients and the expression obtained are sufficiently complicated in the general case (see, e.g., [39, 113, 142]).

There are relatively simple expressions for the second viscosity coefficient only in the approximation where the internal degrees of freedom in a gas are rotational degrees of freedom. The example of such expression for ζ was presented in the previous section.

A more elaborated expression for the second viscosity coefficient depending on rotational degrees of freedom was presented in [212] in the form

$$\zeta = \frac{p\mathcal{R}}{c_V}\gamma_{\rm rot}\tau_{\rm rot}, \tag{3.28}$$

where $\gamma_{\rm rot}$ is the part of the internal energy containing in rotational degrees of freedom.

Let us show that expression (3.28) differs from the formula (3.27) for ζ obtained by using the QGD equation only by a numerical coefficient. For this purpose, we express $\gamma_{\rm rot}$ through γ:

$$\gamma_{\rm rot} = \frac{i}{n},$$

where i is the number of rotational degrees of freedom and n is the total number of degrees of freedom. If we take into account the fact that $n = i + 3$, where 3 is the number of translational degrees of freedom, then the formula for $\gamma_{\rm rot}$ can be rewritten in the form

$$\gamma_{\rm rot} = \frac{n-3}{n}.$$

Taking into account the connection of the number of degrees of freedom with the specific heat ratio γ,

$$n = \frac{2}{\gamma - 1},$$

we obtain

$$\gamma_{\rm rot} = \frac{3}{2}\left(\frac{5}{3} - \gamma\right). \tag{3.29}$$

The relaxation time of rotational degrees of freedom $\tau_{\rm rot}$ is expressed as follows:

$$\tau_{\rm rot} = Z_{\rm rot}\tau_c,$$

where τ_c is the mean collision time of molecules (2.22), $Z_{\rm rot}$ is the coefficient of energy interchange between the rotational and translational degrees of freedom, which is equal to the mean number of molecular collisions necessary for the relaxation to the equilibrium state of the rotational mode. The expressions for $Z_{\rm rot}$ are presented, e.g., in [28, 212]. These expressions for $Z_{\rm rot}$ were used for QGD calculations of nonequilibrium flows in underexpended jets CO_2 [109] and N_2 [146] and also for the numerical modelling of the shock wave structure in nitrogen [89]. For example, for nitrogen, we have

$$Z_{\rm rot} = Z^{\infty}\Bigg/\left[1 + \frac{\pi^{3/2}}{2}\sqrt{\frac{T^*}{T}} + \left(\pi + \frac{\pi^2}{4}\right)\frac{T^*}{T}\right],$$

where $Z^{\infty} = 23$ and $T^* = 91.5$ K, if the gas temperature in the undisturbed flow is equal to 273 K (see [212]). In this case, the value of $Z_{\rm rot}$ varies in the limits 5–16 in the temperature range from 300 to 6000 K (Fig. 3.2).

Let us express τ_c through the dynamical viscosity μ of the gas. The mean free path λ of particles in the gas is determined by the mean collision time of particles and the mean thermal velocity $\langle\bar{c}\rangle$ (see Eq. (2.25)) as follows:

$$\lambda = \tau_c\langle\bar{c}\rangle. \qquad (3.30)$$

The value of the mean thermal velocity of particles in the equilibrium gas is equal to $\langle\bar{c}\rangle = \sqrt{8\mathcal{R}T/\pi}$ (see Sect. 2.5). Therefore,

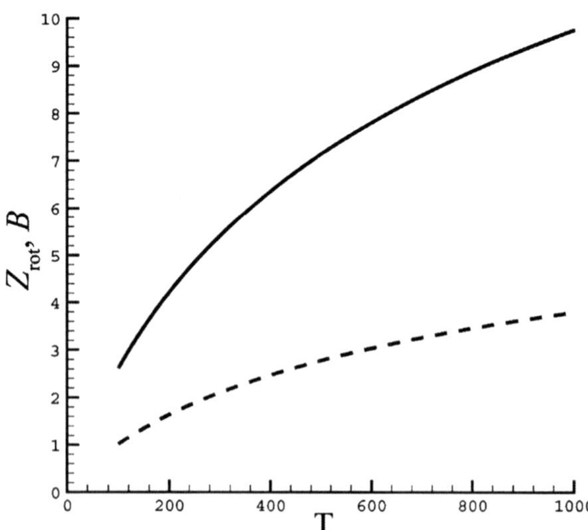

Fig. 3.2 Coefficients B (*dotted line*) and $Z_{\rm rot}$ (*continuous line*) for nitrogen

$$\tau_c = \frac{\lambda}{\sqrt{8\mathcal{R}T/\pi}}.$$

(3.31)

The mean free path can be connected with the viscosity of the gas:

$$\lambda = A\frac{\mu}{p}\sqrt{\mathcal{R}T},$$

(3.32)

where

$$A = \sqrt{\frac{\pi}{2}}$$

the Chapman formula, see [3] or

$$A = \frac{2(7 - 2\omega)(5 - 2\omega)}{15\sqrt{2\pi}}$$

the Bird formula, see [28]. Here ω is the exponent in the viscosity law $\mu \propto T^{\omega}$.

Relations (3.31) and (3.32) allow us to express τ_c through the dynamic viscosity μ of the gas in the form

$$\tau_c = A\frac{\mu}{p}\sqrt{\frac{\pi}{8}}.$$

(3.33)

Substituting the expressions for c_v, γ_{rot} (3.29), and τ_c (3.33) in Eq. (3.28), we write the initial formula for the second viscosity (3.28) in the form

$$\zeta = \mu\left(\frac{5}{3} - \gamma\right)(\gamma - 1)A\frac{3}{2}\sqrt{\frac{\pi}{8}}Z_{\mathrm{rot}} = \mu\left(\frac{5}{3} - \gamma\right)B.$$

(3.34)

The latter formula (3.34) with accuracy up to the numerical factor

$$B = (\gamma - 1)A\frac{3}{2}\sqrt{\frac{\pi}{8}}Z_{\mathrm{rot}}$$

coincides with formula (3.27) for the second viscosity obtained early in analyzing the QGD equations. In Fig. 3.2, we present the values of B (dotted line) and Z_{rot} (continuous line) depending on the temperature for nitrogen. It is seen that the coefficient B varies from 1 to 4.

The crucial role of the volume viscosity coefficient is demonstrated by the calculations of a shock wave in nitrogen presented in Appendix B.

3.5 Navier–Stokes System as an Asymptotic of the QGD System

The dissipative terms Π_{NS} and q_{NS} entering the QGD equations have the order of smallness $O(\mu) \sim O(\tau)$ (or $O(\mathrm{Kn})$ in the dimensionless form) under the corresponding smoothness conditions of the corresponding derivatives. The additional τ-terms to the viscous shear-stress tensor and the vectors of mass flux density and heat flux written in Sect. 3.3 have the same order of smallness. Let us show that for stationary flows, the QGD τ-terms have not the first, but second order of smallness in τ, i.e., $\sim O(\tau^2)$.

In Sect. 3.3, the quantities j_m, Π, A, and q in the QGD system were written in the form of a sum of the corresponding terms of the Navier–Stokes type and QGD additional terms. For convenience of the presentation, we present the form of τ-terms separately:

$$j_{\text{QGD}} = \tau \left[\text{div}(\rho u \otimes u) + \nabla p \right], \tag{3.35}$$

$$\Pi_{\text{QGD}} = \tau u \otimes \left[\rho(u \cdot \nabla)u + \nabla p \right] + \tau \hat{I} \left[(u \cdot \nabla)p + \gamma p \, \text{div}\, u \right], \tag{3.36}$$

$$A_{\text{QGD}} = \Pi_{\text{QGD}} u - \frac{p}{\rho} j_{\text{QGD}}, \tag{3.37}$$

$$q_{\text{QGD}} = \tau \rho u \left[(u \cdot \nabla)\varepsilon + p(u \cdot \nabla)\left(\frac{1}{\rho}\right) \right]; \tag{3.38}$$

here $\hat{I}$ is the identity tensor.

Let us show that in the stationary case, the written quantities formally have the second order of smallness $O(\tau^2)$.

To prove this, we use the approach presented in [184]. Write system (3.8), (3.9), and (3.10) in the stationary case:

$$\text{div}(\rho u) = M, \tag{3.39}$$

$$\text{div}(\rho u \otimes u) + \nabla p = I, \tag{3.40}$$

$$\text{div}\left[\rho u \left(\frac{u^2}{2} + \varepsilon \right) + p u \right] = E. \tag{3.41}$$

The right-hand sides of the corresponding equations are denoted by

$$M = \text{div}\left\{ \tau \left[\text{div}(\rho u \otimes u) + \nabla p \right] \right\},$$

$$I = \text{div}\left\{ \tau \left[\text{div}(\rho u \otimes u \otimes u) + (\nabla \otimes pu) + (\nabla \otimes pu)^T \right] \right\} + \nabla \left\{ \tau [\text{div}(\rho u)] \right\},$$

$$E = \text{div}\left\{ \tau \left\{ \text{div}\left[\rho \left(\frac{u^2}{2} + \varepsilon + 2\frac{p}{\rho} \right) u \otimes u \right] + \nabla \left[p \left(\frac{u^2}{2} + \varepsilon + \frac{p}{\rho} \right) \right] \right\} \right\}.$$

We immediately note that

$$M \sim O(\tau), \quad I \sim O(\tau), \quad E \sim O(\tau). \tag{3.42}$$

3.5.1 QGD Addition to the Vector of Mass Flux Density

It follows from Eqs. (3.35), (3.40), and (3.42) that

$$j_{\text{QGD}} = \tau I \sim O(\tau^2).$$

3.5.2 QGD Addition to the Viscous Shear-Stress Tensor

Applying identity (3.24), we transform Eq. (3.40):

$$u \operatorname{div}(\rho u) + \rho(u \cdot \nabla)u + \nabla p = I. \tag{3.43}$$

Then the first term in Π_{QGD} is written in the form

$$\rho(u \cdot \nabla)u + \nabla p = I - u \operatorname{div}(\rho u) = I - uM. \tag{3.44}$$

Let us calculate the divergence in Eq. (3.41) taking into account the relation $\varepsilon = p/\rho(\gamma - 1)$:

$$\operatorname{div}\left[\rho u \left(\frac{u^2}{2} + \varepsilon\right) + pu\right] = \frac{u^2}{2} \operatorname{div}(\rho u) + \rho u(u \cdot \nabla)u + \frac{\gamma}{\gamma - 1}\left[(u \cdot \nabla)p + p \operatorname{div} u\right].$$

This implies

$$\gamma\left[(u \cdot \nabla)p + p \operatorname{div} u\right] = (\gamma - 1)\left(E - \frac{u^2}{2}M - \rho u(u \cdot \nabla)u\right).$$

Furthermore, we add and subtract $(u \cdot \nabla)p$:

$$(u \cdot \nabla)p + \gamma p \operatorname{div} u = (\gamma - 1)\left(E - \frac{u^2}{2}M - \rho u(u \cdot \nabla)u - (u \cdot \nabla)p\right). \tag{3.45}$$

Let us transform the expression $(u \cdot \nabla)p$. For this purpose, we take the inner product of Eq. (3.43) and u. As a result, we obtain

$$(u \cdot \nabla)p = (u \cdot I) - u^2 M - \rho u(u \cdot \nabla)u.$$

Therefore, taking into account Eq. (3.5), we write Eq. (3.45) as follows:

$$(u \cdot \nabla)p + p \operatorname{div} u = (\gamma - 1)\left(E - \frac{u^2}{2}M - (u \cdot I)\right).$$

Transforming Eq. (3.44) with account for Eq. (3.45), we finally obtain

$$\Pi_{\mathrm{QGD}} = \tau u \otimes (I - uM) + \tau I(\gamma - 1)\left(E - \frac{u^2}{2}M - (u \cdot I)\right).$$

It follows from the last relation that $\Pi_{\mathrm{QGD}} \sim O(\tau^2)$.

3.5.3 QGD Addition to the Work of Pressure and Viscous Friction Forces

It follows from Eqs. (3.35), (3.36), and (3.37) that $A_{\mathrm{QGD}} \sim O(\tau^2)$.

3.5.4 QGD addition to the Heat Flux Vector

Let us transform the left-hand side of expression (3.41):

$$E = \frac{u^2}{2}M + \rho u(u \cdot \nabla)u + \varepsilon M + \rho(u \cdot \nabla)\varepsilon + (u \cdot \nabla)p + p \operatorname{div} u.$$

We have

$$\rho(u \cdot \nabla)\varepsilon + p \operatorname{div} u = E - M\left(\varepsilon + \frac{u^2}{2}\right) - \rho u(u \cdot \nabla)u - (u \cdot \nabla)p.$$

From Eq. (3.39) we express $\operatorname{div} u$:

$$\operatorname{div} u = \frac{M - (u \cdot \nabla)\rho}{\rho} = \frac{M}{\rho} + \rho(u \cdot \nabla)\left(\frac{1}{\rho}\right). \qquad (3.46)$$

Taking Eqs. (3.5) and (3.46) into account, we obtain

$$\rho(u \cdot \nabla)\varepsilon + p\rho(u \cdot \nabla)\left(\frac{1}{\rho}\right) = E - M\left(\varepsilon - \frac{u^2}{2} - \frac{p}{\rho}\right) - (u \cdot I). \qquad (3.47)$$

It follows from Eqs. (3.47) and (3.38) that

$$q_{\mathrm{QGD}} = \tau u\left[E - M\left(\varepsilon - \frac{u^2}{2} - \frac{p}{\rho}\right) - (u \cdot I)\right].$$

From the latter relation, we conclude that $q_{\mathrm{QGD}} \sim O(\tau^2)$.

The results obtained in this section mean that for stationary flows, the QGD equations approximate the Navier–Stokes system with accuracy $O(\tau^2)$, or, in the dimensionless form, with accuracy $O(\mathrm{Kn}^2)$ under a sufficient smoothness of the corresponding partial derivatives. Therefore, as τ tends to zero, we can assume that the solution of the QGD equations with the corresponding order converges to the solution of the Navier–Stokes system. Note that for nonstationary flows, we do not succeed in proving the second-order approximation.

Another important asymptotic of hydrodynamic equations is the boundary layer approximation (see [134, 144]). It was shown (see, e.g., [40, 55, 184]) that the Prandtl equations are an approximation of the boundary layer for QGD equations as for the Navier–Stokes equations. The form of the Prandtl equations for the two-dimensional flows is presented in Appendix B.

3.6 QGD Equations for Gas Flows Under the Existence of Exterior Forces and Heat Sources

In this section, we present another method for constructing the QGD system. Based on this method, we construct the QGD equations for the description of flows under the influence of exterior forces and heat sources. The obtained systems can be used for the numerical modelling of a number of urgent problems, for example, for the combustion problem, for studying the possibility of flow control by using the energy embedding, and for the computation of active media in the resonators of gas lasers and flows of radiating and absorbing gas (see [40]).

To construct this model, following the methodology presented in [184], we write the Euler system of equations for the ideal polytropic gas:

$$\frac{\partial}{\partial t}\rho + \frac{\partial}{\partial x_i}\rho u_i = 0, \tag{3.48}$$

$$\frac{\partial}{\partial t}\rho u_i + \frac{\partial}{\partial x_j}\rho u_i u_j + \frac{\partial}{\partial x_i}p = \rho F_i, \tag{3.49}$$

$$\frac{\partial}{\partial t}\rho\left(\frac{u^2}{2}+\varepsilon\right) + \frac{\partial}{\partial x_i}\rho u_i\left(\frac{u^2}{2}+\varepsilon+\frac{p}{\rho}\right) = \rho u_i F_i + Q, \tag{3.50}$$

with the state equations

$$p = \rho\mathcal{R}T, \quad \varepsilon = \frac{\mathcal{R}}{\gamma-1}T, \tag{3.51}$$

and the following differential identity implied by it:

$$\frac{\partial}{\partial t}\frac{1}{\rho} + u_i\frac{\partial}{\partial x_i}\frac{1}{\rho} - \frac{1}{\rho}\frac{\partial}{\partial x_i}u_i = 0, \tag{3.52}$$

$$\frac{\partial}{\partial t}u_i + u_j\frac{\partial}{\partial x_j}u_i + \frac{1}{\rho}\frac{\partial}{\partial x_i}p - F_i = 0, \tag{3.53}$$

$$\frac{\partial}{\partial t}\varepsilon + u_i\frac{\partial}{\partial x_i}\varepsilon + \frac{p}{\rho}\frac{\partial}{\partial x_i}u_i - \frac{Q}{\rho} = 0, \tag{3.54}$$

$$\frac{\partial}{\partial t}p + u_i\frac{\partial}{\partial x_i}p + \gamma p\frac{\partial}{\partial x_i}u_i - (\gamma-1)Q = 0. \tag{3.55}$$

Here and in what follows, we use the usual notation: ρ is the specific density, u_i is the velocity component, p is the pressure, F_i is the exterior force component, ε is the specific internal energy, and Q is the power of heat sources. In the formulas, the summation with respect to repeating indices is meant. The fulfillment of identities (3.52), (3.53), (3.54), and (3.55) can be verified by a direct substitution of the expressions for the derivatives in time from the Euler equations (3.48), (3.49), and (3.50) in them with the subsequent collection of similar terms.

Let us write the integral conservation laws for a small immovable volume V in the finite-difference form replacing the derivatives in time by the difference quotient for the instants of time t and $t + \Delta t$, where Δt is a small interval of time. Then we can represent the conservation laws of mass, momentum, and total energy in the following form:

$$\int_V \frac{\hat{\rho} - \rho}{\Delta t}\, d^3x + \iint_\Sigma \rho^\star u_i^\star\, d\sigma_i = 0, \tag{3.56}$$

$$\int_V \frac{\hat{\rho}\hat{u}_i - \rho u_i}{\Delta t}\, d^3x + \iint_\Sigma \rho^\star u_i^\star u_j^\star\, d\sigma_j + \iint_\Sigma p^\star\, d\sigma_i$$
$$= \int_V \rho^\star F_i\, d^3x + \iint \Pi^\star_{\mathrm{NS}\, ij}\, d\sigma_j, \tag{3.57}$$

$$\int_V \frac{\hat{\rho}(\hat{u}^2/2 + \hat{\varepsilon}) - \rho(u^2/2 + \varepsilon)}{\Delta t}\, d^3x$$
$$+ \iint_\Sigma \rho^\star u_i^\star \left(\frac{(u^\star)^2}{2} + \varepsilon^\star + \frac{p^\star}{\rho^\star} \right) d\sigma_i + \iint_\Sigma q^\star_{\mathrm{NS}\, i}\, d\sigma_i$$
$$= \int_V \rho^\star u_i^\star F_i\, d^3x + \iint_\Sigma \Pi^\star_{\mathrm{NS}\, ij} u_i^\star\, d\sigma_j + \int_V Q^\star\, d^3x. \tag{3.58}$$

Here, the heat flux vector and the viscous stress tensor are calculated as follows:

$$q_{\mathrm{NS}\, i} = -\varkappa \frac{\partial}{\partial x_i} T, \tag{3.59}$$

$$\Pi_{\mathrm{NS}\, ij} = \mu \left(\frac{\partial}{\partial x_i} u_j + \frac{\partial}{\partial x_j} u_i - \frac{2}{3} \delta_{ij} \frac{\partial}{\partial x_k} u_k \right), \tag{3.60}$$

where μ and $\varkappa$ are the viscosity and heat conduction coefficients, respectively. Quantities denoted by letters with stars in Eqs. (3.56), (3.57), and (3.58) mean the values of the gas-dynamic parameters at the instant of time $t^\star$, where $t < t^\star < t + \Delta t$, and the quantities denoted by letters with hats mean the values of parameters at the instant of time $t + \Delta t$. Introduce the notation $\Delta t/2 = \tau$ and find the gas parameters at a middle point $t = t^\star$ restricting ourselves to the first order of smallness in τ:

$$\rho^\star = \rho + \tau \frac{\partial \rho}{\partial t}, \quad u_i^\star = u_i + \tau \frac{\partial u_i}{\partial t}, \quad p^\star = p + \tau \frac{\partial p}{\partial t}. \tag{3.61}$$

Substitute these expressions in formulas (3.56), (3.57), and (3.58). We assume that the quantities F_i and Q are varied a little for time τ and hence omit the stars. In the formulas obtained, we also neglect the terms of orders $O(\tau^2)$, $O(\tau\mu)$, and $O(\tau\varkappa)$. Again returning to the differential form of the derivative in time in the integrals over the volume, we obtain

$$\int_V \frac{\partial}{\partial t} \rho\, d^3x + \iint_\Sigma \left(\rho u_i + \tau \frac{\partial}{\partial t} \rho u_i \right) d\sigma_i = 0, \tag{3.62}$$

$$\int_V \frac{\partial}{\partial t} \rho u_i\, d^3x + \iint_\Sigma \left(\rho u_i + \tau \frac{\partial}{\partial t} \rho u_i \right) u_j\, d\sigma_j$$
$$+ \iint_\Sigma \tau \rho u_i \frac{\partial}{\partial t} u_j\, d\sigma_j + \iint_\Sigma \left(p + \tau \frac{\partial}{\partial t} p \right) d\sigma_i$$
$$= \int_V \left(\rho + \tau \frac{\partial}{\partial t} \rho \right) F_i\, d^3x + \iint_\Sigma \Pi_{\mathrm{NS}\, ij}\, d\sigma_j, \tag{3.63}$$

$$\int_V \frac{\partial}{\partial t} \rho \left(\frac{u^2}{2} + \varepsilon \right) d^3x + \iint_\Sigma \left(\rho u_i + \tau \frac{\partial}{\partial t} \rho u_i \right) \left(\frac{u^2}{2} + \varepsilon + \frac{p}{\rho} \right) d\sigma_i$$
$$+ \iint_\Sigma \tau \rho u_i \left(u_j \frac{\partial}{\partial t} u_j + \frac{\partial}{\partial t} \varepsilon + p \frac{\partial}{\partial t} \frac{1}{\rho} + \frac{1}{\rho} \frac{\partial}{\partial t} p \right) d\sigma_i$$
$$+ \iint_\Sigma q_{\mathrm{NS}\, i}\, d\sigma_i = \int_V \left(\rho u_i + \tau \frac{\partial}{\partial t} \rho u_i \right) F_i d^3x$$
$$+ \iint_\Sigma \Pi_{\mathrm{NS}\, ij} u_i\, d\sigma_j + \int_V Q\, d^3x. \tag{3.64}$$

We assume that in the zero approximation with respect to τ, the Euler equations (3.48), (3.49), and (3.50) hold for the gas considered. We use these equations and identities (3.52), (3.53), (3.54), and (3.55) following from them in order to exclude the derivatives in time in the terms linear in τ. Let us collect similar terms, group together the terms of the same type, and use the arbitrariness of the volume V in order to pass from the integral form of Eqs. (3.62), (3.63), and (3.64) to their differential form.

As a result, we obtain the system of QGD equations in the form of conservation laws with account for the influence of the exterior forces and heat sources:

$$\frac{\partial}{\partial t} \rho + \frac{\partial}{\partial x_i} j_{mi} = 0, \tag{3.65}$$

$$\frac{\partial}{\partial t} \rho u_i + \frac{\partial}{\partial x_j} j_j u_i + \frac{\partial}{\partial x_i} p = \rho_\star F_i + \frac{\partial}{\partial x_j} \Pi_{ji}, \tag{3.66}$$

$$\frac{\partial}{\partial t} \rho \left(\frac{u^2}{2} + \varepsilon \right) + \frac{\partial}{\partial x_i} j_{mi} \left(\frac{u^2}{2} + \varepsilon + \frac{p}{\rho} \right) + \frac{\partial}{\partial x_i} q_i$$
$$= j_{mi} F_i + \frac{\partial}{\partial x_i} \Pi_{ij} u_j + Q, \tag{3.67}$$

where we have introduced the following notation:

$$j_{mi} = \rho(u_i - w_i), \tag{3.68}$$

$$w_i = \frac{\tau}{\rho} \left(\frac{\partial}{\partial x_j} \rho u_i u_j + \frac{\partial}{\partial x_i} p - \rho F_i \right), \tag{3.69}$$

$$\rho_\star = \left(\rho - \tau \, \frac{\partial}{\partial x_k} \, \rho u_k \right), \tag{3.70}$$

$$\Pi_{ij} = \Pi_{\mathrm{NS}\, ij} + \tau \rho u_i \left(u_k \, \frac{\partial}{\partial x_k} \, u_j + \frac{1}{\rho} \, \frac{\partial}{\partial x_j} \, p - F_j \right)$$

$$+ \tau \delta_{ij} \left(u_k \, \frac{\partial}{\partial x_k} \, p + \gamma p \, \frac{\partial}{\partial x_k} \, u_k - (\gamma - 1) Q \right), \tag{3.71}$$

$$q_i = q_{\mathrm{NS}\, i} - \tau \rho u_i \left(u_j \, \frac{\partial}{\partial x_j} \, \varepsilon + p u_j \, \frac{\partial}{\partial x_j} \, \frac{1}{\rho} - \frac{Q}{\rho} \right). \tag{3.72}$$

3.7 Entropy Balance Equation

In [184], it was shown that in the case where $Q = 0$, the entropy production for system (3.65), (3.66), and (3.67) is nonnegative. Let us generalize this result to the case where $Q \neq 0$.

Denote the total derivative in time by the operator

$$D = \rho \, \frac{\partial}{\partial t} + \rho \, \frac{\partial}{\partial x_i} \, (u_i - w_i) = \frac{\partial}{\partial t} \, \rho + \frac{\partial}{\partial x_i} \, j_{mi}.$$

We assume that the parameters of the gas considered satisfy the Gibbs identity

$$T Ds = D\varepsilon + p D \frac{1}{\rho}.$$

Using system (3.65), (3.66), and (3.67), we obtain the equation for the entropy s from this relation. In Eq. (3.66), we transport all the summands to the left-hand side and multiply the result by u_i:

$$0 = u_i \left(\frac{\partial}{\partial t} \, \rho u_i + \frac{\partial}{\partial x_j} \, j_{mj} u_i + \frac{\partial}{\partial x_i} \, p - \rho_\star F_i - \frac{\partial}{\partial x_j} \, \Pi_{ji} \right)$$

$$= D \frac{u^2}{2} + u_i \, \frac{\partial}{\partial x_i} \, p - \rho_\star F_i u_i - u_i \, \frac{\partial}{\partial x_j} \, \Pi_{ji}.$$

This implies the relation

$$D \frac{u^2}{2} = -u_i \, \frac{\partial}{\partial x_i} \, p + \rho_\star F_i u_i + u_i \, \frac{\partial}{\partial x_j} \, \Pi_{ji}.$$

Equation (3.67) can be written in the form

$$D \left(\frac{u^2}{2} + \varepsilon \right) + \frac{\partial}{\partial x_i} \, j_{mi} \frac{p}{\rho} + \frac{\partial}{\partial x_i} \, q_i = j_{mi} F_i + \frac{\partial}{\partial x_i} \, \Pi_{ij} u_j + Q,$$

whence, using the previous relation, we obtain

$$
D\varepsilon = j_{mi} F_i + \frac{\partial}{\partial x_j} \Pi_{ji} u_i - \frac{\partial}{\partial x_i} q_i - \frac{\partial}{\partial x_i} j_{mi} \frac{p}{\rho}
$$

$$
- \rho_\star F_i u_i - u_i \frac{\partial}{\partial x_j} \Pi_{ji} + u_i \frac{\partial}{\partial x_i} p
$$

$$
= (j_{mi} - \rho_\star u_i) F_i + \Pi_{ji} \frac{\partial}{\partial x_j} u_i - \frac{\partial}{\partial x_i} q_i - \frac{\partial}{\partial x_i} j_{mi} \frac{p}{\rho} + u_i \frac{\partial}{\partial x_i} p + Q.
$$

Note that

$$
D\frac{1}{\rho} = \frac{\partial}{\partial x_i} (u_i - w_i),
$$

and, therefore,

$$
Ds = \frac{1}{T} D\varepsilon + \frac{p}{T} D\frac{1}{\rho}
$$

$$
= \frac{1}{T} \left((j_{mi} - \rho_\star u_i) F_i + \Pi_{ji} \frac{\partial}{\partial x_j} u_i - \frac{\partial}{\partial x_i} q_i - \frac{\partial}{\partial x_i} j_{mi} \frac{p}{\rho} + u_i \frac{\partial}{\partial x_i} p + Q \right)
$$

$$
+ \frac{p}{T} \frac{\partial}{\partial x_i} (u_i - w_i) = - \frac{\partial}{\partial x_i} \frac{q_i}{T} + q_i \frac{\partial}{\partial x_i} \frac{1}{T}
$$

$$
+ \frac{1}{T} \left((j_{mi} - \rho_\star u_i) F_i + \Pi_{ji} \frac{\partial}{\partial x_j} u_i - \right.
$$

$$
\left. - \frac{\partial}{\partial x_i} j_{mi} \frac{p}{\rho} + u_i \frac{\partial}{\partial x_i} p + Q + p \frac{\partial}{\partial x_i} (u_i - w_i) \right).
$$

The latter relation is the entropy balance equation of the form

$$
Ds = - \frac{\partial}{\partial x_i} \frac{q_i}{T} + X,
$$

where the entropy production X is given by the expression

$$
X = q_i \frac{\partial}{\partial x_i} \frac{1}{T} + \frac{1}{T} \left((j_{mi} - \rho_\star u_i) F_i + \Pi_{ji} \frac{\partial}{\partial x_j} u_i \right.
$$

$$
\left. - \frac{\partial}{\partial x_i} j_{mi} \frac{p}{\rho} + u_i \frac{\partial}{\partial x_i} p + Q + p \frac{\partial}{\partial x_i} (u_i - w_i) \right). \tag{3.73}
$$

The criterion of the physical correctness of the hydrodynamic model is the non-negativity of the entropy production: $X \geq 0$ for $Q \geq 0$. Let us show that the above property holds for the QGD model. For this purpose, we transform expression (3.73) taking into account Eqs. (3.68), (3.69), (3.70), (3.71), and (3.72). Let us consider each of the terms separately distinguishing nonnegative combinations of quantities where it is possible.

For the first term, we obtain

$$
q_i \frac{\partial}{\partial x_i} \frac{1}{T} = \left(-\varkappa \frac{\partial}{\partial x_i} T - \tau \rho u_i \left(u_j \frac{\partial}{\partial x_j} \varepsilon + p u_j \frac{\partial}{\partial x_j} \frac{1}{\rho} - \frac{Q}{\rho} \right) \right) \frac{\partial}{\partial x_i} \frac{1}{T}
$$

$$
= \varkappa \left(\frac{\nabla T}{T} \right)^2 + \tau \rho u_i \left(u_j \frac{\partial}{\partial x_j} \varepsilon + p u_j \frac{\partial}{\partial x_j} \frac{1}{\rho} - \frac{Q}{\rho} \right) \frac{1}{T^2} \frac{\partial}{\partial x_i} T
$$

$$
= \varkappa \left(\frac{\nabla T}{T} \right)^2 + \tau \rho \frac{u_i u_j}{T \varepsilon} \left(\frac{\partial}{\partial x_i} \varepsilon \right) \left(\frac{\partial}{\partial x_j} \varepsilon \right)
$$

$$
- \frac{\tau u_i u_j}{T \varepsilon} \frac{p}{\rho} \left(\frac{\partial}{\partial x_i} \varepsilon \right) \left(\frac{\partial}{\partial x_j} \rho \right) - \frac{\tau u_i Q}{T \varepsilon} \frac{\partial}{\partial x_i} \varepsilon.
$$

Let us transform the second summand:

$$
\frac{1}{T} \left((j_{mi} - \rho_\star u_i) F_i \right) = \frac{1}{T} \left(j_{mi} - (\rho - \tau \frac{\partial}{\partial x_j} \rho u_j) u_i \right) F_i
$$

$$
= \frac{F_i}{T} \left(-\rho w_i + \tau u_i \frac{\partial}{\partial x_j} \rho u_j \right)
$$

$$
= \frac{F_i \tau}{T} \left(u_i \frac{\partial}{\partial x_j} \rho u_j - \frac{\partial}{\partial x_j} \rho u_i u_j - \frac{\partial}{\partial x_i} p + \rho F_i \right)
$$

$$
= \frac{\tau \rho}{T} F_i F_i - \frac{\tau \rho F_i}{T} u_j \frac{\partial}{\partial x_j} u_i - \frac{\tau F_i}{T} \frac{\partial}{\partial x_i} p.
$$

Noting that

$$
\frac{\Pi_{\mathrm{NS}\,ij} \Pi_{\mathrm{NS}\,ij}}{2 \mu T} = \frac{\mu}{T} \left(\frac{\partial u_i}{\partial x_j} \frac{\partial u_i}{\partial x_j} - \frac{2}{3} \frac{\partial u_k}{\partial x_k} \frac{\partial u_m}{\partial x_m} + \frac{\partial u_i}{\partial x_j} \frac{\partial u_j}{\partial x_i} \right),
$$

we transform the following fragment in the formula for X:

$$
\frac{1}{T} \Pi_{ji} \frac{\partial}{\partial x_j} u_i = \frac{1}{T} \left(\mu \left(\frac{\partial}{\partial x_j} u_i + \frac{\partial}{\partial x_i} u_j - \frac{2}{3} \delta_{ij} \frac{\partial}{\partial x_k} u_k \right) \right.
$$

$$
+ \tau \rho u_j \left(u_k \frac{\partial}{\partial x_k} u_i + \frac{1}{\rho} \frac{\partial}{\partial x_i} p - F_i \right)
$$

$$
\left. + \tau \delta_{ij} \left(u_k \frac{\partial}{\partial x_k} p + \gamma p \frac{\partial}{\partial x_k} u_k - (\gamma - 1) Q \right) \right) \frac{\partial}{\partial x_j} u_i
$$

$$
= \frac{\Pi_{\mathrm{NS}\,ij} \Pi_{\mathrm{NS}\,ij}}{2 \mu T} + \frac{\tau \rho u_j}{T} \left(u_k \frac{\partial}{\partial x_k} u_i + \frac{1}{\rho} \frac{\partial}{\partial x_i} p - F_i \right) \frac{\partial}{\partial x_j} u_i
$$

$$
+ \frac{\tau}{T} \left(u_k \frac{\partial}{\partial x_k} p + \gamma p \frac{\partial}{\partial x_k} u_k - (\gamma - 1) Q \right) \frac{\partial}{\partial x_i} u_i.
$$

Finally, we consider the remaining terms:

$$-\frac{1}{T}\frac{\partial}{\partial x_i}\, j_{mi}\,\frac{p}{\rho} + \frac{1}{T}u_i\frac{\partial}{\partial x_i}\,p + \frac{p}{T}\frac{\partial}{\partial x_i}(u_i - w_i)$$

$$= \frac{\tau}{\rho T}\left(\frac{\partial}{\partial x_j}\,\rho u_i u_j + \frac{\partial}{\partial x_i}\,p - \rho F_i\right)\frac{\partial}{\partial x_i}\,p.$$

Collecting together all the fragments of the initial formula, we obtain

$$X = \varkappa\left(\frac{\nabla T}{T}\right)^2 + \frac{\Pi_{\mathrm{NS}\,ij}\Pi_{\mathrm{NS}\,ij}}{2\mu T} + \frac{\tau\rho}{T}\left(u_k\frac{\partial}{\partial x_k}u_i + \frac{1}{\rho}\frac{\partial}{\partial x_i}p - F_i\right)^2$$

$$+ \frac{\tau\rho u_i u_j}{T\varepsilon}\left(\frac{\partial}{\partial x_i}\varepsilon\right)\left(\frac{\partial}{\partial x_j}\varepsilon\right) - \frac{\tau p u_i u_j}{T\varepsilon\rho}\left(\frac{\partial}{\partial x_i}\varepsilon\right)\left(\frac{\partial}{\partial x_j}\rho\right)$$

$$- \frac{\tau u_i Q}{T\varepsilon}\frac{\partial}{\partial x_i}\varepsilon + \frac{\tau}{T}\left(u_k\frac{\partial}{\partial x_k}p + \gamma p\frac{\partial}{\partial x_k}u_k - (\gamma - 1)Q\right)\frac{\partial}{\partial x_i}u_i$$

$$+ \frac{\tau}{T}u_i\left(\frac{\partial}{\partial x_i}p\right)\left(\frac{\partial}{\partial x_j}u_j\right) + \frac{\tau}{\rho T}u_i u_j\left(\frac{\partial}{\partial x_i}p\right)\left(\frac{\partial}{\partial x_j}\rho\right) + \frac{Q}{T}.$$

For brevity, here and in what follows, we use the notation $(a_i)^2 \equiv \delta_{ij}a_i a_j$.
Using the relation

$$p = (\gamma - 1)\varepsilon\rho,$$

we can write the previous relation in the form

$$X = \varkappa\left(\frac{\nabla T}{T}\right)^2 + \frac{\Pi_{\mathrm{NS}\,ij}\Pi_{\mathrm{NS}\,ij}}{2\mu T} + \frac{\tau\rho}{T}\left(u_k\frac{\partial}{\partial x_k}u_i + \frac{1}{\rho}\frac{\partial}{\partial x_i}p - F_i\right)^2$$

$$+ \frac{\tau p}{\rho^2 T}\left(\frac{\partial}{\partial x_i}\rho u_i\right)^2 + \frac{\tau\rho}{T\varepsilon}\left(u_i\frac{\partial}{\partial x_i}\varepsilon + \frac{p}{\rho}\frac{\partial}{\partial x_i}u_i - \frac{Q}{\rho}\right)^2 + \frac{Q}{T_\star}, \qquad (3.74)$$

where

$$T_\star = T\left(1 + \tau\left(\frac{1}{\varepsilon}u_j\frac{\partial\varepsilon}{\partial x_i} + (\gamma - 1)\frac{\partial u_i}{\partial x_i} - \frac{Q}{\varepsilon\rho}\right)\right)^{-1}.$$

The last two terms in Eq. (3.74) containing Q can be rewritten in the form[1]

$$\frac{\tau\rho}{T\varepsilon}\left(u_i\frac{\partial}{\partial x_i}\varepsilon + \frac{p}{\rho}\frac{\partial}{\partial x_i}u_i - \frac{1}{2}\frac{Q}{\rho}\right)^2 + \frac{Q}{T}\left(1 - \frac{(\gamma - 1)\tau Q}{4p}\right).$$

---

[1] This transformation was performed by A. A. Zlotnik.

Therefore, the entropy production is nonnegative if

$$\frac{(\gamma - 1)Q\tau}{4p} \leq 1.$$

As is seen from the obtained expression, for $Q > 0$ (for $Q < 0$, the entropy can decrease), the entropy production is nonnegative if the quantity τ is sufficiently small. For $\tau = 0$, the entropy production (3.74) coincides with the corresponding quantity calculated for the Navier–Stokes equations.

Chapter 4
Quasi-gas-dynamic Equations and Coordinate Systems

In this chapter, the quasi-gas-dynamic equations are written in the tensor representations in arbitrary orthogonal coordinates, which allow one to use the QGD system in a coordinate system convenient for solving the problem considered. The written form of equations enlarges the class of problems solved to complicated computational domains, including three-dimensional domains, in which the coordinate grid can be given analytically or numerically. This allows one to construct homogeneous finite-difference schemes on quasi-orthogonal spatial grids in the transformed coordinate space (see [201]). In the last two sections, the QGD equations are written in Cartesian and cylindrical coordinate systems (see [80]).

4.1 Quasi-gas-dynamic Equations in an Arbitrary Coordinate System

In accordance with [49, 125, 158], we present the formulas of tensor analysis, which are needed in what follows.

In the Euclidean space, let us consider a certain basis e_j in which any vector is given by its coordinates (x^1, x^2, x^3). Let r_i be a new basis in which any vector is given by the coordinates $(\widehat{x}^1, \widehat{x}^2, \widehat{x}^3)$.

Define the passage from the basis e_j to the new basis r_i by using the following transformation (here and in what follows, the summation with respect to repeating indices is meant):

$$r_i = b_i^j e_j, \quad b_i^j = \frac{\partial x^j}{\partial \widehat{x}^i}, \tag{4.1}$$

where b_i^j is the transition matrix from the basis e_j to the basis r_i.

In differential geometry and tensor analysis, the concept of the metric tensor is introduced; this tensor reflects the properties of the intrinsic geometry of the domain considered. The components of the metric tensor are defined as follows:

$$g_{ij} = \frac{\partial x^k}{\partial \widehat{x}^i} \frac{\partial x^k}{\partial \widehat{x}^j}, \tag{4.2}$$

where $g \equiv \det(g_{ij})$ is the determinant of the metric tensor.

T.G. Elizarova, *Quasi-Gas Dynamic Equations*, Computational Fluid and Solid Mechanics, DOI 10.1007/978-3-642-00292-2_4,
© Springer-Verlag Berlin Heidelberg 2009

Consider a certain vector $U = U^i e_i$ in the basis e_i, where U^i are the coordinates of the vector in this basis. When passing to a new orthonormal basis $(\widehat{e}_i = r_i / \sqrt{|g_{ii}|})$, the coordinates of the vector U are transformed by the formula

$$\widehat{U}^j = U^i \, \bar{b}_i^j \, \sqrt{|g_{jj}|}, \quad U^i = \frac{\widehat{U}^j}{\sqrt{|g_{jj}|}} \, b_j^i, \tag{4.3}$$

where $\bar{b} = b^{-1}$ is the matrix of the inverse passage from the basis r_i to the basis e_j.

The coordinates of the vector $\widehat{U}^i$ in the normed basis $\widehat{e}_j$ are connected with the coordinates U^i of the vector in the basis r_i by the relation

$$\widehat{U}^i = \sqrt{|g_{ii}|}U^i. \tag{4.4}$$

A similar formula for the components of the tensor Π^{ik} has the form

$$\widehat{\Pi}^{ik} = \sqrt{|g_{ii}| \cdot |g_{kk}|}\Pi^{ik}. \tag{4.5}$$

In formulas (4.4) and (4.5) the summation with respect to repeating indices is not performed.

Now let us write the formulas of tensor analysis necessary for the representation of the QGD system of equations in the tensor form.

The inner product of vectors:

$$(A \cdot B) = g_{kl} A^k B^l. \tag{4.6}$$

The covariant and contravariant derivatives of a scalar-valued function:

$$\nabla_i A = \frac{\partial A}{\partial x^i}, \quad \nabla^i A = g^{ij} \nabla_j A. \tag{4.7}$$

The covariant derivatives of tensors of the first and second rank:

$$\nabla_j U^i = \frac{\partial U^i}{\partial x^j} + U^m \Gamma^i_{mj}, \tag{4.8}$$

$$\nabla_i A^{ij} = \frac{\partial A^{ij}}{\partial x^i} + \Gamma^i_{li} A^{lj} + \Gamma^j_{mi} A^{im}, \tag{4.9}$$

where

$$\Gamma^k_{ij} = \frac{1}{2} g^{kl} \left(\frac{\partial g_{lj}}{\partial x^i} + \frac{\partial g_{il}}{\partial x^j} - \frac{\partial g_{ij}}{\partial x^l} \right) \tag{4.10}$$

are the Christoffel symbols of the second kind.

The divergence of a vector (a particular case of Eq. (4.8) for $i = j$):

$$\nabla_i A^i = \frac{1}{\sqrt{|g|}} \frac{\partial(\sqrt{|g|}A^i)}{\partial x^i}. \tag{4.11}$$

In what follows, we consider orthogonal coordinate systems in which elements of metric tensors satisfy the condition $g_{ik} = g_{ii}\delta^i_k$, where δ^i_k is the Kronecker symbol:

$$\delta^i_k = \begin{cases} 1 & \text{if } i = k, \\ 0 & \text{if } i \neq k. \end{cases}$$

In an orthogonal coordinate system, the following relations for calculating the Christoffel symbols Γ^k_{ij} hold:

$$\Gamma^k_{ij} = 0 \quad \text{for} \quad i \neq j, \quad j \neq k, \quad k \neq i, \tag{4.12}$$

$$\Gamma^k_{ii} = -\frac{1}{2g_{kk}}\frac{\partial g_{ii}}{\partial x^k} \quad \text{for} \quad i \neq k, \tag{4.13}$$

$$\Gamma^k_{ik} = \Gamma^k_{ki} = \frac{1}{2g_{kk}\dfrac{\partial g_{kk}}{\partial x^i}} = \frac{1}{2}\frac{\partial(\ln g_{kk})}{\partial x^i}. \tag{4.14}$$

Let us write the QGD system in the tensor representation.
The mass-balance equation has the form

$$\frac{\partial \rho}{\partial t} + \frac{1}{\sqrt{|g|}}\frac{\partial}{\partial x^i}(\sqrt{|g|}\,j^i_m) = 0, \tag{4.15}$$

where the vector of mass flux density is defined as follows:

$$j^i_m = \rho\left(\frac{\widehat{U}^i}{\sqrt{|g_{ii}|}} - w^i\right),$$

where

$$w^i = \frac{\tau}{\rho}\left[\frac{\partial}{\partial x^j}\left(\rho\frac{\widehat{U}^j}{\sqrt{|g_{jj}|}}\frac{\widehat{U}^i}{\sqrt{|g_{ii}|}}\right) + \Gamma^j_{lj}\rho\frac{\widehat{U}^l}{\sqrt{|g_{ll}|}}\frac{\widehat{U}^i}{\sqrt{|g_{ii}|}}\right]$$
$$+ \frac{\tau}{\rho}\left[\Gamma^i_{mj}\rho\frac{\widehat{U}^m}{\sqrt{|g_{mm}|}}\frac{\widehat{U}^j}{\sqrt{|g_{jj}|}} + g^{ij}\frac{\partial p}{\partial x^j}\right].$$

The impulse balance equation has the form

$$\frac{\partial}{\partial t}\left(\rho\frac{\widehat{U}^i}{\sqrt{|g_{ii}|}}\right) + \frac{\partial}{\partial x^j}\left(j^j_m\frac{\widehat{U}^i}{\sqrt{|g_{ii}|}}\right) + \Gamma^j_{lj}\left(j^l_m\frac{\widehat{U}^i}{\sqrt{|g_{ii}|}}\right)$$
$$+ \Gamma^i_{mj}\left(j^j_m\frac{\widehat{U}^m}{\sqrt{|g_{mm}|}}\right) + g^{ij}\frac{\partial p}{\partial x^j} = \frac{\partial \Pi^{ji}}{\partial x^j} + \Gamma^j_{lj}\Pi^{li} + \Gamma^i_{mj}\Pi^{jm}, \tag{4.16}$$

and the components of the viscous stress tensor are defined by the formulas

$$\Pi^{ji} = \tau \frac{\widehat{U}^j}{\sqrt{|g_{jj}|}} \left\{ \rho \frac{\widehat{U}^k}{\sqrt{|g_{kk}|}} \left[\frac{\partial \left(\widehat{U}^i / \sqrt{|g_{ii}|} \right)}{\partial x^k} + \frac{\widehat{U}^m}{\sqrt{|g_{mm}|}} \Gamma^i_{mk} \right] + g^{ik} \frac{\partial p}{\partial x^k} \right\}$$

$$+ \tau g^{ji} \left[\frac{\widehat{U}^k}{\sqrt{|g_{kk}|}} \frac{\partial p}{\partial x^k} + \gamma p \frac{1}{\sqrt{|g|}} \frac{\partial}{\partial x^m} \left(\sqrt{|g|} \frac{\widehat{U}^m}{\sqrt{|g_{mm}|}} \right) \right] + \Pi^{ji}_{\mathrm{NS}},$$

$$\Pi^{ji}_{\mathrm{NS}} = \mu \left\{ g^{jk} \left[\frac{\partial \left(\widehat{U}^i / \sqrt{|g_{ii}|} \right)}{\partial x^k} + \frac{\widehat{U}^m}{\sqrt{|g_{mm}|}} \Gamma^i_{mk} \right] \right\}$$

$$+ \mu \left\{ g^{ik} \left[\frac{\partial \left(\widehat{U}^j / \sqrt{|g_{jj}|} \right)}{\partial x^k} + \frac{\widehat{U}^m}{\sqrt{|g_{mm}|}} \Gamma^j_{mk} \right] \right.$$

$$\left. - \frac{2}{3} g^{ji} \frac{1}{\sqrt{|g|}} \frac{\partial}{\partial x^m} \left(\sqrt{|g|} \frac{\widehat{U}^m}{\sqrt{|g_{mm}|}} \right) \right\}.$$

The energy balance equation has the form

$$\frac{\partial E}{\partial t} + \frac{1}{\sqrt{|g|}} \frac{\partial}{\partial x^j} \left(\sqrt{|g|} j^j_m \left(\frac{E+p}{\rho} \right) \right) + \frac{1}{\sqrt{|g|}} \frac{\partial}{\partial x^j} \left(\sqrt{|g|} q^j \right)$$

$$= \frac{1}{\sqrt{|g|}} \frac{\partial}{\partial x^j} \left(\sqrt{|g|} g_{kl} \Pi^{jk} \frac{\widehat{U}^l}{\sqrt{|g_{ll}|}} \right), \tag{4.17}$$

where the energy and the heat flux are written as follows:

$$E = \rho \left(\frac{g_{kl}}{2} \frac{\widehat{U}^k}{\sqrt{|g_{kk}|}} \frac{\widehat{U}^l}{\sqrt{|g_{ll}|}} + \frac{1}{\gamma - 1} \frac{p}{\rho} \right),$$

$$q^j = - \frac{\mu}{\mathrm{Pr}} \frac{\gamma}{\gamma - 1} g^{jk} \frac{\partial}{\partial x^k} \left(\frac{p}{\rho} \right)$$

$$- \tau \rho \frac{\widehat{U}^j}{\sqrt{|g_{jj}|}} \left[\frac{1}{\gamma - 1} \frac{\widehat{U}^k}{\sqrt{|g_{kk}|}} \frac{\partial}{\partial x^k} \left(\frac{p}{\rho} \right) + p \frac{\widehat{U}^k}{\sqrt{|g_{kk}|}} \frac{\partial}{\partial x^k} \left(\frac{1}{\rho} \right) \right].$$

The tensor form (4.15), (4.16), and (4.17) allows one to adopt the QGD system to arbitrary orthogonal coordinate systems: Cartesian, cylindrical, spherical, ellipsoidal, and also to coordinate systems constructed numerically.

4.2 Cartesian Coordinate System

In the Cartesian coordinate system, we set

$$\widehat{x}^1 = x, \quad \widehat{x}^2 = y, \quad \widehat{x}^3 = z.$$

The metric tensor is diagonal:

$$g_{ij} = \begin{pmatrix} 1 & 0 & 0 \\ 0 & 1 & 0 \\ 0 & 0 & 1 \end{pmatrix}, \qquad g \equiv \det g_{ij} = 1. \tag{4.18}$$

Formulas (4.6), (4.7), (4.8), (4.9), (4.10), and (4.11) take a simpler form.
   The inner product of vectors:

$$(A \cdot B) = A^i B_i. \tag{4.19}$$

The direct tensor product of vectors:

$$(A \otimes B) = A^i B^j. \tag{4.20}$$

The derivative (gradient) of a scalar-valued function:

$$\operatorname{grad} p = \nabla_i p = \nabla^i p = \frac{\partial p}{\partial \hat{x}^i}. \tag{4.21}$$

The derivatives of tensors of the first and second rank:

$$\nabla_j A^i = \frac{\partial A^i}{\partial \hat{x}^j}, \tag{4.22}$$

$$\nabla_i \Pi^{ij} = \frac{\partial \Pi^{ij}}{\partial \hat{x}^i}. \tag{4.23}$$

The divergence:

$$\operatorname{div} A = (\nabla \cdot A) = \nabla_i A^j = \frac{\partial A_x}{\partial x} + \frac{\partial A_y}{\partial y} + \frac{\partial A_z}{\partial z}, \tag{4.24}$$

$$\operatorname{div} \Pi = (\nabla \cdot \Pi) = \nabla_i \Pi^{ij}. \tag{4.25}$$

The divergence of the direct tensor product:

$$\operatorname{div}(A \otimes B) = \nabla_i(A^i B^j). \tag{4.26}$$

The divergence of the inner product of a tensor and a vector:

$$\operatorname{div}(\Pi \cdot A) = (\nabla \cdot (\Pi \cdot A)) = \nabla_i(\Pi^{ij} A_j). \tag{4.27}$$

The QGD system (4.15), (4.16), and (4.17) is written in the form

$$\frac{\partial}{\partial t}\rho + \nabla_i j_m^i = 0, \tag{4.28}$$

$$\frac{\partial}{\partial t}(\rho\, u^j) + \nabla_i\left(j_m^i u^j\right) + \nabla^j p = \nabla_i \Pi^{ij}, \tag{4.29}$$

$$\frac{\partial}{\partial t}\left[\rho\left(\frac{u^2}{2}+\varepsilon\right)\right]+\nabla_i\left\{j_m^i\left[\left(\frac{u^2}{2}+\varepsilon\right)+\frac{p}{\rho}\right]\right\}+\nabla_i q^i=\nabla_i(\Pi^{ij}u_j). \quad (4.30)$$

Expressions (4.4) for the desired components of the vectors j_m^i and q^i are calculated as follows:

$$j_{mx}=j_m^1,\quad j_{my}=j_m^2,\quad j_{mz}=j_m^3,$$
$$q_x=q^1,\quad q_y=q^2,\quad q_z=q^3.$$

In the variables x, y, and z, system (4.28), (4.29), and (4.30) is written as follows:

$$\frac{\partial\rho}{\partial t}+\frac{\partial j_{mx}}{\partial x}+\frac{\partial j_{my}}{\partial y}+\frac{\partial j_{mz}}{\partial z}=0, \quad (4.31)$$

$$\frac{\partial(\rho u_x)}{\partial t}+\frac{\partial(j_{mx}u_x)}{\partial x}+\frac{\partial(j_{my}u_x)}{\partial y}+\frac{\partial(j_{mz}u_x)}{\partial z}+\frac{\partial p}{\partial x}=\frac{\partial\Pi_{xx}}{\partial x}+\frac{\partial\Pi_{yx}}{\partial y}+\frac{\partial\Pi_{zx}}{\partial z}, \quad (4.32)$$

$$\frac{\partial(\rho u_y)}{\partial t}+\frac{\partial(j_{mx}u_y)}{\partial x}+\frac{\partial(j_{my}u_y)}{\partial y}+\frac{\partial(j_{mz}u_y)}{\partial z}+\frac{\partial p}{\partial y}=\frac{\partial\Pi_{xy}}{\partial x}+\frac{\partial\Pi_{yy}}{\partial y}+\frac{\partial\Pi_{zy}}{\partial z}, \quad (4.33)$$

$$\frac{\partial(\rho u_z)}{\partial t}+\frac{\partial(j_{mx}u_z)}{\partial x}+\frac{\partial(j_{my}u_z)}{\partial y}+\frac{\partial(j_{mz}u_z)}{\partial z}+\frac{\partial p}{\partial z}=\frac{\partial\Pi_{xz}}{\partial x}+\frac{\partial\Pi_{yz}}{\partial y}+\frac{\partial\Pi_{zz}}{\partial z}, \quad (4.34)$$

$$\frac{\partial E}{\partial t}+\frac{\partial(j_{mx}H)}{\partial x}+\frac{\partial(j_{my}H)}{\partial y}+\frac{\partial(j_{mz}H)}{\partial z}+\frac{\partial q_x}{\partial x}+\frac{\partial q_y}{\partial y}+\frac{\partial q_z}{\partial z}$$
$$=\frac{\partial}{\partial x}\left(\Pi_{xx}u_x+\Pi_{xy}u_y+\Pi_{xz}u_z\right)+\frac{\partial}{\partial y}\left(\Pi_{yx}u_x+\Pi_{yy}u_y+\Pi_{yz}u_z\right)$$
$$+\frac{\partial}{\partial z}\left(\Pi_{zx}u_x+\Pi_{zy}u_y+\Pi_{zz}u_z\right). \quad (4.35)$$

Here, $H=(E+p)/\rho$, and the components of the vector of mass flux density j_m has the form

$$j_{mx}=j_m^1=\rho u_x-\tau\left(\frac{\partial(\rho u_x^2)}{\partial x}+\frac{\partial(\rho u_x u_y)}{\partial y}+\frac{\partial(\rho u_x u_z)}{\partial z}+\frac{\partial p}{\partial x}\right),$$
$$j_{my}=j_m^2=\rho u_y-\tau\left(\frac{\partial(\rho u_x u_y)}{\partial x}+\frac{\partial(\rho u_y^2)}{\partial y}+\frac{\partial(\rho u_y u_z)}{\partial z}+\frac{\partial p}{\partial y}\right), \quad (4.36)$$
$$j_{mz}=j_m^3=\rho u_z-\tau\left(\frac{\partial(\rho u_x u_z)}{\partial x}+\frac{\partial(\rho u_y u_z)}{\partial y}+\frac{\partial(\rho u_z^2)}{\partial z}+\frac{\partial p}{\partial z}\right).$$

In formulas (4.36), the first terms correspond to the vector of Navier–Stokes mass flux density, and the subsequent summands (with the factor τ as a coefficient) are QGD additional terms.

The elements of the viscous stress tensor Π have the form

$$\Pi_{xx} = \Pi^{11}, \qquad \Pi_{xy} = \Pi^{12}, \qquad \Pi_{xz} = \Pi^{13},$$
$$\Pi_{yx} = \Pi^{21}, \qquad \Pi_{yy} = \Pi^{22}, \qquad \Pi_{yz} = \Pi^{23}, \qquad (4.37)$$
$$\Pi_{zx} = \Pi^{31}, \qquad \Pi_{zy} = \Pi^{32}, \qquad \Pi_{zz} = \Pi^{33},$$

$$\Pi_{xx} = \mu\left(2\frac{\partial u_x}{\partial x} - \left[\frac{2}{3} - \frac{\zeta}{\mu}\right]\operatorname{div} u\right) + \tau\gamma p \operatorname{div} u$$
$$+ \tau\rho u_x\left(u_x\frac{\partial u_x}{\partial x} + u_y\frac{\partial u_x}{\partial y} + u_z\frac{\partial u_x}{\partial z}\right) + \tau\left(2u_x\frac{\partial p}{\partial x} + u_y\frac{\partial p}{\partial y} + u_z\frac{\partial p}{\partial z}\right),$$

$$\Pi_{xy} = \mu\left(\frac{\partial u_y}{\partial x} + \frac{\partial u_x}{\partial y}\right) + \tau\rho u_x\left(u_x\frac{\partial u_y}{\partial x} + u_y\frac{\partial u_y}{\partial y} + u_z\frac{\partial u_y}{\partial z} + \frac{1}{\rho}\frac{\partial p}{\partial y}\right),$$

$$\Pi_{xz} = \mu\left(\frac{\partial u_z}{\partial x} + \frac{\partial u_x}{\partial z}\right) + \tau\rho u_x\left(u_x\frac{\partial u_z}{\partial x} + u_y\frac{\partial u_z}{\partial y} + u_z\frac{\partial u_z}{\partial z} + \frac{1}{\rho}\frac{\partial p}{\partial z}\right),$$

$$\Pi_{yx} = \mu\left(\frac{\partial u_x}{\partial y} + \frac{\partial u_y}{\partial x}\right) + \tau\rho u_y\left(u_x\frac{\partial u_x}{\partial x} + u_y\frac{\partial u_x}{\partial y} + u_z\frac{\partial u_x}{\partial z} + \frac{1}{\rho}\frac{\partial p}{\partial x}\right),$$

$$\Pi_{yy} = \mu\left(2\frac{\partial u_y}{\partial y} - \left[\frac{2}{3} - \frac{\zeta}{\mu}\right]\operatorname{div} u\right) + \tau\gamma p \operatorname{div} u$$
$$+ \tau\rho u_y\left(u_x\frac{\partial u_y}{\partial x} + u_y\frac{\partial u_y}{\partial y} + u_z\frac{\partial u_y}{\partial z}\right) + \tau\left(u_x\frac{\partial p}{\partial x} + 2u_y\frac{\partial p}{\partial y} + u_z\frac{\partial p}{\partial z}\right),$$

$$\Pi_{yz} = \mu\left(\frac{\partial u_z}{\partial y} + \frac{\partial u_y}{\partial z}\right) + \tau\rho u_y\left(u_x\frac{\partial u_z}{\partial x} + u_y\frac{\partial u_z}{\partial y} + u_z\frac{\partial u_z}{\partial z} + \frac{1}{\rho}\frac{\partial p}{\partial z}\right),$$

$$\Pi_{zx} = \mu\left(\frac{\partial u_x}{\partial z} + \frac{\partial u_z}{\partial x}\right) + \tau\rho u_z\left(u_x\frac{\partial u_x}{\partial x} + u_y\frac{\partial u_x}{\partial y} + u_z\frac{\partial u_x}{\partial z} + \frac{1}{\rho}\frac{\partial p}{\partial x}\right),$$

$$\Pi_{zy} = \mu\left(\frac{\partial u_y}{\partial z} + \frac{\partial u_z}{\partial y}\right) + \tau\rho u_z\left(u_x\frac{\partial u_y}{\partial x} + u_y\frac{\partial u_y}{\partial y} + u_z\frac{\partial u_y}{\partial z} + \frac{1}{\rho}\frac{\partial p}{\partial y}\right),$$

$$\Pi_{zz} = \mu\left(2\frac{\partial u_z}{\partial z} - \left[\frac{2}{3} - \frac{\zeta}{\mu}\right]\operatorname{div} u\right) + \tau\gamma p \operatorname{div} u$$
$$+ \tau\rho u_z\left(u_x\frac{\partial u_z}{\partial x} + u_y\frac{\partial u_z}{\partial y} + u_z\frac{\partial u_z}{\partial z}\right) + \tau\left(u_x\frac{\partial p}{\partial x} + u_y\frac{\partial p}{\partial y} + 2u_z\frac{\partial p}{\partial z}\right).$$

In expressions (4.37) for the components of the viscous stress tensor, the terms with the factor μ correspond to the Navier–Stokes tensor, and the components with the factor τ are QGD additions.

Furthermore, we present a more convenient and economical numerical simulation form of the components of the mass flux density and those of the viscous stress tensor for the case $\zeta = 0$:

$$j_{mx} = \rho(u_x - w_x), \quad j_{my} = \rho(u_y - w_y), \quad j_{mz} = \rho(u_z - w_z),$$

where

$$w_x = \frac{\tau}{\rho}\left[\frac{\partial(\rho u_x^2)}{\partial x} + \frac{\partial(\rho u_x u_y)}{\partial y} + \frac{\partial(\rho u_x u_z)}{\partial z} + \frac{\partial p}{\partial x}\right], \tag{4.38}$$

$$w_y = \frac{\tau}{\rho}\left[\frac{\partial(\rho u_x u_y)}{\partial x} + \frac{\partial(\rho u_y^2)}{\partial y} + \frac{\partial(\rho u_y u_z)}{\partial z} + \frac{\partial p}{\partial y}\right], \tag{4.39}$$

$$w_z = \frac{\tau}{\rho}\left[\frac{\partial(\rho u_x u_z)}{\partial x} + \frac{\partial(\rho u_y u_z)}{\partial y} + \frac{\partial(\rho u_z^2)}{\partial z} + \frac{\partial p}{\partial z}\right]. \tag{4.40}$$

The components of the viscous stress tensor Π are written by using auxiliary quantities denoted by the superscript (*) in the form

$$\Pi_{xx} = \Pi_{xx}^{NS} + u_x \cdot w_x^* + R^*, \qquad \Pi_{xx}^{NS} = 2\mu \cdot \frac{\partial u_x}{\partial x} - \frac{2}{3}\mu \cdot \mathrm{div}\, u,$$

$$\Pi_{xy} = \Pi_{xy}^{NS} + u_x \cdot w_y^*, \qquad \Pi_{xy}^{NS} = \Pi_{yx}^{NS} = \mu \cdot \left(\frac{\partial u_y}{\partial x} + \frac{\partial u_x}{\partial y}\right),$$

$$\Pi_{xz} = \Pi_{xz}^{NS} + u_x \cdot w_z^*, \qquad \Pi_{xz}^{NS} = \Pi_{zx}^{NS} = \mu \cdot \left(\frac{\partial u_z}{\partial x} + \frac{\partial u_x}{\partial z}\right),$$

$$\Pi_{yx} = \Pi_{yx}^{NS} + u_y \cdot w_x^*,$$

$$\Pi_{yy} = \Pi_{yy}^{NS} + u_y \cdot w_y^* + R^*, \qquad \Pi_{yy}^{NS} = 2\mu \cdot \frac{\partial u_y}{\partial y} - \frac{2}{3}\mu \cdot \mathrm{div}\, u,$$

$$\Pi_{yz} = \Pi_{yz}^{NS} + u_y \cdot w_z^*, \qquad \Pi_{yz}^{NS} = \Pi_{zy}^{NS} = \mu \cdot \left(\frac{\partial u_z}{\partial y} + \frac{\partial u_y}{\partial z}\right),$$

$$\Pi_{zx} = \Pi_{zx}^{NS} + u_z \cdot w_x^*,$$

$$\Pi_{zy} = \Pi_{zy}^{NS} + u_z \cdot w_y^*,$$

$$\Pi_{zz} = \Pi_{zz}^{NS} + u_z \cdot w_z^* + R^*, \qquad \Pi_{zz}^{NS} = 2\mu \cdot \frac{\partial u_z}{\partial z} - \frac{2}{3}\mu \cdot \mathrm{div}\, u,$$

and the expressions for the auxiliary quantities w_x^*, w_y^*, w_z^*, and R^* are given by the formulas

$$
\begin{aligned}
w_x^* &= \tau \left[\rho \left(u_x \frac{\partial u_x}{\partial x} + u_y \frac{\partial u_x}{\partial y} + u_z \frac{\partial u_x}{\partial z} \right) + \frac{\partial p}{\partial x} \right], \\
w_y^* &= \tau \left[\rho \left(u_x \frac{\partial u_y}{\partial x} + u_y \frac{\partial u_y}{\partial y} + u_z \frac{\partial u_y}{\partial z} \right) + \frac{\partial p}{\partial y} \right], \\
w_z^* &= \tau \left[\rho \left(u_x \frac{\partial u_z}{\partial x} + u_y \frac{\partial u_z}{\partial y} + u_z \frac{\partial u_z}{\partial z} \right) + \frac{\partial p}{\partial z} \right], \\
R^* &= \tau \left[u_x \frac{\partial p}{\partial x} + u_y \frac{\partial p}{\partial y} + u_z \frac{\partial p}{\partial z} + \gamma \cdot p \cdot \operatorname{div} u \right].
\end{aligned}
\tag{4.41}
$$

The components of the heat flux vector q are calculated as follows:

$$
\begin{aligned}
q_x &= q^1 = q_x^{\mathrm{NS}} - u_x \cdot R^q, \\
q_y &= q^2 = q_y^{\mathrm{NS}} - u_y \cdot R^q, \\
q_z &= q^3 = q_z^{\mathrm{NS}} - u_z \cdot R^q,
\end{aligned}
\tag{4.42}
$$

where

$$
\begin{aligned}
R^q &= \tau \rho \frac{1}{\gamma - 1} \left[u_x \frac{\partial}{\partial x} \left(\frac{p}{\rho} \right) + u_y \frac{\partial}{\partial y} \left(\frac{p}{\rho} \right) + u_z \frac{\partial}{\partial z} \left(\frac{p}{\rho} \right) \right] \\
&\quad + \tau \rho p \left[u_x \frac{\partial}{\partial x} \left(\frac{1}{\rho} \right) + u_y \frac{\partial}{\partial y} \left(\frac{1}{\rho} \right) + u_z \frac{\partial}{\partial z} \left(\frac{1}{\rho} \right) \right].
\end{aligned}
\tag{4.43}
$$

The Navier–Stokes terms q_x^{NS}, q_y^{NS}, and q_z^{NS} are found by the formulas

$$
q_x^{\mathrm{NS}} = -\varkappa \frac{\partial T}{\partial x}, \quad
q_y^{\mathrm{NS}} = -\varkappa \frac{\partial T}{\partial y}, \quad
q_z^{\mathrm{NS}} = -\varkappa \frac{\partial T}{\partial z}.
$$

4.3 Cylindrical Coordinate System

In the cylindrical coordinate system, we set

$$
\widehat{x}^1 = r, \quad \widehat{x}^2 = \varphi, \quad \widehat{x}^3 = z.
$$

The metric tensor has the diagonal form

$$
g_{ij} = \begin{pmatrix} 1 & 0 & 0 \\ 0 & r^2 & 0 \\ 0 & 0 & 1 \end{pmatrix},
\tag{4.44}
$$

and its determinant is equal to $g \equiv \det g_{ij} = r^2$.

Let us use formulas (4.4), (4.5), (4.6), (4.7), (4.8), (4.9), (4.10), and (4.11) presented above and write the r-, φ-, and z-components of the vector j_m of the mass flux density, the viscous stress tensor Π, and the heat flux vector q.

The components of the vector j_m of the mass flux density are

$$j_{mr} = j_m^1 = \rho u_r - \tau \left(\frac{1}{r} \frac{\partial (r\rho u_r^2)}{\partial r} + \frac{1}{r} \frac{\partial (\rho u_r u_\varphi)}{\partial \varphi} + \frac{\partial (\rho u_r u_z)}{\partial z} + \frac{\partial p}{\partial r} \right),$$

$$j_{m\varphi} = r j_m^2 = \frac{\rho u_\varphi}{r} - \tau \left(\frac{1}{r} \frac{\partial (r\rho u_\varphi u_r)}{\partial r} + \frac{1}{r^2} \frac{\partial (\rho u_\varphi^2)}{\partial \varphi} + \frac{1}{r} \frac{\partial (\rho u_\varphi u_z)}{\partial z} + \frac{1}{r^2} \frac{\partial p}{\partial \varphi} \right),$$

$$\text{(4.45)}$$

$$j_{mz} = j_m^3 = \rho u_z - \tau \left(\frac{1}{r} \frac{\partial (r\rho u_z u_r)}{\partial r} + \frac{1}{r} \frac{\partial (\rho u_z u_\varphi)}{\partial \varphi} + \frac{\partial (\rho u_z^2)}{\partial z} + \frac{\partial p}{\partial z} \right).$$

In each of the formulas presented above, the first term of the right-hand side corresponds to the Navier–Stokes vector j_m^{NS} of the mass flux density and the subsequent terms (with the factor τ as the coefficient) are QGD additions j_m^{QGD}.

The elements of the viscous stress tensor Π have the form

$$\Pi_{rr} = \Pi^{11} = \mu \left(2\frac{\partial u_r}{\partial r} - \left[\frac{2}{3} - \frac{\zeta}{\mu} \right] \operatorname{div} u \right)$$

$$+ \tau \rho u_r \left(u_r \frac{\partial u_r}{\partial r} + \frac{u_\varphi}{r} \frac{\partial u_r}{\partial \varphi} - \frac{u_\varphi}{r} + u_z \frac{\partial u_r}{\partial z} + \frac{1}{\rho} \frac{\partial p}{\partial r} \right)$$

$$+ \tau \left(u_r \frac{\partial p}{\partial r} + \frac{u_\varphi}{r^3} \frac{\partial p}{\partial \varphi} + u_z \frac{\partial p}{\partial z} + \gamma p \operatorname{div} u \right),$$

$$\Pi_{r\varphi} = r \Pi^{12} = \mu \left(\frac{1}{r^2} \frac{\partial u_r}{\partial \varphi} - \frac{u_\varphi}{r^2} + \frac{1}{r} \frac{\partial u_\varphi}{\partial r} \right)$$

$$+ \tau \rho \frac{u_r}{r} \left(r u_r \frac{\partial u_\varphi}{\partial r} - r^2 u_r u_\varphi + u_\varphi \frac{\partial u_\varphi}{\partial \varphi} + r u_z \frac{\partial u_\varphi}{\partial z} + \frac{1}{\rho} \frac{\partial p}{\partial \varphi} \right),$$

$$\Pi_{rz} = \Pi^{13} = \mu \left(\frac{\partial u_r}{\partial z} + \frac{\partial u_z}{\partial r} \right) + \tau \rho u_r \left(u_r \frac{\partial u_z}{\partial r} + \frac{u_\varphi}{r} \frac{\partial u_z}{\partial \varphi} + u_z \frac{\partial u_z}{\partial z} + \frac{1}{\rho} \frac{\partial p}{\partial z} \right),$$

$$\Pi_{\varphi r} = r \Pi^{21} = \mu \left(\frac{1}{r^2} \frac{\partial u_r}{\partial \varphi} - \frac{u_\varphi}{r^2} + \frac{1}{r} \frac{\partial u_\varphi}{\partial r} \right)$$

$$+ \tau \rho u_\varphi \left(u_r \frac{\partial u_r}{\partial r} + \frac{u_\varphi}{r} \frac{\partial u_r}{\partial \varphi} - \frac{u_\varphi^2}{r} + u_z \frac{\partial u_r}{\partial z} + \frac{1}{\rho} \frac{\partial p}{\partial r} \right),$$

$$
\Pi_{\varphi\varphi} = r^2 \Pi^{22} = \frac{\mu}{r^2}\left(2\frac{u_r}{r} + 2\frac{1}{r}\frac{\partial u_\varphi}{\partial\varphi} - \left[\frac{2}{3} - \zeta\frac{r^2}{\mu}\right]\operatorname{div} u\right)
$$

$$
+ \tau\rho\frac{u_\varphi}{r}\left(ru_r\frac{\partial u_\varphi}{\partial r} + u_\varphi\frac{\partial u_\varphi}{\partial\varphi} + u_r u_\varphi + ru_z\frac{\partial u_\varphi}{\partial z} + \frac{1}{\rho}\frac{\partial p}{\partial\varphi}\right)
$$

$$
+ \tau\left(u_r\frac{\partial p}{\partial r} + \frac{u_\varphi}{r^3}\frac{\partial p}{\partial\varphi} + u_z\frac{\partial p}{\partial z} + \gamma p\operatorname{div} u\right),
$$

$$
\Pi_{\varphi z} = r\Pi^{23} = \mu\left(\frac{1}{r}\frac{\partial u_\varphi}{\partial z} + \frac{1}{r^2}\frac{\partial u_z}{\partial\varphi}\right)
$$

$$
+ \tau\rho u_\varphi\left(u_r\frac{\partial u_z}{\partial r} + \frac{u_\varphi}{r}\frac{\partial u_z}{\partial\varphi} + u_z\frac{\partial u_z}{\partial z} + \frac{1}{\rho}\frac{\partial p}{\partial z}\right),
$$

$$
\Pi_{zr} = \Pi^{31} = \mu\left(\frac{\partial u_r}{\partial z} + \frac{\partial u_z}{\partial r}\right)
$$

$$
+ \tau\rho u_z\left(u_r\frac{\partial u_r}{\partial r} + \frac{u_\varphi}{r}\frac{\partial u_r}{\partial\varphi} - \frac{u_\varphi^2}{r} + u_z\frac{\partial u_r}{\partial z} + \frac{1}{\rho}\frac{\partial p}{\partial r}\right),
$$

$$
\Pi_{z\varphi} = r\Pi^{32} = \mu\left(\frac{1}{r}\frac{\partial u_\varphi}{\partial z} + \frac{1}{r^2}\frac{\partial u_z}{\partial\varphi}\right)
$$

$$
+ \tau\rho\frac{u_z}{r}\left(ru_r\frac{\partial u_\varphi}{\partial r} + u_\varphi\frac{\partial u_\varphi}{\partial\varphi} + u_r u_\varphi + ru_z\frac{\partial u_\varphi}{\partial z} + \frac{1}{\rho}\frac{\partial p}{\partial\varphi}\right),
$$

$$
\Pi_{zz} = \Pi^{33} = \mu\left(2\frac{\partial u_z}{\partial z} - \left[\frac{2}{3} - \frac{\zeta}{\mu}\right]\operatorname{div} u\right)
$$

$$
+ \tau\rho u_z\left(u_r\frac{\partial u_z}{\partial r} + \frac{u_\varphi}{r}\frac{\partial u_z}{\partial\varphi} + u_z\frac{\partial u_z}{\partial z} + \frac{1}{\rho}\frac{\partial p}{\partial z}\right)
$$

$$
+ \tau\left(u_r\frac{\partial p}{\partial r} + \frac{u_\varphi}{r}\frac{\partial p}{\partial\varphi} + u_z\frac{\partial p}{\partial z} + \gamma p\operatorname{div} u\right).
$$

In the expressions for the components of the viscous stress tensor, the terms with the factor μ correspond to the Navier–Stokes tensor and the terms with the factor τ are QGD additions.

Finally, we write the components of the heat flux vector q:

$$
\begin{aligned}
q_r &= q^1 = q_r^{\mathrm{NS}} - u_r \cdot R^q, \\
q_z &= q^2 = q_z^{\mathrm{NS}} - u_z \cdot R^q, \\
q_\varphi &= q^3 = q_\varphi^{\mathrm{NS}} - u_\varphi \cdot R^q,
\end{aligned}
\tag{4.46}
$$

where

$$R^q = \tau\rho\frac{1}{\gamma-1}\left[u_r\frac{\partial}{\partial r}\left(\frac{p}{\rho}\right) + u_z\frac{\partial}{\partial z}\left(\frac{p}{\rho}\right) + \frac{u_\varphi}{r}\frac{\partial}{\partial\varphi}\left(\frac{p}{\rho}\right)\right]$$

$$+ \tau\rho p\left[u_r\frac{\partial}{\partial r}\left(\frac{1}{\rho}\right) + u_z\frac{\partial}{\partial z}\left(\frac{1}{\rho}\right) + \frac{u_\varphi}{r}\frac{\partial}{\partial\varphi}\left(\frac{1}{\rho}\right)\right]. \qquad (4.47)$$

The Navier–Stokes summands q_r^{NS}, q_z^{NS}, and q_φ^{NS} are found by the formulas

$$q_r^{\mathrm{NS}} = -\varkappa\frac{\partial T}{\partial r}, \quad q_z^{\mathrm{NS}} = -\varkappa\frac{\partial T}{\partial z}, \quad q_\varphi^{\mathrm{NS}} = -\frac{\varkappa}{r^2}\frac{\partial T}{\partial\varphi}.$$

Chapter 5
Numerical Algorithms for Solving Gas-Dynamic Problems

In this chapter, we present numerical algorithms based on the quasi-gas-dynamic equations for solving gas-dynamic problems.

We construct finite-difference approximations in the flux form directly for the vector j_m of the mass flux density, the heat flux vector q, and the viscous stress tensor Π. This corresponds to the writing of the gas-dynamic equations in the form of conservation laws and makes the algorithm sufficiently compact and efficient. The stability of the finite-difference algorithm is ensured by introducing an artificial dissipation whose form is determined by the QGD additions. For clearness of presentation, we construct the algorithms on the rectangular computation grids. We consider one- and two-dimensional flows.

The algorithm can be used for computing the viscous, as well as nonviscous, supersonic nonstationary flows. We describe a modification of the numerical algorithm for computing the subsonic flows.

For illustration of the work of numerical algorithms, we present the results of computing the one-dimensional problem on the strong discontinuity step evolution and two-dimensional problems on the axially symmetric flow in a neighborhood of a cylinder and the planar flow near forward-facing and backward-facing steps. The latter problem is described in Appendix D.

The content of this chapter is based on the results of [71, 81, 83, 84, 184, 186].

5.1 System for Planar Two-Dimensional Flows

Let us present the system of the quasi-gas-dynamic equations in the Cartesian coordinates for two-dimensional flows:

$$\frac{\partial \rho}{\partial t} + \frac{\partial j_{mx}}{\partial x} + \frac{\partial j_{my}}{\partial y} = 0, \tag{5.1}$$

$$\frac{\partial (\rho u_x)}{\partial t} + \frac{\partial (j_{mx} u_x)}{\partial x} + \frac{\partial (j_{my} u_x)}{\partial y} + \frac{\partial p}{\partial x} = \frac{\partial \Pi_{xx}}{\partial x} + \frac{\partial \Pi_{yx}}{\partial y}, \tag{5.2}$$

$$\frac{\partial (\rho u_y)}{\partial t} + \frac{\partial (j_{mx} u_y)}{\partial x} + \frac{\partial (j_{my} u_y)}{\partial y} + \frac{\partial p}{\partial y} = \frac{\partial \Pi_{xy}}{\partial x} + \frac{\partial \Pi_{yy}}{\partial y}, \tag{5.3}$$

T.G. Elizarova, *Quasi-Gas Dynamic Equations*, Computational Fluid and Solid Mechanics, DOI 10.1007/978-3-642-00292-2_5,
© Springer-Verlag Berlin Heidelberg 2009

$$\frac{\partial E}{\partial t} + \frac{\partial(j_{mx}H)}{\partial x} + \frac{\partial(j_{my}H)}{\partial y} + \frac{\partial q_x}{\partial x} + \frac{\partial q_y}{\partial y}$$

$$= \frac{\partial}{\partial x}\left(\Pi_{xx}u_x + \Pi_{xy}u_y\right) + \frac{\partial}{\partial y}\left(\Pi_{yx}u_x + \Pi_{yy}u_y\right), \qquad (5.4)$$

where E is the total energy of the volume unit and H is the total specific enthalpy:

$$E = \rho\frac{u_x^2 + u_y^2}{2} + \frac{p}{\gamma - 1}, \quad H = \frac{(E + p)}{\rho}, \qquad p = \rho\mathcal{R}T. \qquad (5.5)$$

The components of the vector j_m of the mass flux density are calculated by the formulas

$$j_{mx} = \rho(u_x - w_x), \quad j_{my} = \rho(u_y - w_y), \qquad (5.6)$$

where

$$w_x = \frac{\tau}{\rho}\left[\frac{\partial(\rho u_x^2)}{\partial x} + \frac{\partial(\rho u_x u_y)}{\partial y} + \frac{\partial p}{\partial x}\right],$$

$$\qquad (5.7)$$

$$w_y = \frac{\tau}{\rho}\left[\frac{\partial(\rho u_x u_y)}{\partial x} + \frac{\partial(\rho u_y^2)}{\partial y} + \frac{\partial p}{\partial y}\right].$$

The components of the viscous shear-stress tensor Π are defined by using the following expressions efficient from the viewpoint of the program implementation:

$$\Pi_{xx} = \Pi_{xx}^{NS} + u_x w_x^* + R^*, \quad \Pi_{xx}^{NS} = 2\mu\frac{\partial u_x}{\partial x} - \frac{2}{3}\mu\,\text{div}\,u,$$

$$\Pi_{xy} = \Pi_{xy}^{NS} + u_x w_y^*, \qquad \Pi_{xy}^{NS} = \Pi_{yx}^{NS} = \mu\left(\frac{\partial u_y}{\partial x} + \frac{\partial u_x}{\partial y}\right),$$

$$\qquad (5.8)$$

$$\Pi_{yx} = \Pi_{yx}^{NS} + u_y w_x^*,$$

$$\Pi_{yy} = \Pi_{yy}^{NS} + u_y w_y^* + R^*, \quad \Pi_{yy}^{NS} = 2\mu\frac{\partial u_y}{\partial y} - \frac{2}{3}\mu\,\text{div}\,u;$$

here, $\Pi_{xx}^{NS}, \Pi_{xy}^{NS}, \Pi_{yx}^{NS}$, and Π_{yy}^{NS} are components of the Navier–Stokes viscous shear-stress tensor.

The quantities w_x^*, w_y^*, and R^* are calculated by the formulas

$$w_x^* = \tau \left[\rho u_x \frac{\partial u_x}{\partial x} + \rho u_y \frac{\partial u_x}{\partial y} + \frac{\partial p}{\partial x} \right],$$

$$w_y^* = \tau \left[\rho u_x \frac{\partial u_y}{\partial x} + \rho u_y \frac{\partial u_y}{\partial y} + \frac{\partial p}{\partial y} \right], \tag{5.9}$$

$$R^* = \tau \left[u_x \frac{\partial p}{\partial x} + u_y \frac{\partial p}{\partial y} + \gamma p \, div \, u \right],$$

and the divergence div u of the velocity vector is given by

$$div \, u = \frac{\partial u_x}{\partial x} + \frac{\partial u_y}{\partial y}. \tag{5.10}$$

The components of the heat flux vector q have the form

$$q_x = q_x^{NS} - u_x R^q, \quad q_y = q_y^{NS} - u_y R^q, \tag{5.11}$$

$$R^q = \tau \rho \left[\frac{u_x}{\gamma - 1} \frac{\partial}{\partial x} \left(\frac{p}{\rho} \right) + \frac{u_y}{\gamma - 1} \frac{\partial}{\partial y} \left(\frac{p}{\rho} \right) + p u_x \frac{\partial}{\partial x} \left(\frac{1}{\rho} \right) + p u_y \frac{\partial}{\partial y} \left(\frac{1}{\rho} \right) \right],$$

where the Navier–Stokes terms q_x^{NS} and q_y^{NS} are calculated as follows:

$$q_x^{NS} = -\varkappa \frac{\partial T}{\partial x}, \quad q_y^{NS} = -\varkappa \frac{\partial T}{\partial y}. \tag{5.12}$$

The dependence of the dynamical viscosity coefficient μ on the temperature is chosen in the following form:

$$\mu = \mu_\infty \left(\frac{T}{T_\infty} \right)^\omega, \tag{5.13}$$

where $\mu_\infty = \mu(T_\infty)$ is the known value of μ for the temperature T_∞.

The heat conduction coefficient $\varkappa$ and the relaxation parameter τ are connected with the dynamical viscosity coefficient μ by the relations

$$\varkappa = \frac{\gamma \mathcal{R}}{(\gamma - 1) \Pr} \mu, \quad \tau = \frac{1}{p \, Sc} \mu, \tag{5.14}$$

where Pr is the Prandtl number and Sc is the Schmidt number.

5.2 System for Cylindrical Two-Dimensional Flows

In the cylindrical coordinates and in the two-dimensional case, the QGD system of equations has the form

$$\frac{\partial \rho}{\partial t} + \frac{1}{r}\frac{\partial(r j_{mr})}{\partial r} + \frac{\partial j_{mz}}{\partial z} = 0, \tag{5.15}$$

$$\frac{\partial(\rho u_r)}{\partial t} + \frac{1}{r}\frac{\partial(r j_{mr} u_r)}{\partial r} + \frac{\partial(j_{mz} u_r)}{\partial z} + \frac{\partial p}{\partial r} = \frac{1}{r}\frac{\partial(r \Pi_{rr})}{\partial r} + \frac{\partial \Pi_{zr}}{\partial z} - \frac{\Pi_{\phi\phi}}{r}, \tag{5.16}$$

$$\frac{\partial(\rho u_z)}{\partial t} + \frac{1}{r}\frac{\partial(r j_{mr} u_z)}{\partial r} + \frac{\partial(j_{mz} u_z)}{\partial z} + \frac{\partial p}{\partial z} = \frac{1}{r}\frac{\partial(r \Pi_{rz})}{\partial r} + \frac{\partial \Pi_{zz}}{\partial z}, \tag{5.17}$$

$$\frac{\partial E}{\partial t} + \frac{1}{r}\frac{\partial(r j_{mr} H)}{\partial r} + \frac{\partial(j_{mz} H)}{\partial z} + \frac{1}{r}\frac{\partial(r q_r)}{\partial r} + \frac{\partial q_z}{\partial z}$$
$$= \frac{1}{r}\frac{\partial}{\partial r}\left[r\left(\Pi_{rr} u_r + \Pi_{rz} u_z\right)\right] + \frac{\partial}{\partial z}\left(\Pi_{zr} u_r + \Pi_{zz} u_z\right), \tag{5.18}$$

where E is the total energy of the volume unit and H is the total specific enthalpy:

$$E = \rho \frac{u_r^2 + u_z^2}{2} + \frac{p}{\gamma - 1}, \quad H = \frac{(E + p)}{\rho}, \quad p = \rho \mathcal{R} T. \tag{5.19}$$

Here, u_r and u_z are the projections of the velocity vector u on the axes r and z, respectively. The influence of the exterior forces is not taken into account.

The components of the vector j_m of the mass flux density are calculated by the formulas

$$j_{mr} = \rho(u_r - w_r), \quad j_{mz} = \rho(u_z - w_z), \tag{5.20}$$

where

$$w_r = \frac{\tau}{\rho}\left[\frac{1}{r}\frac{\partial}{\partial r}(r \rho u_r^2) + \frac{\partial}{\partial z}(\rho u_r u_z) + \frac{\partial p}{\partial r}\right],$$

$$w_z = \frac{\tau}{\rho}\left[\frac{1}{r}\frac{\partial}{\partial r}(r \rho u_r u_z) + \frac{\partial}{\partial z}(\rho u_z^2) + \frac{\partial p}{\partial z}\right].$$

We write the components of the viscous shear-stress tensor Π in the following form convenient for program implementation:

$$\Pi_{rr} = \Pi_{rr}^{\text{NS}} + u_r w_r^* + R^*, \qquad \Pi_{rr}^{\text{NS}} = 2\mu \frac{\partial u_r}{\partial r} - \frac{2}{3}\mu \operatorname{div} u,$$

$$\Pi_{rz} = \Pi_{rz}^{\text{NS}} + u_r w_z^*, \qquad \Pi_{rz}^{\text{NS}} = \Pi_{zr}^{\text{NS}} = \mu\left(\frac{\partial u_r}{\partial z} + \frac{\partial u_z}{\partial r}\right),$$

$$\Pi_{zr} = \Pi_{zr}^{\text{NS}} + u_z w_r^*, \tag{5.21}$$

$$\Pi_{\phi\phi} = \Pi_{\phi\phi}^{\text{NS}} + R^*, \qquad \Pi_{\phi\phi}^{\text{NS}} = 2\mu \frac{u_r}{r} - \frac{2}{3}\mu \operatorname{div} u,$$

$$\Pi_{zz} = \Pi_{zz}^{\text{NS}} + u_z w_z^* + R^*, \qquad \Pi_{zz}^{\text{NS}} = 2\mu \frac{\partial u_z}{\partial z} - \frac{2}{3}\mu \operatorname{div} u;$$

here, Π_{rr}^{NS}, Π_{rz}^{NS}, Π_{zr}^{NS}, $\Pi_{\phi\phi}^{NS}$, and Π_{zz}^{NS} are the components of the Navier–Stokes viscous shear-stress tensor.

The quantities w_r^*, w_z^*, and R^* are calculated by the formulas

$$w_r^* = \tau \left[\rho u_r \frac{\partial u_r}{\partial r} + \rho u_z \frac{\partial u_r}{\partial z} + \frac{\partial p}{\partial r} \right],$$

$$w_z^* = \tau \left[\rho u_r \frac{\partial u_z}{\partial r} + \rho u_z \frac{\partial u_z}{\partial z} + \frac{\partial p}{\partial z} \right],$$

$$R^* = \tau \left[u_r \frac{\partial p}{\partial r} + u_z \frac{\partial p}{\partial z} + \gamma p \operatorname{div} u \right],$$

and the divergence $\operatorname{div} u$ of the velocity vector is given by

$$\operatorname{div} u = \frac{1}{r} \frac{\partial (r u_r)}{\partial r} + \frac{\partial u_z}{\partial z}.$$

The components of the heat flux vector q are found by the formulas

$$q_r = q_r^{NS} - u_r R^q, \qquad q_z = q_z^{NS} - u_z R^q, \tag{5.22}$$

$$R^q = \tau \rho \left[\frac{u_r}{\gamma - 1} \frac{\partial}{\partial r} \left(\frac{p}{\rho} \right) + \frac{u_z}{\gamma - 1} \frac{\partial}{\partial z} \left(\frac{p}{\rho} \right) + p u_r \frac{\partial}{\partial r} \left(\frac{1}{\rho} \right) + p u_z \frac{\partial}{\partial z} \left(\frac{1}{\rho} \right) \right],$$

where the Navier–Stokes terms q_r^{NS} and q_z^{NS} are given by the formulas

$$q_r^{NS} = -\varkappa \frac{\partial T}{\partial r}, \qquad q_z^{NS} = -\varkappa \frac{\partial T}{\partial z}. \tag{5.23}$$

5.3 Boundary Conditions

The systems presented must be complemented by initial and boundary conditions. The statement of these conditions is determined by a concrete problem being solved. As an example, we present a variant of statement of boundary conditions for the problem of supersonic gas flow in a neighborhood of a cylinder in the axially symmetric geometry. The scheme of the computation domain for this flow is presented in Fig. 5.7 (see p. 95).

The profile of the flow on the input boundary located in the plane $z = L$ is given in the form

$$\rho = \rho_\infty, \quad u_z = u_{z\infty}, \quad u_r = 0, \quad p = p_\infty. \tag{5.24}$$

On the symmetry axis coinciding with the axis z, we pose the symmetry conditions

$$\frac{\partial \rho}{\partial r} = 0, \quad \frac{\partial u_z}{\partial r} = 0, \quad u_r = 0, \quad \frac{\partial p}{\partial r} = 0. \tag{5.25}$$

On free boundaries where it is assumed that the gas flows out from the domain considered, we pose the so-called "soft" boundary conditions or the "drift" conditions. Under these conditions, we assume that the normal derivatives of the density, the pressure, and the components of the velocity vanish. For example, if the boundary lies in the plane $z = 0$, then these conditions are written as follows:

$$\frac{\partial \rho}{\partial z} = 0, \quad \frac{\partial u_z}{\partial z} = 0, \quad \frac{\partial u_r}{\partial z} = 0, \quad \frac{\partial p}{\partial z} = 0. \tag{5.26}$$

Let an impermeable rigid wall be one of the boundaries. On the rigid wall, we can pose the non-leakage condition $u_n = 0$ and the adhesion (or no-slip) condition $u_t = 0$ or the sliding condition $\partial u_t / \partial n = 0$ for the velocity, where u_n and u_t are the normal and tangential components of the flow velocity, respectively. These conditions must be complemented by the boundary condition for the temperature, which is determined by physical conditions on the rigid wall.

In the QGD system of equations, the vector of the mass flux density is calculated as follows:

$$j_m^{\text{QGD}} = \rho u - \rho w = \rho u - \tau(\text{div}(\rho u \otimes u) + \nabla p). \tag{5.27}$$

For the normal component of the mass flux flowing through the rigid boundary and calculated by formula (5.27) to be equal to zero, it is necessary to complement the non-leakage condition for the velocity by the condition for the pressure of the form $\partial p / \partial n = 0$. This relation is an additional condition necessary for closing the QGD system, by which the statement of the problem in the framework of the QGD system differs from the statement of the same problem in the framework of the Navier–Stokes model.

For definiteness, on the lateral boundary $r = R_c$ of the cylinder (see Fig. 5.7, p. 95), as the boundary conditions, we pose the adhesion condition for the velocity, assign a constant temperature, and assume that the surface is impermeable. Then the boundary conditions are written as follows:

$$u_z = 0, \quad u_r = 0, \quad T = T_0, \quad \frac{\partial p}{\partial r} = 0. \tag{5.28}$$

Let us write the expressions for the heat flux and the friction force on the boundary with the non-leakage condition $u_n = 0$. Let the boundary lie in the plane $z = L_c$ (the butt-end of the cylinder). The friction force on this boundary is determined by the component of the viscous shear-stress tensor Π_{zr}, and the heat flux q_z is given by

$$q_z = q_z^{\text{NS}} - u_z R^q,$$

$$\Pi_{zr} = \Pi_{zr}^{\text{NS}} + \tau u_z \left[\rho u_z \frac{\partial u_r}{\partial z} + \rho u_r \frac{\partial u_r}{\partial r} + \frac{\partial p}{\partial r} \right]. \tag{5.29}$$

Taking into account the non-leakage conditions on the wall ($u_z = 0$), we immediately obtain the expressions for the heat flux and the friction force on the wall in the form

$$q_z = q_z^{\text{NS}} = -\varkappa \frac{\partial T}{\partial z}, \quad \Pi_{zr} = \Pi_{zr}^{\text{NS}} = \mu \frac{\partial u_r}{\partial z}. \tag{5.30}$$

In other words, for the QGD equations, the expressions for the heat flux and the friction force on the rigid wall (5.30) coincide with the corresponding quantities for the Navier–Stokes equations.

5.4 Dimensionless Form of the Equations

For numerically solving the gas-dynamic equations, it is convenient to represent them in the dimensionless form. First, this allows one to operate with quantities of order unity in computations, and, second, this allows one to distinguish dimensionless coefficients in the equations on which the solution of the problem depends, the so-called similarity parameters.

As the main dimensional parameters, we choose the characteristic linear size R_c (for example, the radius of the cylinder), the density ρ_∞, and the sound speed c_∞ of the incident unperturbed flow.

Let us write the relations between the dimensional and dimensionless quantities denoting the dimension-free quantities by the "tilde" sign $\sim$:

$$\rho = \tilde{\rho} \rho_\infty, \quad u = \tilde{u} c_\infty, \quad c = \tilde{c} c_\infty, \quad p = \tilde{p} \rho_\infty c_\infty^2,$$

$$x = \tilde{x} R_c, \quad t = \tilde{t} \frac{R_c}{c_\infty},$$

$$T = \frac{p}{\mathcal{R}\rho} = \frac{\tilde{p} \rho_\infty c_\infty^2}{\mathcal{R} \tilde{\rho} \rho_\infty} = \frac{\tilde{p} \gamma}{\tilde{\rho}} \frac{1}{\mathcal{R}\gamma} c_\infty^2 = \tilde{T} \frac{c_\infty^2}{\gamma \mathcal{R}}. \tag{5.31}$$

Introduce the Mach and Reynolds numbers:

$$\text{Ma}_\infty = \frac{u_\infty}{c_\infty}, \quad \text{Re}_\infty = \frac{\rho_\infty u_\infty R_c}{\mu_\infty}. \tag{5.32}$$

The total energy equation does not change the form

$$E = \frac{\rho u^2}{2} + \frac{p}{(\gamma - 1)} \quad \Rightarrow \quad \tilde{E} = \frac{\tilde{\rho} \tilde{u}^2}{2} + \frac{\tilde{p}}{(\gamma - 1)}. \tag{5.33}$$

In passing to dimensionless quantities, the sound speed transforms as follows:

$$c = \sqrt{\gamma \mathcal{R} T} \quad \Rightarrow \quad \tilde{c} = \sqrt{\tilde{T}}. \tag{5.34}$$

The state equation becomes

$$p = \rho \mathcal{R} T, \quad \tilde{p}\rho_\infty c_\infty^2 = \tilde{\rho}\rho_\infty \tilde{T}\frac{c_\infty^2}{\gamma \mathcal{R}} \quad \Rightarrow \quad \tilde{p} = \frac{\tilde{\rho}\tilde{T}}{\gamma}. \tag{5.35}$$

Therefore, the connection equations (5.34) and (5.35) change their form after passing to dimensionless quantities.

Substituting relations (5.31) in the equations of the QGD system, we can verify that the reduction to the dimensionless form does not change the form of the initial equations. Moreover, the dimensionless viscosity coefficients, heat conduction coefficients, and the parameter τ are calculated as follows:

$$\tilde{\mu} = \frac{\mathrm{Ma}}{\mathrm{Re}}\tilde{T}^\omega, \tag{5.36}$$

$$\tilde{\varkappa} = \frac{1}{(\gamma - 1)\,\mathrm{Pr}}\frac{\mathrm{Ma}}{\mathrm{Re}}\tilde{T}^\omega, \tag{5.37}$$

$$\tilde{\tau} = \frac{\mathrm{Ma}}{\mathrm{Re}\,\mathrm{Sc}}\frac{\tilde{T}^\omega}{\tilde{p}}. \tag{5.38}$$

Furthermore, using the dimensionless QGD equations and the connection equations, we omit the tilde sign $\sim$.

5.5 Finite-Difference Approximation

For definiteness, let us consider the approximation of equations in the cylindrical geometry. Denote by ω_h the set of nodes (i, j) of the spatial grid. For simplicity, we consider the uniform grid in which the spatial steps are denoted by h_z and h_r.

We refer all gas-dynamic quantities (the density ρ, the velocity components u_z and u_r, and the pressure p) to the nodes of the grid ω_h. We calculate the values of the gas-dynamic quantities at the half-integer nodes

$$(z_{i\pm0.5}, r_j), \quad (z_i, r_{j\pm0.5}), \quad (z_{i\pm0.5}, r_{j\pm0.5}) \tag{5.39}$$

as the arithmetical mean of their values at the adjacent nodes, i.e., the value of a function ψ from the set $\{\rho, u_r, u_z, p\}$ at the half-integer nodes is calculated as follows:

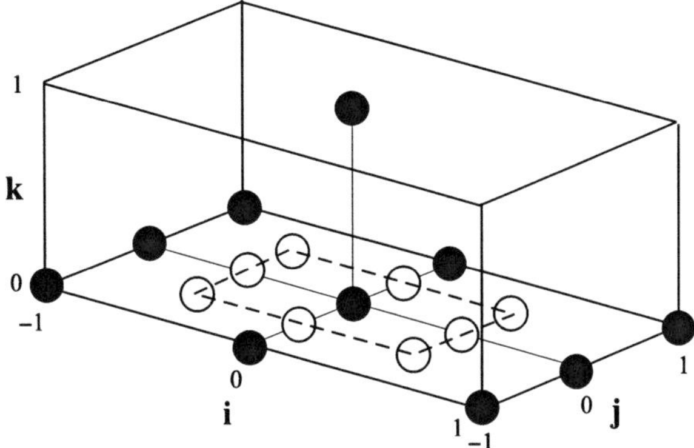

Fig. 5.1 Stencil for space approximating the finite-difference derivatives

$$\psi_{i\pm0.5,j} = 0.5(\psi_{i\pm1,j} + \psi_{i,j}),$$
$$\psi_{i,j\pm0.5} = 0.5(\psi_{i,j\pm1} + \psi_{i,j}), \tag{5.40}$$
$$\psi_{i\pm0.5,j\pm0.5} = 0.25(\psi_{i\pm1,j\pm1} + \psi_{i,j\pm1} + \psi_{i\pm1,j} + \psi_{i,j}).$$

Therefore, to compute the spatial derivatives, we use the nine-point stencil (see Fig. 5.1).

For functions $f = f(\rho, u_z, u_r, p)$ of gas-dynamic parameters, we set

$$f_{i,j} = f(\rho_{i,j}, (u_r)_{i,j}, (u_z)_{i,j}, p_{i,j}),$$
$$f_{i\pm0.5,j} = f(\rho_{i\pm0.5,j}, (u_r)_{i\pm0.5,j}, (u_z)_{i\pm0.5,j}, p_{i\pm0.5,j}), \tag{5.41}$$
$$f_{i,j\pm0.5} = f(\rho_{i,j\pm0.5}, (u_r)_{i,j\pm0.5}, (u_z)_{i,j\pm0.5}, p_{i,j\pm0.5}), \text{ etc.}$$

For numerical solution of system (5.15), (5.16), (5.17), and (5.18), we use the finite-difference scheme explicit in time. The derivatives in time are approximated by forward differences with first-order accuracy. The spatial derivatives are approximated by central differences with second-order accuracy.

Let us approximate the differential equation (5.15) by the finite-difference equation

$$\frac{\widehat{\rho}_{ij} - \rho_{ij}}{\Delta t} + \frac{1}{r_j h_r}\left[(rj_{mr})_{i,j+0.5} - (rj_{mr})_{i,j-0.5}\right]$$
$$+ \frac{1}{h_z}[(j_{mz})_{i+0.5,j} - (j_{mr})_{i-0.5,j}] = 0, \tag{5.42}$$

where Δt is the step in time. The quantity with the superscript $\widehat{\rho}_{i,j}$ is calculated on the next time layer.

The other equations of system (5.15), (5.16), (5.17), and (5.18) are approximated similarly:

$$\frac{(\widehat{\rho u_r})_{ij} - (\rho u_r)_{ij}}{\Delta t} + \frac{1}{r_j h_r}\left[(r j_{mr})_{i,j+0.5}(u_r)_{i,j+0.5} - (r j_{mr})_{i,j-0.5}(u_r)_{i,j-0.5}\right]$$

$$+ \frac{1}{h_z}\left[(j_{mz}u_r)_{i+0.5,j} - (j_{mz}u_r)_{i-0.5,j}\right] + \frac{1}{h_r}\left[p_{i,j+0.5} - p_{i,j-0.5}\right]$$

$$= \frac{1}{r_j h_r}\left[(r\Pi_{rr})_{i,j+0.5} - (r\Pi_{rr})_{i,j-0.5}\right]$$

$$+ \frac{1}{h_z}\left[(\Pi_{zr})_{i+0.5,j} - (\Pi_{zr})_{i-0.5,j}\right] - \frac{(\Pi_{\varphi\varphi})_{ij}}{r_j}, \qquad (5.43)$$

$$\frac{(\widehat{\rho u_z})_{ij} - (\rho u_z)_{ij}}{\Delta t} + \frac{1}{r_j h_r}\left[(r j_{mr})_{i,j+0.5}(u_z)_{i,j+0.5} - (r j_{mr})_{i,j-0.5}(u_z)_{i,j-0.5}\right]$$

$$+ \frac{1}{h_z}\left[(j_{mz}u_z)_{i+0.5,j} - (j_{mz}u_z)_{i-0.5,j}\right] + \frac{1}{h_z}\left[p_{i+0.5,j} - p_{i-0.5,j}\right]$$

$$= \frac{1}{r_j h_r}\left[r_{j+0.5}(\Pi_{rz})_{i,j+0.5} - r_{j-0.5}(\Pi_{rz})_{i,j-0.5}\right]$$

$$+ \frac{1}{h_z}\left[(\Pi_{zz})_{i+0.5,j} - (\Pi_{zz})_{i-0.5,j}\right], \qquad (5.44)$$

$$\frac{\widehat{E}_{ij} - E_{ij}}{\Delta t} + \frac{1}{r_j h_r}\left[(r j_{mr})_{i,j+0.5}H_{i,j+0.5} - (r j_{mr})_{i,j-0.5}H_{i,j-0.5}\right]$$

$$+ \frac{1}{h_z}\left[(j_{mz}H)_{i+0.5,j} - (j_{mz}H)_{i-0.5,j}\right]$$

$$+ \frac{1}{r_j h_r}\left[r_{j+0.5}(q_r)_{i,j+0.5} - r_{j-0.5}(q_r)_{i,j-0.5}\right]$$

$$+ \frac{1}{h_z}\left[(q_z)_{i+0.5,j} - (q_z)_{i-0.5,j}\right]$$

$$= \frac{1}{r_j h_r}\left[(r\Pi_{rr})_{i,j+0.5}(u_r)_{i,j+0.5} - (r\Pi_{rr})_{i,j-0.5}(u_r)_{i,j-0.5}\right]$$

$$+ \frac{1}{r_j h_r}\left[r_{j+0.5}(\Pi_{rz}u_z)_{i,j+0.5} - r_{j-0.5}(\Pi_{rz}u_z)_{i,j-0.5}\right]$$

$$+ \frac{1}{h_z}\left[(\Pi_{zr}u_r)_{i+0.5,j} - (\Pi_{zr}u_r)_{i-0.5,j}\right]$$

$$+ \frac{1}{h_z}\left[(\Pi_{zz}u_z)_{i+0.5,j} - (\Pi_{zz}u_z)_{i-0.5,j}\right], \qquad (z_i, r_j) \in \omega_h. \, (5.45)$$

Further, we construct the finite-difference approximations directly for the components of the mass flux density vector j_m, for those of the heat flux q, and for the components of the viscous shear-stress tensor Π written in the form (5.20), (5.21), (5.22), and (5.23). These quantities are approximated at half-integer points. For example, the quantities in Eq. (5.42) are approximated as follows:

$$(r j_{mr})_{i,j\pm0.5} = \rho_{i,j\pm0.5}[r_{j\pm0.5}(u_r)_{i,j\pm0.5} - (r w_r)_{i,j\pm0.5}],$$
$$(j_{mz})_{i\pm0.5,j} = \rho_{i\pm0.5,j}[(u_z)_{i\pm0.5,j} - (w_z)_{i\pm0.5,j}].$$

We construct the finite-difference approximations of the other terms of the equations similarly.

The numerical solution of the system of finite-difference equations (5.42), (5.43), (5.44), and (5.45) on each time level is sequentially performed as follows. By the gas-dynamic parameters ρ, u_z, u_r, and p known from the previous time level, we compute the values of quantities at the half-integer points (5.40). Then we calculate the values of the fluxes j_m, q, and Π at the corresponding points; after that, these values are substituted in the finite-difference expressions (5.42), (5.43), (5.44), and (5.45).

We add the initial and boundary conditions to the system of finite-difference equations (5.42), (5.43), (5.44), and (5.45).

For the uniform computation of the gas-dynamic quantities at all interior points of the computation domain, including boundary nodes, we introduce a system of fictitious cells adjacent to each of the boundaries. In this case, the boundary of the domain lies at the half-integer points (see Fig. 5.8, p. 97). The values of the density, velocity components, and pressure in fictitious cells are given in a way that allows us to approximate the corresponding boundary condition.

For example, on the boundary lying at the point $i = 1/2$, let the value of a function f_w be given. The point $i = 0$ is fictitious, and the point $i = 1$ is the nearest interior point adjacent to the boundary, the value at which is calculated in each time layer. Then f_0 at the fictitious point is chosen from the condition

$$f_w = \frac{f_0 + f_1}{2} \quad \rightarrow \quad f_0 = 2f_w - f_1.$$

If the condition for the derivative of the form $\partial f/\partial n = 0$ is given on the boundary, then f_0 at the fictitious point is given by

$$\frac{f_0 - f_1}{h} = 0 \quad \rightarrow \quad f_0 = f_1.$$

Therefore, the algorithm for finding the density, velocity components, and pressure in each time level consists of two stages. First, on the basis of already calculated values, we fill in the fictitious cells according to the rule indicated above, and then we calculate the values of $\widehat{\rho}_{ij}$, $(\widehat{u}_r)_{ij}$, $(\widehat{u}_z)_{ij}$, and $\widehat{E}_{ij}$ at all interior points in the next time layer.

The stationary solution is found by using the ascertainment method, as it is considered as attained under the fulfillment of the following criterion:

$$\frac{1}{N_r N_z} \sum_{(z_i,r_j)\in\omega_h} \left| \frac{\widehat{\rho}_{ij} - \rho_{ij}}{\rho_{ij}\,\Delta t} \right| \le \epsilon, \tag{5.46}$$

in which the discrepancy ϵ can be varied depending on the requirements for the accuracy of solutions. In the computations presented below, the value of ϵ is chosen in the range from 10^{-3} up to 10^{-7}. Here, N_r and N_z are the numbers of nodes of the grid in r and z, respectively.

Other criteria for ascertainment of the stationary solution are possible, for example,

$$\max_{(z_i,r_j)\in\omega_h} \left| \frac{\widehat{\rho}_{ij} - \rho_{ij}}{\Delta t} \right| \le \epsilon.$$

5.6 Introducing the Artificial Dissipation

In numerical modelling of the gas-dynamic flows, various singularities of solution arise, for example, solution discontinuities, the shock waves and contact discontinuities, boundary layers, and other zones with large gradients of parameters. In these cases, it is impossible to directly perform the computation according to finite-difference schemes obtained by central finite-difference approximations of the initial equations.

A known method of computation of a shock wave without explicit distinguishing of its front on the grid is the method of the smearing of the front that accounts for introducing in the system of difference equations some dissipative terms called the pseudo-viscosity or the artificial viscosity. Such approaches to solving the gas-dynamic problems are elaborated in detail and are presented, e.g., in [7, 131, 163, 167, 171].

The additional terms introduced model the action of the real viscosity, i.e., they lead to the dissipation of the kinetic energy. For example, the following additions to the pressure are well known:

$$p \to p + \omega,$$

where ω is interpreted as the artificial viscosity. Most often, one considers the linear artificial viscosity

$$\omega = -\nu\rho\frac{\partial u}{\partial x}$$

or the quadratic artificial viscosity

$$\omega = -\nu\rho\left(\frac{\partial u}{\partial x}\right)^2.$$

Here, ν is a certain coefficient proportional to the step h of the spatial grid. In solving gas-dynamic problems for a polytropic gas whose heat conductivity is equal to zero, one uses the Neumann–Richtmyer viscosity (see, e.g., [167]), where

$$\nu = h\rho \left| \frac{\partial u}{\partial x} \right|.$$

Therefore, the artificial viscosity is a certain regularizer that allows one to perform the uniform computation of shock waves or other solution singularities without their explicit isolation on the grid. In this case, by the front domain one means the zone of sharp variation of the flow parameters. In computing of shock waves, the width of the shock profile stipulated by the action of the pseudo-viscosity is proportional to h and has no relation to the real width of the wave front which is equal to several mean free paths of molecules (see Appendix C).

In algorithms based on the QGD equations, as the regularizer, one takes terms proportional to the parameter τ.

To ensure the stability of the QGD algorithm, we introduce an additional term proportional to the step of the spatial grid in formula (5.38) for calculation of the relaxation parameter τ. This term determines an additional artificial dissipation. The dimensionless coefficient τ is calculated as follows:

$$\tau = \frac{\mathrm{Ma}}{\mathrm{Re}} \frac{T^\omega}{p\,\mathrm{Sc}} + \alpha \frac{h}{c}, \tag{5.47}$$

where α is a numerical coefficient of the order of unity, which is found by fitting. For the finite-difference scheme described above, $h = \sqrt{h_z^2 + h_r^2}$. The dynamical viscosity coefficient μ and the heat conductivity coefficient $\varkappa$ are calculated through the relaxation parameter τ by the formula

$$\mu = \tau p\,\mathrm{Sc}, \quad \varkappa = \tau \frac{p\,\mathrm{Sc}}{\mathrm{Pr}(\gamma - 1)}. \tag{5.48}$$

Therefore, the stabilizing term $\alpha h/c$ is included in μ and $\varkappa$ and hence in the expressions for the heat flux and the friction force on the boundary (5.30).

Starting from formula (5.47), we see that the grid viscosity can be formally considered as small if

$$\alpha h < \frac{\mathrm{Ma}}{\mathrm{Re}} \frac{T^{\omega+1/2}}{p}. \tag{5.49}$$

The introduction of the artificial dissipation proportional to the grid step h makes the resulting finite-difference scheme the first-order accuracy scheme with respect to the space.

Therefore, the finite-difference schemes based on the QGD equations of the form (5.42), (5.43), (5.44), and (5.45) approximate the initial-boundary-value problem with the first-order accuracy in time and space $O(\alpha h + \Delta t)$. The scheme is explicit, uniform, and conservative. The additional terms proportional to $\alpha h/c$ are interpreted as artificial regularizers.

5.7 Problem on the Strong Discontinuity Step Evolution

As an example of the use of the algorithm constructed above, we consider the one-dimensional, strong discontinuity step evolution problem in the approximation of a nonviscous, non-heat-conducting gas. In other words, we solve the problem in the framework of the Euler equation. This problem is a known test for estimating the stability and the accuracy of numerical algorithms, and various approaches to the consideration of this flow are presented, in particular, in [103, 167, 198].

In this case, the parameter τ in Eq. (5.47) determines the value of the artificial dissipation and is calculated as follows:

$$\tau = \alpha \frac{h}{c}. \tag{5.50}$$

The QGD equations for the one-dimensional planar flow have the form

$$\frac{\partial \rho}{\partial t} + \frac{\partial j_m}{\partial x} = 0, \tag{5.51}$$

$$\frac{\partial (\rho u)}{\partial t} + \frac{\partial (j_m u)}{\partial x} + \frac{\partial p}{\partial x} = \frac{\partial \Pi_{xx}}{\partial x}, \tag{5.52}$$

$$\frac{\partial E}{\partial t} + \frac{\partial (j_m H)}{\partial x} + \frac{\partial q}{\partial x} = \frac{\partial (\Pi_{xx} u)}{\partial x}. \tag{5.53}$$

Here, E and H are the total energy of the volume unit and the specific enthalpy, respectively, which are calculated by the formulas

$$E = \frac{1}{2}\rho u^2 + \frac{p}{\gamma - 1}, \quad H = \frac{E + p}{\rho}.$$

The vector of the mass flux density is calculated as follows:

$$j_m = \rho(u - w),$$

where

$$w = \frac{\tau}{\rho}\frac{\partial}{\partial x}(\rho u^2 + p).$$

The component of the viscous shear-stress tensor entering system (5.51), (5.52), and (5.53) is

$$\Pi_{xx} = \frac{4}{3}\mu \frac{\partial u}{\partial x} + u\tau \left(\rho u \frac{\partial u}{\partial x} + \frac{\partial p}{\partial x} \right) + \tau \left(u \frac{\partial p}{\partial x} + \gamma p \frac{\partial u}{\partial x} \right).$$

The heat flux vector q is given by

$$q = -\varkappa \frac{\partial T}{\partial x} - \tau \rho u \left[\frac{u}{\gamma - 1} \frac{\partial}{\partial x} \left(\frac{p}{\rho} \right) + pu \frac{\partial}{\partial x} \left(\frac{1}{\rho} \right) \right].$$

The passage to dimensionless quantities does not change the form of the equations. The relaxation parameter and the coefficients of viscosity and heat conduction in the dimensionless form are calculated as follows:

$$\tau = \alpha \frac{h}{c}, \quad \mu = \tau p \,\mathrm{Sc}, \quad \varkappa = \frac{\tau p \,\mathrm{Sc}}{\Pr(\gamma - 1)}.$$

Introduce the uniform grid in the coordinate x with step h, node coordinates x_i, and index values $i = 0, \ldots, N_x - 1$ and also the grid in time with step Δt. The values of all gas-dynamic quantities—the velocity, the density, and the pressure—are found at the nodes of the grid. The values of fluxes are found at the half-integer nodes. To solve problem (5.51), (5.52), and (5.53), we use the following finite-difference scheme explicit in time:

$$\widehat{\rho_i} = \rho_i - \frac{\Delta t}{h} \left(j_{m,i+1/2} - j_{m,i-1/2} \right),$$

$$\widehat{\rho_i u_i} = \rho_i u_i + \frac{\Delta t}{h} \Big[\left(\Pi_{xx\,i+1/2} - \Pi_{xx\,i-1/2} \right) - \left(p_{i+1/2} - p_{i-1/2} \right)$$
$$- \left(j_{m,i+1/2} u_{i+1/2} - j_{m,i-1/2} u_{i-1/2} \right) \Big],$$

$$\widehat{E_i} = E_i + \frac{\Delta t}{h} \Big[\left(\Pi_{xx\,i+1/2} u_{i+1/2} - \Pi_{xx\,i-1/2} u_{i-1/2} \right)$$
$$- \left(q_{i+1/2} - q_{i-1/2} \right) - \left(\frac{j_{m,i+1/2}}{\rho_{i+1/2}} \left(E_{i+1/2} + p_{i+1/2} \right) \right.$$
$$\left. - \frac{j_{m,i-1/2}}{\rho_{i-1/2}} \left(E_{i-1/2} + p_{i-1/2} \right) \right) \Big],$$

$$p_i = (\gamma - 1) \left(E_i - \frac{\rho_i u_i^2}{2} \right).$$

The discrete analog of the mass flux j_m has the form

$$j_{m,i+1/2} = \rho_{i+1/2}(u_{i+1/2} - w_{i+1/2}),$$

$$w_{i+1/2} = \frac{\tau_{i+1/2}}{\rho_{i+1/2}} \frac{1}{h} \left(\rho_{i+1} u_{i+1}^2 + p_{i+1} - \rho_i u_i^2 - p_i \right).$$

The discrete expressions for Π_{xx} and q are written similarly.

Note that the finite-difference equations for the density coincide with the corresponding equation in the Lax–Wendroff scheme (see, e.g., [7]).

The problem is solved on the closed interval $0 \leq x \leq 200$. The initial conditions are a discontinuity at the point $x = 100$. The values of the gas-dynamic parameters to the left and to the right from the discontinuity are as follows:

$$\rho(x, 0) = \begin{cases} 8, & x \leq 99, \\ 1, & x > 99, \end{cases} \quad p(x, 0) = \begin{cases} 480, & x \leq 99, \\ 1, & x > 99, \end{cases} \quad u(x, 0) = 0.$$

In spite of the simplicity of the statement, this problem reflects the main peculiarities of nonstationary gas-dynamic flows, since in the flow domain, a shock wave, a contact discontinuity, and a rarefaction wave are formed, and it is often used for debugging the numerical algorithm for computing supersonic flows.

The computations were carried out for $\gamma = 5/3$, $\mathrm{Pr} = 2/3$, and $\mathrm{Sc} = 1$. The parameters of the above computations are systematized in Table 5.1, where we indicate the number of the computation variant, the value of the spatial step h, the number N_x of grid points, the step in time Δt, and the number N_t of step in time necessary for attaining the dimensionless time $t_0 = 4$. This instant of time corresponds to a number of tests presented, e.g., in [56, 103, 213].

The results of computations up to the instant of time $t_0 = 4$ are presented in Figs. 5.2, 5.3, 5.4, 5.5, and 5.6. In Figs. 5.2, 5.3, and 5.4, we show the distributions of density, pressure, and velocity carried out on condensing spatial grids. In the same figures, we present the self-similar solution of the problem according to [167]. We present the distribution of gas-dynamic quantities in the whole computational domain and also represent the fragments of solution in the shock wave zone. The dotted lines correspond to variants from 1 to 6 in accordance with the table. To variant 1, we put in correspondence the dotted line which smoothens the solution in the highest degree. From the presented pictures, the convergence of the numerical solution to the self-similar solution as the spatial grid condenses is clearly seen.

In Figs. 5.5 and 5.6, we show the influence of the choice of the coefficient α determining the value of the regularization parameter on the accuracy of the numerical solutions for $h = 0.03125$. It is seen that when α decreases, the accuracy of the numerical solution increases. However, for very small values $\alpha = 0.02$ and 0.1, in the shock wave front, oscillations arise whose values unboundedly increase when

Table 5.1 Variants of computing the discontinuity step evolution problem, $\alpha = 0.5$

No.	h	N_x	Δt	N_t	t_0
1	1	200	2×10^{-3}	2×10^3	4
2	0.5	400	2×10^{-3}	2×10^3	4
3	0.25	800	2×10^{-3}	2×10^3	4
4	0.125	1600	2×10^{-4}	2×10^4	4
5	0.0625	3200	2×10^{-4}	2×10^4	4
6	0.03125	6400	2×10^{-4}	2×10^4	4

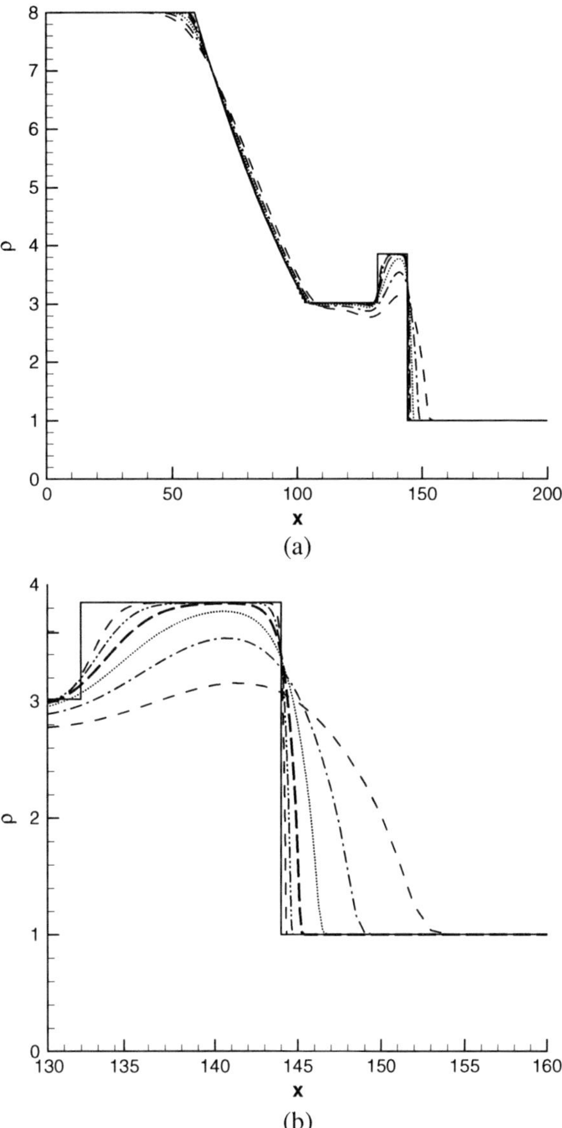

Fig. 5.2 Distribution of the density along the axis (in the enlarged scale in (**b**))

α decreases. In a number of cases, the suppression of this instability is possible by diminishing the step in time Δt.

To improve the influence of the coefficient α, in Fig. 5.6, we depict a fragment of the graph of the density depending on the quantity α for $\alpha = 0.5, 0.4, 0.3$, and 0.2 for $h = 0.03125$ ($h/c = 0.003125$). We see the rapid convergence of solution when the coefficient α decreases.

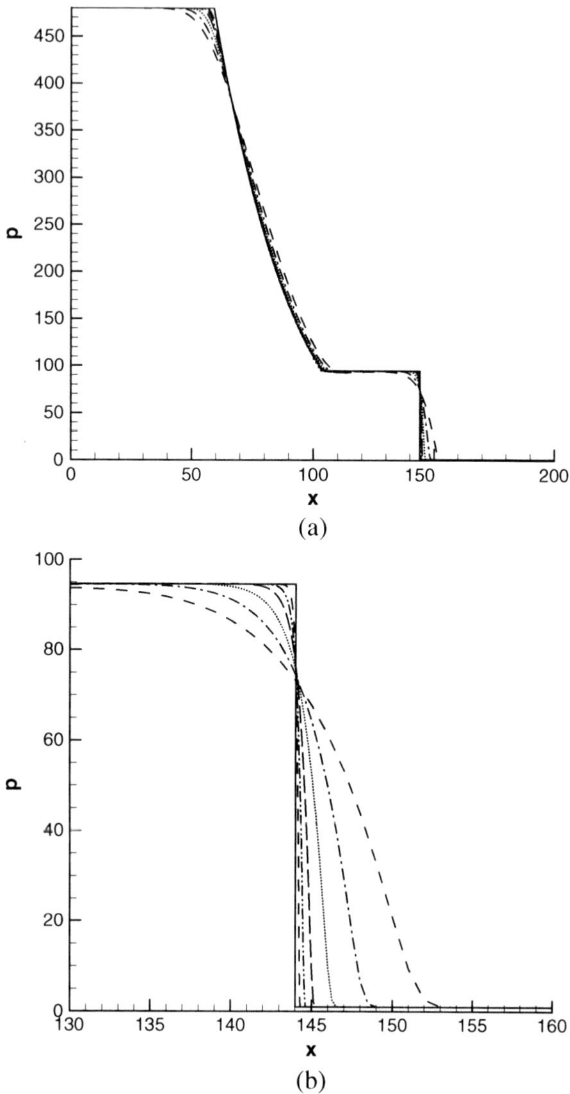

Fig. 5.3 Distribution of the pressure along the axis (in the enlarged scale in (**b**))

For $\alpha = 0.2$, the shock wave smears approximately by four steps of the grid, which corresponds to the uniform calculation algorithms of the advanced order of accuracy for such problems. However, for $\alpha = 0.1$, the diminishing of the step in time even by 100 times leads to an oscillating solution. The optimal value of the regularization parameter ensuring the maximal step in time and the best accuracy in this problem is attained for $\alpha = 0.2$–0.3 (see Sect. 5.11).

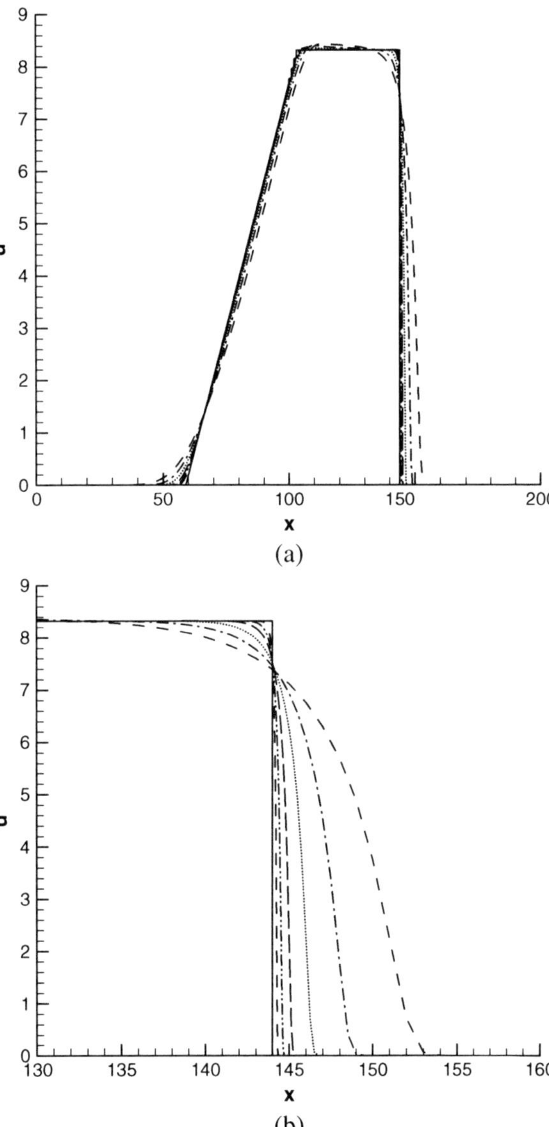

Fig. 5.4 Distribution of the velocity along the axis (in the enlarged scale in (**b**))

(a)

(b)

5.8 Flow Around a Cylindrical Obstacle

5.8.1 Problem Formulation and the Numerical Algorithm

As an example of the implementation of the two-dimensional finite-difference algorithm, we present the results of computing the problems on the flow in a neighborhood of a cylindrical butt-end of radius R_c, which is placed in the homogeneous

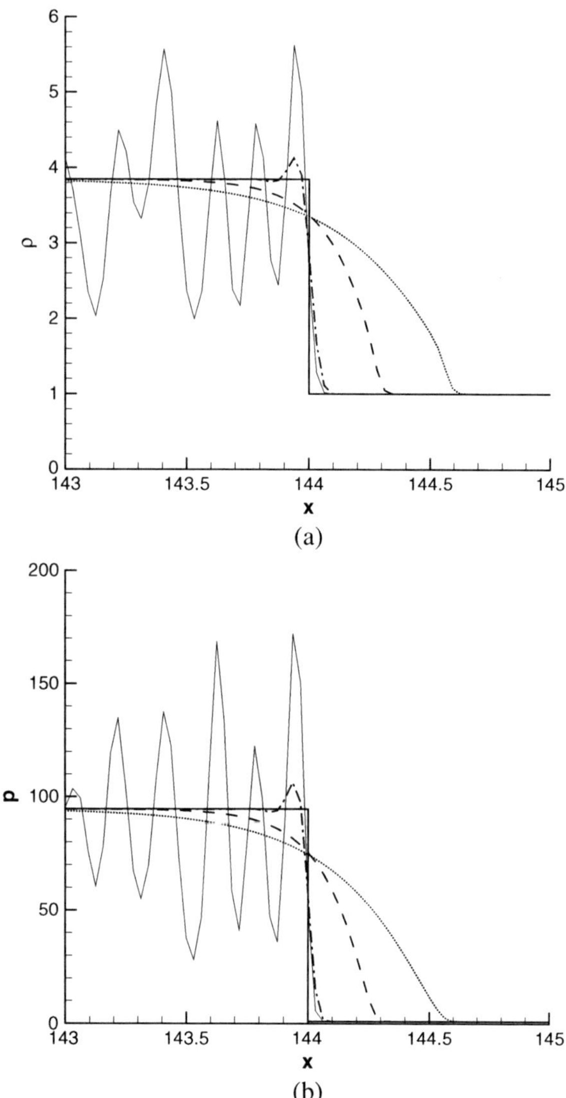

Fig. 5.5 Graphs of the density (**a**) and the pressure (**b**) for $\alpha = 0.02$ (*thin line*), $\alpha = 0.1$ (*dash-dotted line*), $\alpha = 0.5$ (*dotted line*), and $\alpha = 1$ (*continuous line*)

supersonic flow of a viscous, compressible, heat-conducting gas parallel to the axis of the cylinder. The computation domain for this problem is presented in Fig. 5.7.

Introduce the cylindrical coordinate system directing the axis z along the axis of the cylinder and the axis r along the surface of the butt-end (Fig. 5.7). The size of the computational domain is $L \times H$ and the size of the cylinder is $L_c \times R_c$, where R_c is the radius of the cylinder.

We consider the flow of a one-atom rigid sphere gas with the parameters $\gamma = 5/3$, Pr $= 2/3$, Re $= 1000$, Sc $= 0.77$, and $\omega = 0.5$. As the initial conditions for

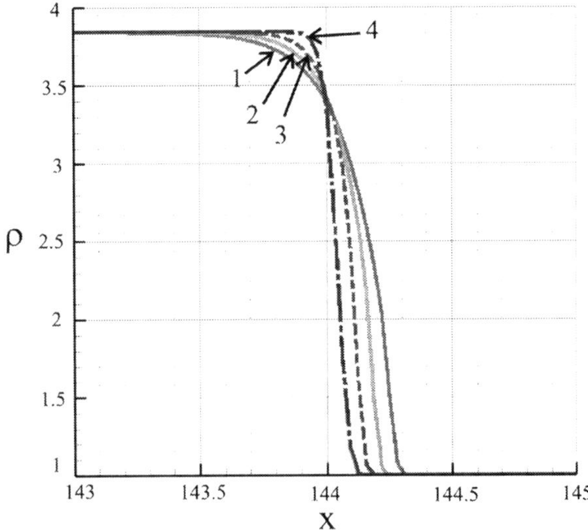

Fig. 5.6 Graphs of the density. *Line 1* corresponds to $\alpha = 0.5$, *line 2* corresponds to $\alpha = 0.4$, *line 3* corresponds to $\alpha = 0.3$, and *line 4* corresponds to $\alpha = 0.2$

system (5.15), (5.16), (5.17), and (5.18), we use the parameters of the incident flow:

$$\rho|_{t=0} = \rho_\infty, \quad u_z|_{t=0} = -U_\infty, \quad u_r|_{t=0} = 0, \quad p|_{t=0} = p_\infty. \tag{5.54}$$

After introducing the dimensionless quantities, these parameters are equal to $\rho_\infty = 1$, $U_\infty = -\,\mathrm{Ma}$, and $p_\infty = 1/\gamma$.

On the right input boundary ($z = L, 0 < r < H$), we give the parameters of the incident flow (5.54):

$$\rho = 1, \quad u_z = -\,\mathrm{Ma}, \quad u_r = 0, \quad p = \frac{1}{\gamma}; \tag{5.55}$$

on the symmetry axis ($L_c < z < L, r = 0$), we pose the "symmetry conditions":

$$\frac{\partial \rho}{\partial r} = 0, \quad \frac{\partial u_z}{\partial r} = 0, \quad u_r = 0, \quad \frac{\partial p}{\partial r} = 0; \tag{5.56}$$

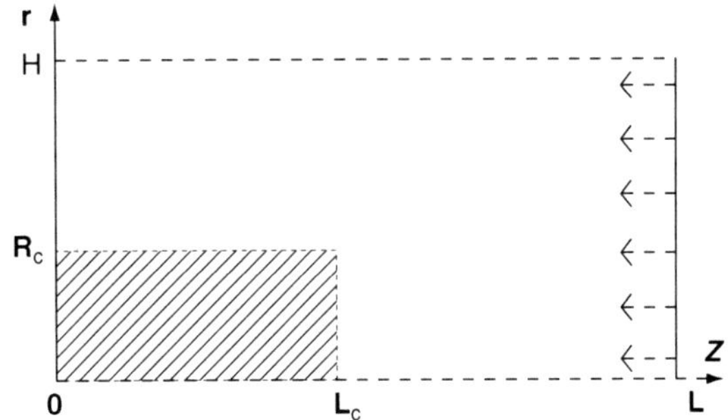

Fig. 5.7 Scheme of the computational domain in the problem of flow past cylindrical butt-end

on the butt-end of the cylinder ($z = L_c, 0 < r < R_c$), we pose the "slip conditions" for the velocity:

$$\frac{\partial \rho}{\partial z} = 0, \quad u_z = 0, \quad \frac{\partial u_r}{\partial z} = 0, \quad \frac{\partial p}{\partial z} = 0; \tag{5.57}$$

and on the lateral surface of the cylinder ($0 < z < L_c, r = R_c$), we pose the "adhesion conditions" for the velocity:

$$\frac{\partial \rho}{\partial r} = 0, \quad u_z = 0, \quad u_r = 0, \quad \frac{\partial p}{\partial r} = 0. \tag{5.58}$$

The surface of the cylinder is assumed to be adiabatic.

On the left output and upper free boundaries, we pose the so-called "soft" boundary conditions, i.e., we assume that the normal derivatives of the density, pressure, and velocity components vanish.

The output boundary ($z = 0, R_c < r < H$):

$$\frac{\partial \rho}{\partial z} = 0, \quad \frac{\partial u_z}{\partial z} = 0, \quad \frac{\partial u_r}{\partial z} = 0, \quad \frac{\partial p}{\partial z} = 0, \tag{5.59}$$

and the free boundary ($0 < z < L, r = H$):

$$\frac{\partial \rho}{\partial r} = 0, \quad \frac{\partial u_z}{\partial r} = 0, \quad \frac{\partial u_r}{\partial r} = 0, \quad \frac{\partial p}{\partial r} = 0. \tag{5.60}$$

When approximating the boundary conditions by using fictitious nodes, the cell corresponding to the corner point with coordinates (L_c, R_c) turns out to be not unique (Fig. 5.7). Precisely, in calculating the values of functions at the fictitious node ($i - 1, j - 1$), it is required to use the value of the function at the node ($i - 1, j$) in one case, and in the other case, it is required to use the value of the function at the computational node ($i, j - 1$) on the previous time layer (Fig. 5.8).

In the computations for solving this problem presented below, the numerical algorithm was modified in the following way. We partition the computational domain $L \times H$ into two subdomains Π_1 and Π_2,

$$\Pi_1 = \{(z, r) : \ L_c < z < L, \ 0 < r < H\},$$
$$\Pi_2 = \{(z, r) : \ 0 < z < L_c, \ R_c < r < H\},$$

and seek for the numerical solution in the domain $G = \Pi_1 \cup \Pi_2$.

The algorithm for finding basic gas-dynamic quantities ρ, u_z, u_r, and p on each time layer consists of two stages. At the first stage, we fill in all fictitious cells by using the boundary conditions. At the second stage, we compute the flow parameters $\widehat{\rho}, \widehat{u}_z, \widehat{u}_r$, and $\widehat{p}$ on the next time layer at all interior points of the domain by using the same formulas. The system of equations on each time layer is solved by

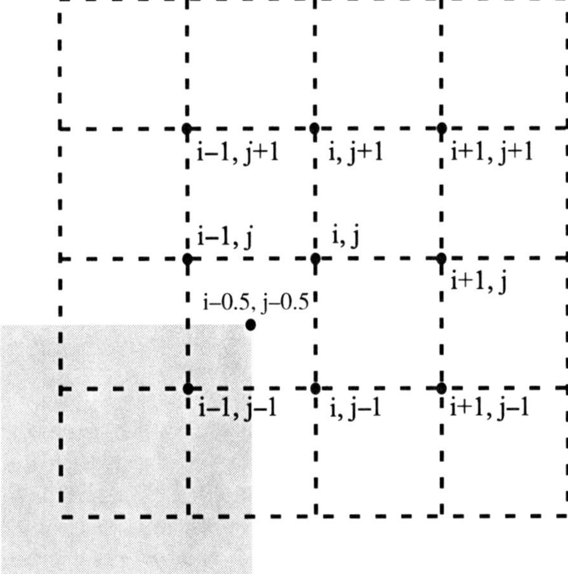

Fig. 5.8 Structure of the grid in a neighborhood of a corner point

turns: first, in the right domain Π_1 and then in the left domain Π_2 (for this problem, when flow comes from the right). After that, we pass to the next step in time.

5.8.2 Results of Computation

In this problem, the Reynolds number is chosen to be sufficiently large, and therefore, the accuracy of computations presented here can be estimated by comparing the gas parameters on the butt-end of the cylinder with the stagnation parameters.

The stagnation parameters are the values of the density ρ_s, the pressure p_s, and the temperature T_s at a drag point calculated by using the Bernoulli theorem (see, e.g., [134]):

$$
p_s = p_2 \left(1 + \frac{\gamma - 1}{2} \frac{u_{z2}^2}{c_2^2} \right)^{\gamma/(\gamma-1)} ,
$$

$$
T_s = T_2 \left(1 + \frac{\gamma - 1}{2} \frac{u_{z2}^2}{c_2^2} \right) ,
$$

$$
\rho_s = \frac{\gamma p_s}{T_s} ,
$$

(5.61)

where p_2, u_{z2}, T_2, and c_2 are the pressure, the velocity, the temperature, and the sound speed after the shock wave, respectively, which are obtained from the Renkin–Hugoniot conditions (see [134]):

$$\rho_2 = \rho_1 \frac{(\gamma + 1)\,\mathrm{Ma}_1^2}{2 + (\gamma - 1)\,\mathrm{Ma}_1^2},$$

$$p_2 = p_1 \frac{2\gamma\,\mathrm{Ma}_1^2 - (\gamma - 1)}{\gamma + 1}, \tag{5.62}$$

$$u_{z2} = u_{z1} \frac{2 + (\gamma - 1)\,\mathrm{Ma}_1^2}{(\gamma + 1)\,\mathrm{Ma}_1^2}, \quad T_2 = \gamma \frac{p_2}{\rho_2}, \quad c_2 = \sqrt{T_2},$$

where ρ_1, u_{z1}, p_1, and T_1 are the known parameters of the incident flow. According to Eq. (5.62), the maximum value of the density in the shock wave is defined as $\rho_{\max} = (\gamma + 1)/(\gamma - 1)$ and for $\gamma = 5/3$, it is equal to $\rho_{\max} = 4$.

The distance S between the shock wave and the butt-end of the cylinder can be estimated on the basis of Lunev's approximative formula (see [22]):

$$S = R_c \sqrt{k}(1 + 0.6\,k), \quad k = \frac{\rho_1}{\rho_2}, \tag{5.63}$$

where $\rho_1 = \rho_\infty$ is the density on the input boundary and ρ_2 is the density after the shock wave calculated by the Renkin-Hugoniot conditions (5.62). The values of S thus computed are presented in Table 5.2. The position of the shock wave in the computation can be found by the position of the sound line, i.e., the line on which the local Mach number is $\mathrm{Ma} = 1$.

The computations were carried out on uniform spatial grids with steps $h_r = h_z = 0.05$ and 0.025 for the Mach numbers $\mathrm{Ma} = 1.5, 2, 3, 5$, and 50 and various values of the parameter α. In Table 5.3, we present the characteristics of computations and the comparison of stagnation parameters obtained in computations with the theoretical values (they are distinguished by boldface).

For all Mach numbers, including $\mathrm{Ma} = 50$, the values of the stagnation parameters obtained in the computation correspond to the theoretical values calculated by formulas (5.62).

The distribution of the density and the streamlines for the stationary flow for the Mach numbers 1.5, 2, 3, 5, and 50 ($\alpha = 0.5$) are presented in Fig. 5.9. In the same figure, we show the change of the form and position of the shock wave and the flow structure with the growth of the incident flow velocity (Mach number). The values

Table 5.2 Position of the shock wave

Ma	1.5	2	3	5	50
S	1.03	0.84	0.69	0.62	0.58

Table 5.3 Computation of the supersonic axially symmetric flow

Ma	α	Grid	$h_r = h_z$	Δt	Discrepancy ε	Number of steps	Stagnation parameters ρ_s	p_s	T_s
Theoretical values for Ma $= 1.5^\dagger$							**2.172**	**2.280**	**1.750**
1.5	**1.0**	120 × 120	0.05	10^{-3}	10^{-5}	28 034	2.112	2.248	1.774
1.5	**0.5**	120 × 120	0.05	10^{-3}	10^{-5}	27 363	2.156	2.280	1.763
1.5	**0.5**	160 × 120	0.025	$5 \cdot 10^{-4}$	10^{-4}	126 000	2.186	2.305	1.757
1.5	**0.2**	120 × 120	0.05	10^{-3}	10^{-5}	27 637	2.189	2.303	1.753
Theoretical values for Ma $= 2^\dagger$							**2.719**	**3.807**	**2.333**
2	**1.0**	80 × 60	0.05	10^{-3}	10^{-5}	16 255	2.648	3.704	2.331
2	**0.5**	80 × 60	0.05	10^{-3}	10^{-5}	15 473	2.720	3.781	2.317
2	**0.2**	80 × 60	0.05	10^{-3}	10^{-5}	15 961	2.781	3.843	2.303
2	**1.0**	160 × 120	0.025	$5 \cdot 10^{-4}$	10^{-5}	30 903	2.717	3.777	2.317
2	**0.5**	160 × 120	0.025	$5 \cdot 10^{-4}$	10^{-5}	30 774	2.760	3.818	2.306
2	**0.2**	160 × 120	0.025	$5 \cdot 10^{-4}$	10^{-5}	31 494	2.791	3.848	2.298
Theoretical values for Ma $= 3^\dagger$							**3.418**	**8.204**	**4.000**
3	**1.0**	160 × 120	0.025	10^{-4}	10^{-5}	90 107	3.457	8.089	3.900
3	**0.5**	160 × 120	0.025	10^{-4}	10^{-5}	91 504	3.525	8.204	3.879
Theoretical values for Ma $= 5^\dagger$							**3.982**	**22.330**	**9.333**
5	**1.0**	160 × 120	0.025	10^{-4}	10^{-5}	57 436	4.076	21.914	8.961
5	**0.5**	160 × 120	0.025	10^{-4}	10^{-5}	57 505	4.167	22.287	8.914
Theoretical values for Ma $= 50^\dagger$							**4.402**	**2203.6**	**834.33**
50	**1.0**	80 × 60	0.05	10^{-6}	10^{-4}	125 000	4.756	2181.1	764.35
50	**0.5**	80 × 60	0.05	$5 \cdot 10^{-6}$	10^{-4}	119 637	4.739	2211.8	777.87

† Theoretical values are calculated by formulas (5.62).

of position of shock waves obtained in computation qualitatively coincide with the theoretical values from Table 5.2.

On the presented graphs, we see a high accuracy of computing the shock waves and the absence of oscillations of solution for large Mach numbers Ma. The optimal value of the numerical coefficient from the viewpoint of accuracy and efficiency is $\alpha = 0.5$. In these calculations, the rate of convergence is practically independent of the parameter α. For large Mach numbers, the condensation of the spatial grid leads to the increase of accuracy of the numerical solution.

5.9 Nonviscid Flow in a Channel with a Forward-Facing Step

Let us present an example of a nonviscous supersonic flow in a planar channel with a ledge. A complicated configuration of shock waves serves as a known test for estimating the validity of higher-order accuracy methods for solving the Euler and Navier–Stokes equations (see, e.g., [101, 208]).

The problem is solved in the dimensionless form in the following statement: the length of the channel is 3, its width is 1, the height of the ledge is 0.2, and its length is 2.4. We consider the flow of a nonviscous, non-heat-conducting gas $(1/\mathrm{Re} = 0)$

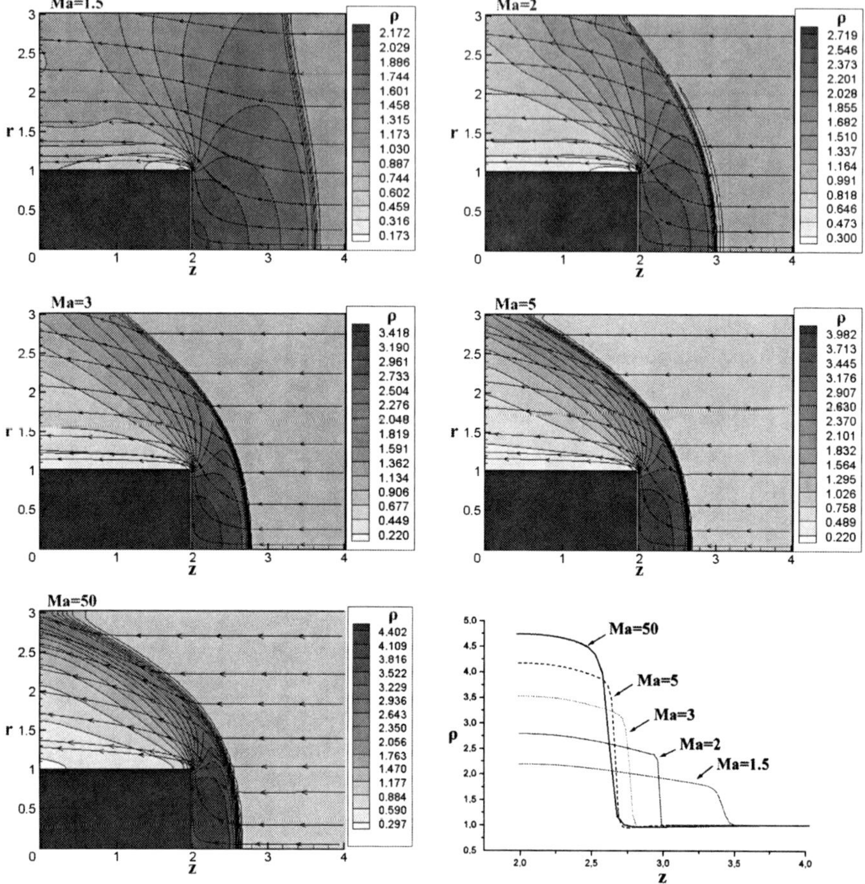

Fig. 5.9 Distribution of the density in the axially symmetric flow

with specific heat ratio $\gamma = 1.4$ and Ma $= 3$. The gas enters from the right (see Fig. 5.10).

The flow is described by the QGD equations (5.1), (5.2), (5.3), and (5.4) written in the Cartesian coordinate system. As the initial conditions, we use the parameters of the incident flow.

The boundary conditions are given as follows. On the input boundary, the values of gas-dynamic parameters are assumed to be equal to the values of the incident flow, i.e., $\rho = 1$, $u_x = -\,\text{Ma}$, $u_y = 0$, and $p = 1/\gamma$.

On the output boundary, we pose the "soft" boundary conditions $\partial f/\partial x = 0$, where $f = (\rho, p, u_x, u_y)$. On the rigid walls of the channel and ledge, we pose the "symmetry" boundary conditions:

$$\frac{\partial p}{\partial n} = 0, \quad \frac{\partial \rho}{\partial n} = 0, \quad \frac{\partial u_s}{\partial n} = 0, \quad u_n = 0,$$

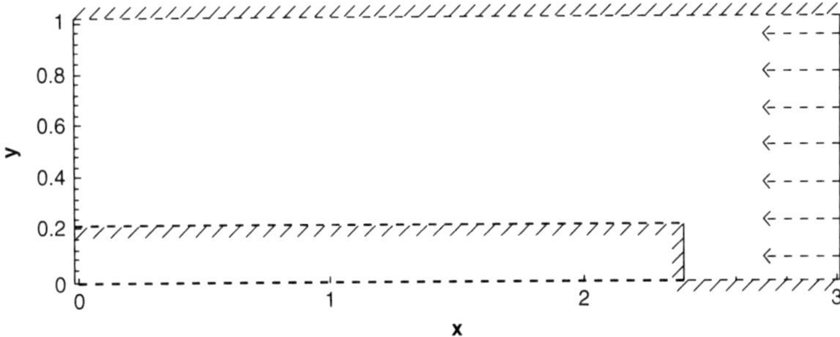

Fig. 5.10 Scheme of the computational domain in the flow problem in a channel with a ledge

where n is the normal, and s is the tangent to the corresponding boundary. The main parameters of computations are presented in Table 5.4. In computing the relaxation parameter τ, we take $\alpha = 0.3$.

In Fig. 5.11, we present the distribution of density at the instant of time $t = 4$ (50 isolines are equidistantly located) obtained in computing on sequentially condensing uniform grids: 120×40, 240×80, and 480×160.

We clearly trace the formation of the secondary waves reflecting from the upper wall of the channel and upper surface of the ledge. After the rarefaction wave, over the corner of the ledge, the gas density is minimal, and near the contact discontinuity, after the triple point over the ledge, the gas density is maximal. The pictures presented clearly demonstrate the convergence of the numerical solution when the spatial grid condenses.

The picture of the flow forming at the instant of time $t = 4$ corresponds to the data of [208] obtained by using piecewise-parabolic schemes of the third-order accuracy in the space and to the results of [101], where the splitting methods of the first, second, and third order of accuracy were applied.

The minimal and maximal values of the density obtained in computing on different grids are presented in Table 5.4. According to the paper [208], which is standard for this problem, $\rho_{\max}/\rho_\infty = 4.541$ and $\rho_{\min}/\rho_\infty = 0.181$. We see that the maximal values of the density well coincide. In condensing the spatial grid, the quantity $\rho_{\min}$ is improved.

In Fig. 5.12, we present the process of ascertainment of the flow. We depict the distribution of density obtained at the instants of time $t = 0.5, 1, 2, 4, 7$, and 15 on the grid 240×80.

Table 5.4 Computation of a flow in a channel with ledge

Grid	$h_x = h_y$	Δt	Number of steps	t	Time of comput.	$\rho_{\min}$	$\rho_{\max}$
120×40	0.025	0.001	4 000	4	4 min	0.551	4.464
240×80	0.0125	0.0005	8 000	4	40 min	0.377	4.553
480×160	0.00625	0.0001	40 000	4	14 h	0.247	4.595

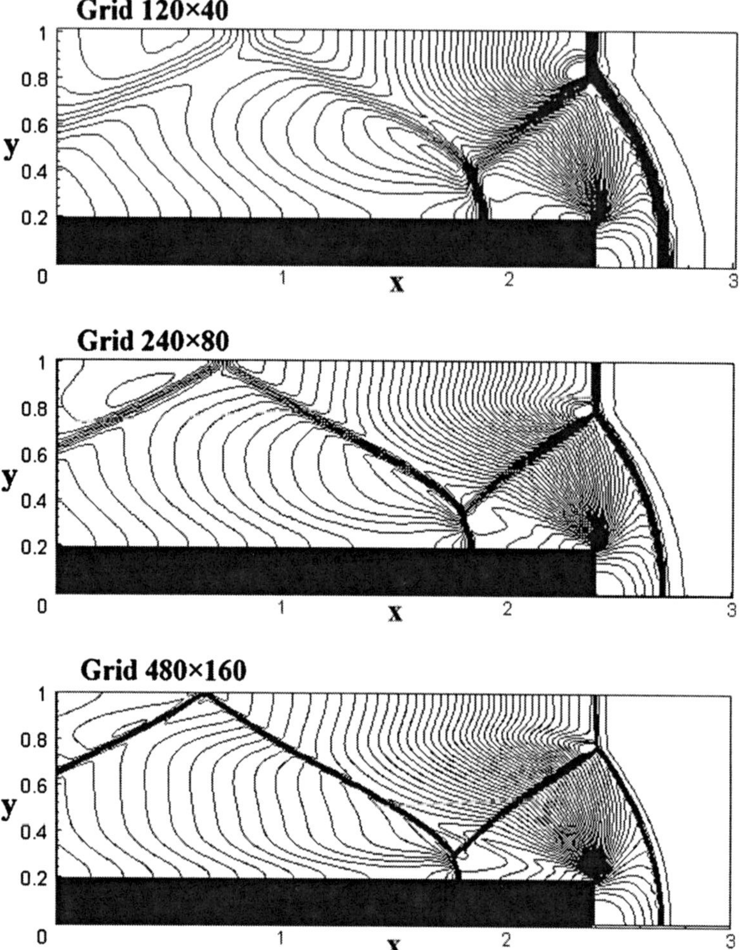

Fig. 5.11 Distribution of the density for a flow in the channel for $t = 4$. Computations on three condensing grids

5.10 Numerical Algorithm for Computing Subsonic Flows

If the gas velocity in the problem of our interest does not exceed the sound speed, then such flows are not accompanied by forming strong shock waves, although in these flows, local supersonic zones can exist. The numerical modelling of such subsonic flows has its own peculiarities, which refer to the regularization method of the numerical solution and to the statement of conditions on free boundaries. In contrast to the supersonic flows, the statement of nonreflecting boundary conditions causes additional difficulties.

In what follows, we present the peculiarities of the numerical algorithm for computing subsonic flows, by which this differs method from the algorithms for computing the supersonic flows described above.

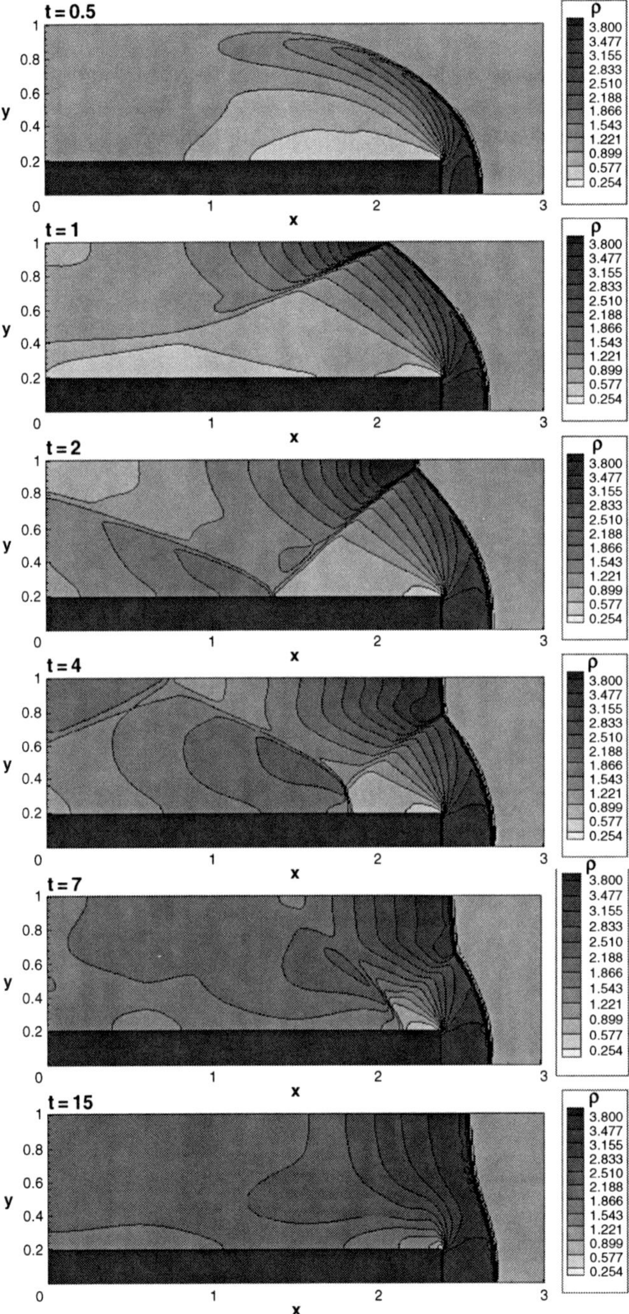

Fig. 5.12 The process of ascertainment of the density field in the flow problem in a channel with a ledge

5.10.1 Dimensionless Form of Equations and the Regularization

In contrast to supersonic problems, in computing the subsonic flows, as a velocity unit, we choose the velocity of the incident flow u_∞. Thus, the main dimensional parameters are the density ρ_∞ of the incident flow, the velocity u_∞ of the incident flow, and the characteristic length denoted by H.

The relations between the dimensional and dimensionless quantities (the sign $\sim$ over a variable refers to the dimensionless parameters) have the form

$$\rho = \tilde{\rho}\rho_\infty, \quad u = \tilde{u}u_\infty, \quad p = \tilde{p}\rho_\infty u_\infty^2,$$

$$x = \tilde{x}H, \quad t = \tilde{t}\frac{H}{u_\infty}, \quad T = \tilde{T}\frac{u_\infty^2}{\gamma\mathcal{R}}. \tag{5.64}$$

In this case, the dimensionless viscosity and heat conduction coefficients and the parameter τ are calculated as follows:

$$\mu = \frac{1}{\mathrm{Re}}(\mathrm{Ma}^2\,T)^\omega, \quad \varkappa = \frac{\mu}{\mathrm{Pr}(\gamma-1)}, \quad \tau = \frac{\mu}{p\,\mathrm{Sc}}. \tag{5.65}$$

Here, in contrast to the algorithm for computing supersonic flows, it suffices to introduce an artificial additional term proportional to the step h of the spatial grid only in additional QGD terms in the form

$$\tau = \frac{1}{\mathrm{Re}}\frac{1}{p\,\mathrm{Sc}}\left(\mathrm{Ma}^2\,T\right)^\omega + \frac{\alpha h}{c}, \tag{5.66}$$

where α is the numerical coefficient of the order of unity, which is found by fitting.

The dimensionless values of the pressure and the temperature can be estimated as follows:

$$T \sim T_\infty = c_\infty^2 = \left(\frac{u_\infty}{\mathrm{Ma}}\right)^2 = \frac{1}{\mathrm{Ma}^2},$$

$$p \sim \frac{\rho_\infty T_\infty}{\gamma} = \frac{1}{\gamma\cdot\mathrm{Ma}^2},$$

where the dimensionless parameters of the incident flow are equal to unity ($\rho_\infty = 1$ and $u_\infty = 1$).

As a result, the estimation formula for τ has the form

$$\tau = \frac{1}{\mathrm{Re}}\frac{\gamma\cdot\mathrm{Ma}^2}{\mathrm{Sc}} + \alpha h\,\mathrm{Ma}\,.$$

Formally, the grid addition can be assumed small if

$$\alpha h < \frac{\mathrm{Ma}}{\mathrm{Re}}\,.$$

We stress that here, in contrast to the algorithm for computing supersonic flows, an additional dissipation proportional to $\alpha h/c$ is included only in the expression for τ and does not enter the expressions for the coefficients μ and $\varkappa$, and hence it does not enter the formulas for calculating the heat flux and the friction force on the wall.

5.10.2 Nonreflecting Boundary Conditions

In computing the subsonic flows, there arises the problem of constructing and numerically implementing the boundary conditions on the free boundaries of the computation domain, the so-called "nonreflecting" boundary conditions. These conditions must not distort the flow field inside the computational domain ensuring the absorption or transmission of the perturbations coming to the boundary, which arise inside the computational domain.

As a rule, as the conditions on the free boundary, one uses the boundary conditions based on the Riemann invariants for the Euler equations or the characteristic boundary conditions. These conditions are applied in computations of viscous and nonviscous flows with subsonic speeds on boundaries. Many variants of the formulation and numerical implementation of conditions of such a type are suggested (see, e.g., [42, 96, 114]). However, their use is referred to considerable difficulties, which are stipulated by a large number of variants for their construction in the differential or finite-difference form, as well as a nonsatisfactory mathematical justification of these condition for viscous gas flows.

In contrast to the algorithms based on the Navier–Stokes equations, in the QGD algorithms, for the statement of conditions on the free subsonic boundaries, in a number of problems, it is possible to use simple boundary conditions analogous to the conditions for viscous incompressible fluid flows (see [65, 163]). Such conditions for incoming and outgoing boundaries have the following form: on the input boundary, one assigns the components of the velocity, the density, and the gradient of the pressure:

$$\frac{\partial p}{\partial n} = \alpha_p, \quad u = u_\infty, \quad \rho = \rho_\infty, \tag{5.67}$$

where $\alpha_p \sim 1/\mathrm{Re}$ is a certain constant and n is the outward normal to the boundary. On the output boundary, we pose the "soft" boundary conditions, except for the pressure, which is assumed to be constant:

$$\frac{\partial \rho}{\partial n} = 0, \quad \frac{\partial u}{\partial n} = 0, \quad p = p_\infty. \tag{5.68}$$

The algorithm described is used in computations of the air flow in a channel with a sudden expansion and contraction. The results of computations are presented in Appendix D. Particularly, we obtain that the optimal range for the regularizing parameter is $0.2 < \alpha < 0.4$. This range corresponds to the best accuracy of solution and the minimum number of steps up to the convergence.

When the Mach number decreases, the rate of convergence of the constructed method also decreases, which is natural in using the complete gas-dynamic equations for computing subsonic flows. For the numerical modelling of flows with lower Mach numbers, the so-called weakly compressible flows, special methods exist (see [136]).

5.11 Stability and Accuracy of QGD Algorithms

The algorithms based on the QGD equations and described above are conditionally stable finite-difference schemes explicit in time. As the practice of numerical computations and theoretical estimates shows, the restriction on the step in time for the QGD algorithms is determined by the Courant condition, which has the form

$$\Delta t \le \beta \left(\frac{h}{c}\right)_{\min}, \tag{5.69}$$

where $\beta = \beta(\alpha)$ is a numerical coefficient, h is the minimal step of the spatial grid, and c is the maximum sound speed in the computation domain.

By using the energy inequalities method, Yu. V. Sheretov [189] obtained a sufficient stability condition of the Courant type for the QGD algorithm. He considered one-dimensional flows in the framework of the Euler equations in the acoustic approximation. For the finite-difference scheme, which is one-dimensional in space and has a constant step in space (see Sect. 5.7), the condition obtained has the form

$$\Delta t \le \beta \frac{h}{c_*}, \tag{5.70}$$

where $c_* = \sqrt{\gamma \mathcal{R} T_*}$ is the sound speed mean in the space at the initial instant of time. The coefficient β is given by

$$\beta = \min(\beta_A, \beta_B, \beta_C), \tag{5.71}$$

where

$$\beta_A = \frac{A}{A^*}, \quad \beta_B = \frac{B}{B^*}, \quad \beta_C = \frac{C}{C^*}.$$

The values of A, B, and C and A^*, B^*, and C^* are determined by the quantities γ and Pr and the value of the coefficient α entering the formula for calculating the artificial dissipation (5.50):

$$A = \frac{\alpha}{\gamma}, \quad B = \frac{\alpha}{\gamma}\left(\frac{4}{3} + \gamma\right), \quad C = \frac{\alpha}{\mathrm{Pr}(\gamma - 1)}, \tag{5.72}$$

$$A^* = 2\gamma A^2 + 2(\gamma - 1)AC + \gamma A + B + \frac{1}{2\gamma}, \tag{5.73}$$

$$B^* = 2B^2 + A + \frac{B}{\gamma} + \frac{(\gamma - 1)C}{\gamma} + \frac{1}{2}, \tag{5.74}$$

$$C^* = (\gamma - 1)C(2A + 2C + 1). \tag{5.75}$$

In Fig. 5.13, we present the following graphs: β_A (the dashed line), β_B (the dash-dotted line), and β_C (the full line) depending on α for air ($\mathrm{Pr} = 0.74$, $\gamma = 1.4$). It is seen that the strongest condition is the restriction connected with β_A. For example, if $\alpha = 0.5$, the finite-difference algorithm is stable for $\beta < 0.15$. In the same graph, we present two dependencies $\beta(\alpha)$ obtained directly in the numerical experiments. The graphs are obtained as follows: for each given α, we find the maximum value of the coefficient β that ensures the stable solution of the problem. Line 1 corresponds to the computation of the subsonic flow in the planar channel for $\mathrm{Re} = 1000$, $\mathrm{Ma} = 0.1$, and $h = 0.075$ (see Appendix D). Line 2 corresponds to the computation of the strong discontinuity step evolution problem (see Sect. 5.7) for $h = 0.03125$.

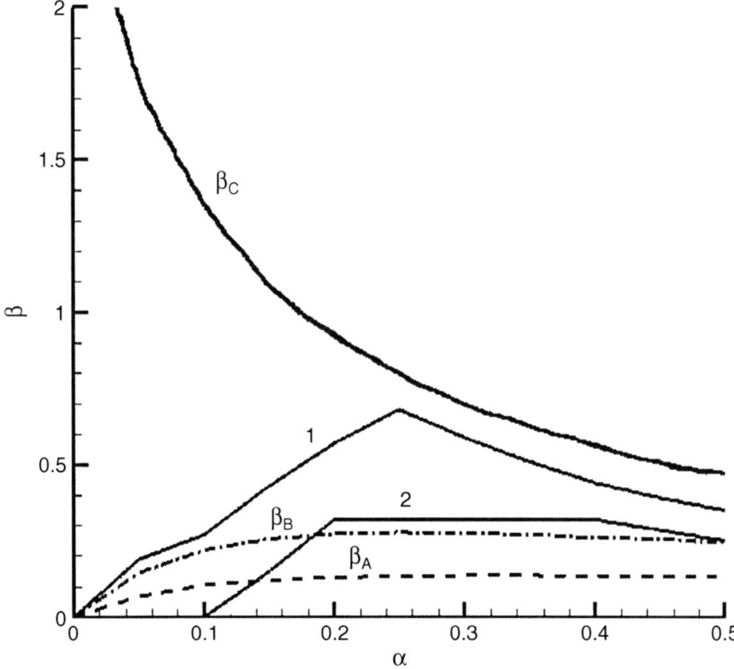

Fig. 5.13 Dependence $\beta(\alpha)$

In both cases, the quantity β exceeds the theoretical estimate obtained as a sufficient condition for the linearized system. For the problem of the flow in the channel, the viscosity coefficient is chosen in the form (5.66), and for small α, the condition on the coefficient β turns out to be weaker than that for the supersonic flow in the discontinuity step evolution problem. The experimental curves attain the maximum in the domain $\alpha \sim 0.2$–0.4, which corresponds to the maximal possible values of the step in time. Precisely these values of the coefficient α are most effective for numerical calculations.

Without use of the artificial viscosity ($\alpha = 0$), the QGD algorithms are finite-difference schemes of the second-order accuracy in the space. For numerical modelling of flows with moderate Reynolds numbers, such an approximation of the QGD equations turns out to be stable. In this case, the step in time ensuring the stability of the QGD algorithms exceeds by an order the step in time necessary for the stable computation of the same flows on the basis of the Navier–Stokes model with the same approximation in space and time. The comparison of the above models for the flow problem in a shock wave is presented in Appendix C.

When introducing the artificial viscosity of the form (5.47) or (5.66), the QGD schemes constructed above formally become first-order schemes. However, the accuracy of the obtained numerical solution rapidly increases when the step of the spatial grid decreases and the coefficient α decreases. This is clearly demonstrated in Sect. 5.7 (the discontinuity step evolution problem) and in Sect. 5.9 (the flow problem in a channel with a ledge).

Therefore, in spite of a sufficiently complicated form of additional terms, a high accuracy and a rapid convergence make the QGD algorithms compete with the existing methods for solving the Navier–Stokes equations.

As the Navier–Stokes equations, the QGD equations allow one to construct other finite-difference algorithms different from those described above. The construction of an implicit finite-difference QGD algorithm is presented in [100]. Examples of schemes constructed by using the method of flux splitting are presented in [102, 103]. Second- and third-order accuracy algorithms for the QGD equations are studied in [101]. In these works, it was noted that in the examples considered, the decrease of the coefficient α in the formula for calculating τ is equivalent to increasing the accuracy order of the scheme.

Chapter 6
Algorithms for Solving Quasi-gas-dynamic Equations on Nonstructured Grids

In this chapter, we generalize the proposed numerical algorithms to the case of nonstructured or irregular two-dimensional spatial grids. The use of irregular grids seems to be perspective for computing flows in domains with complicated boundaries. Moreover, the freedom of the calculator in choosing the location of nodes of the spatial grid allows him/her to approximate the flow zones with strong gradients in detail and diminish the dependence of the numerical solution from the direction of grid lines given a priori. This chapter is mainly based on the results of [71, 78].

6.1 Choice of the Grid and Constructing the Control Volume

In numerical modelling of flows in domains with complicated boundaries, it is not always convenient to use rectangular spatial computation grids. A more general form of grids is the so-called nonstructured, or irregular, grids, in which the computation nodes can be placed arbitrarily. The topology of regular grids is uniquely determined by indices of grid points. For irregular grids, there is no such correspondence. Irregular grids allow one to approximate the boundaries of the computational domains and the characteristic peculiarities of the flow with a needed accuracy. For two-dimensional computational domains, irregular grids can be associated with triangles, and for three-dimensional domains, they can be associated with tetrahedrons. In what follows, for simplicity of presentation, we restrict ourselves to the consideration of two-dimensional computation domains.

Let us define a grid by the set of nodes $M = \{M_i \in \mathbb{R}^2, i = 1, \ldots, n\}$. Then the grid can be represented as a set of triangles with vertices $M_0, M_1, \ldots, M_n$ (see Fig. 6.1).

To construct the finite-difference scheme, we use the integro-interpolation method, or the finite-volume method (see [170]). This method presupposes that the finite-difference approximation of the equations is constructed on the basis of approximate integration of equations written in the flux form with respect to a certain volume, which is said to be control. The control volume is constructed around points at which the computation of the gas-dynamic quantities is performed. For

T.G. Elizarova, *Quasi-Gas Dynamic Equations*, Computational Fluid and Solid Mechanics, DOI 10.1007/978-3-642-00292-2_6,
© Springer-Verlag Berlin Heidelberg 2009

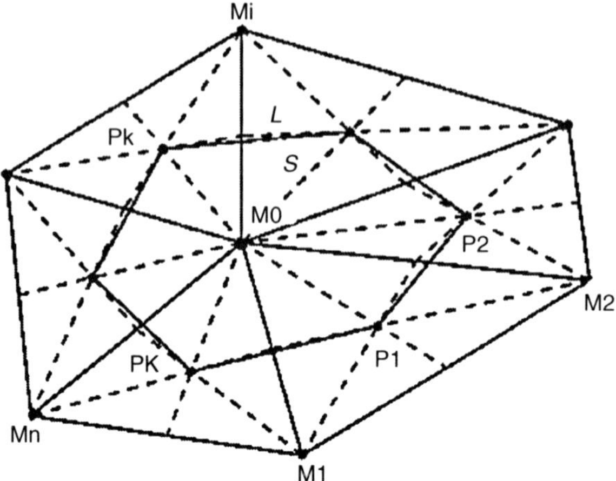

Fig. 6.1 Choice of the control volume

rectangular grids, it is natural to choose the control volume in the form of a rectangle, which leads to the construction of the finite-difference approximations written in the previous chapter.

On the grid consisting of triangles, the control volume can be chosen in various ways. To a considerable extent, the accuracy of the obtained finite-difference scheme is determined by the form of triangles and the method for choosing the control volume. In particular, on the grids consisting of equilateral triangles, we can construct the schemes of the second and higher orders of accuracy. In the construction of the finite-difference approximations, it turns out to be convenient to use grids satisfying the Delone triangulation principle (see [157, 159]). In these grids, triangles are chosen in such a way that the disc circumscribed around any triangle contains no node different from the vertices of the triangle.

If the gas-dynamic quantities u, ρ, p, and E are given at the centers of the inscribed circles, then the triangles themselves can be used as control volumes.

Let the gas-dynamic quantities u, ρ, p, and E be given at the nodes M_i of the grid. Then it is convenient to choose the control volumes in the form of polygons that can be constructed in various ways. Let us present two methods for constructing the control volume, which ensure a high accuracy of the finite-difference approximation of the equations for a sufficiently regular triangular grid (see [2, 40, 94, 157]).

For each node M_i of a triangular grid, we construct the contour L consisting of intersection points P_k of medians of the triangles containing this node. Denote these points by P_0, P_1, ..., P_K (see Fig. 6.1). Denote the number of nodes of the contour by K. The domain bounded by this contour is a computational cell—the control volume. The computation cell is irregular and, therefore, the number of nodes of the contour K is not fixed. The neighboring nodes compose the stencil of the point M_0 on which the approximation of equations is performed. In Fig. 6.1, we present the control volume with boundary L centered at the point M_0.

The method for constructing the control volume chosen above imposes a number of restrictions on the grid structure. Precisely, the node for which the control volume is constructed must not leave the boundaries of the contour L. If the grid does not satisfy this condition, then the accuracy of approximation of the spatial derivatives decreases.

The second variant for choosing the control volume in the form of the so-called Dirichlet cell helps us to avoid this problem. The Dirichlet cell for a point M_0 is defined as the set of points of the computation domain that are nearer to the node M_0 than to any other node from M_i (see [161]). For a Dirichlet cell, the contour L is the broken line connecting the intersection points of the middle perpendiculars of the corresponding triangles.

In what follows, in constructing a finite-difference algorithm for the QGD equations, we use the control volume constructed on the basis of intersection points of medians.

6.2 Approximation of the System

For convenience of the further presentation, we write the system of the QGD equations (3.13), (3.14), and (3.15) for a planar two-dimensional flow in the form of a single vector equation:

$$\frac{\partial U}{\partial t} - \operatorname{div} W = 0, \tag{6.1}$$

where we have introduced the notation

$$U = \begin{pmatrix} \rho \\ \rho u_x \\ \rho u_y \\ E \end{pmatrix}, \quad W = \begin{pmatrix} W_1 \\ W_2 \\ W_3 \\ W_4 \end{pmatrix}, \quad \operatorname{div} W = \begin{pmatrix} \operatorname{div} W_1 \\ \operatorname{div} W_2 \\ \operatorname{div} W_3 \\ \operatorname{div} W_4 \end{pmatrix},$$

$$W_1 = -j_m, \quad \begin{pmatrix} W_2 \\ W_3 \end{pmatrix} = \Pi - j_m \otimes u - pe, \quad W_4 = \Pi u - q - \frac{E + p}{\rho} j_m.$$

Here, $E = \frac{1}{2}\rho u^2 + \varepsilon \rho$ is the total energy. In what follows, all operations with W are also performed componentwise.

In accordance with the finite-volume method, we integrate Eq. (6.1) over the control volume or the cell S:

$$\frac{\partial}{\partial t} \int_S U \, dS - \oint_L (W \cdot n) dl = 0. \tag{6.2}$$

Here, for the integral of W, we have used the Gauss formula, which connects the integral over volume with the integral over the surface in the three-dimensional case, or the Green formula in the two-dimensional case (see [130]), S is the cell area, L is the cell contour, and n is the outward normal to the contour.

Furthermore, applying the mean-value formula to the first term in Eq. (6.2), we obtain

$$\frac{\partial \bar{U}}{\partial t} = \frac{1}{S} \oint_{L} (W \cdot n)\,dl, \qquad (6.3)$$

where

$$\bar{U} = S^{-1} \int_{S} U \, dS$$

is the mean value of the velocity vector U relative to the center of the computation cell S. In what follows, we denote $\bar{U}$ by U.

Let us approximate the contour integral as follows:

$$\oint_{L} (W \cdot n)\,dl = \sum_{k}(W_x(P_{k+1/2})n_x(P_{k+1/2}) + W_y(P_{k+1/2})n_y(P_{k+1/2}))L_k. \quad (6.4)$$

Here, the values of the fluxes W_x and W_y and the projections of the normals n_x and n_y are calculated at the middle points $P_{k+1/2}$ of the segments $L_k = L(P_k, P_{k+1})$ composing the contour L.

Therefore, we obtain the finite-difference formula for the finite-volume method:

$$\widehat{U}_i = U_i - \frac{\Delta t}{S} \sum_{k}(W_x(P_{k+1/2})n_x(P_{k+1/2}) + W_y(P_{k+1/2})n_y(P_{k+1/2}))L_k. \quad (6.5)$$

In computing by formula (6.5), we need the values of the partial derivatives in x and y of the density, the velocity, and the pressure at the middle points of the segments L_k. The values of the gas-dynamic quantities are found as the arithmetical means of the values at the endpoints.

In approximating Eq. (6.5), we also need the value of the gas-dynamic quantities at the centers of triangles, the intersection points of their medians (the points $P_1, \ldots, P_K$; see Fig. 6.1). We define the gas-dynamic quantities at the centers of triangles as the arithmetical means of the values at the vertices of triangles.

In the cylindrical geometry, because of the absence of dependence on the angle ϕ, the form of integral (6.4) does not change, and only the change of coordinates (x, y) by the coordinates (r, z) is needed.

6.3 Approximation of Partial Derivatives

Let us consider methods for calculating the partial derivatives needed for the calculation of the contour integral. For this purpose, we write the Green formula:

$$\iint_{G} \left(\frac{\partial Q}{\partial x} - \frac{\partial P}{\partial y} \right) dx\,dy = \int_{\partial G} P\,dx + Q\,dy, \qquad (6.6)$$

where $P = P(x, y)$ and $Q = Q(x, y)$ are certain scalar-valued functions given in the domain G and ∂G is the boundary of this domain. Also, let us present the formula connecting the curvilinear integrals of the first and second kind:

$$\int_{\partial G} P\, dx + Q\, dy = \int_{\partial G} (P \cos \alpha + Q \cos \beta)\, dl. \tag{6.7}$$

The connection between the tangent l and the normal n to the contour ∂G is presented in Fig. 6.2 and is explained by the following formulas:

$$n = \{n_x, n_y\} = \{\cos \alpha_x, \cos \alpha_y\}, \tag{6.8}$$

$$\frac{\partial x}{\partial l} = \cos \beta = \sin \gamma = -\cos \alpha_y = -n_y, \tag{6.9}$$

$$\frac{\partial y}{\partial l} = \cos \gamma = \sin \beta = \cos \alpha_x = n_x. \tag{6.10}$$

Using these formulas, we find the partial derivatives $\partial P/\partial x$ and $\partial Q/\partial y$. For this purpose, let us consider the divergence of the vector $A = \{P, Q\}$. Using formulas (6.6), (6.7), (6.8), (6.9), and (6.10), we obtain

$$\iint_G \left(\frac{\partial P}{\partial x} + \frac{\partial Q}{\partial y} \right) dx dy = \int_{\partial G} -Q\, dx + P\, dy = \int_{\partial G} (-Q \cos \beta + P \cos \gamma)\, dl$$

$$= \int_{\partial G} \left(Q n_y + P n_x \right) dl = \int_{\partial G} (A \cdot n) dl. \tag{6.11}$$

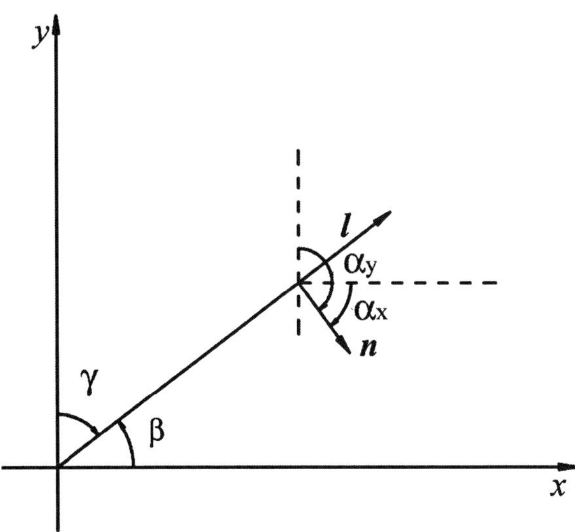

Fig. 6.2 Connection of the angles between the tangent and the normal

Furthermore,

$$\iint\limits_{G} \frac{\partial P}{\partial x} dx dy = \int\limits_{\partial G} P n_x \, dl, \tag{6.12}$$

$$\iint\limits_{G} \frac{\partial Q}{\partial y} dx dy = \int\limits_{\partial G} Q n_y \, dl, \tag{6.13}$$

or, using the mean-value formula, we obtain

$$\frac{\partial P}{\partial x} = \frac{1}{S} \int\limits_{\partial G} P n_x \, dl, \tag{6.14}$$

$$\frac{\partial Q}{\partial y} = \frac{1}{S} \int\limits_{\partial G} Q n_y \, dl, \tag{6.15}$$

where S is the area of the domain G.

Let the domain G be the quadrangle with vertices M_0, P_2, M_2, and P_1 (see Figs. 6.1 and 6.3). Consider methods for expressing the partial derivatives. Consider a quantity f defined at the vertices of the quadrangle and find its partial derivatives $\partial f / \partial x$ and $\partial f / \partial y$ at the point $P_{3/2}$.

We can seek for the partial derivatives by using the following two methods: by using the Green formulas (6.14) and (6.15), i.e., through the integral over the contour 0123:

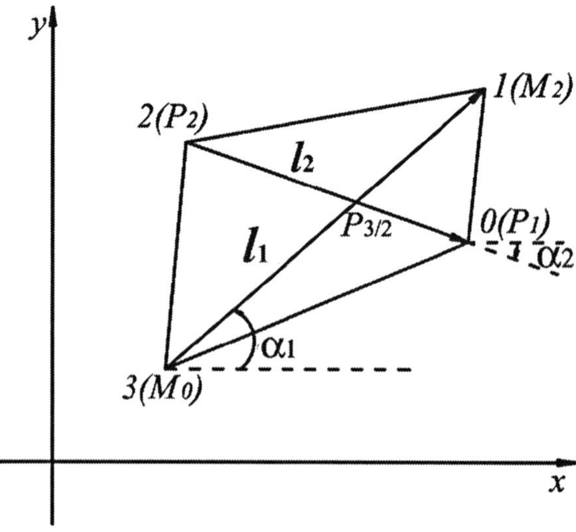

Fig. 6.3 Contour ∂G for the calculation of the partial derivatives

$$\frac{\partial f}{\partial x} = \frac{1}{S} \int_{\partial G} f n_x \, dl, \tag{6.16}$$

$$\frac{\partial f}{\partial y} = \frac{1}{S} \int_{\partial G} f n_y \, dl, \tag{6.17}$$

or using the derivatives in directions l_1 and l_2:

$$\frac{\partial f}{\partial l_1} = \frac{\partial f}{\partial x}\frac{\partial x}{\partial l_1} + \frac{\partial f}{\partial y}\frac{\partial y}{\partial l_1}, \tag{6.18}$$

$$\frac{\partial f}{\partial l_2} = \frac{\partial f}{\partial x}\frac{\partial x}{\partial l_2} + \frac{\partial f}{\partial y}\frac{\partial y}{\partial l_2}. \tag{6.19}$$

Let us show that both methods for calculating the difference derivatives lead to the same result.

Let us use the first method. For this purpose, we convert the integrals in formulas (6.16) and (6.17) similar to Eq. (6.4). We can express the components of the unit normal vector through the coordinates of points as follows:

$$n(x_1, y_1, x_2, y_2) = \{\cos\alpha, \sin\alpha\}$$
$$= \{y_2 - y_1, x_1 - x_2\} \cdot \frac{1}{\sqrt{(x_2 - x_1)^2 + (y_2 - y_1)^2}}, \tag{6.20}$$

and the area can be expressed as

$$S = \frac{1}{2}\left[(x_3 - x_1)(y_0 - y_2) + (x_0 - x_2)(y_1 - y_3)\right]. \tag{6.21}$$

Further,

$$S\frac{\partial f}{\partial x} = \int_{0123} f n_x \, dl = \sum_{i=0}^{3} \frac{f_{i+1} + f_i}{2}(y_{i+1} - y_i)$$
$$= \frac{f_1 + f_0}{2}(y_1 - y_0) + \frac{f_2 + f_1}{2}(y_2 - y_1)$$
$$+ \frac{f_3 + f_2}{2}(y_3 - y_2) + \frac{f_0 + f_3}{2}(y_0 - y_3)$$
$$= \frac{1}{2}[y_1(f_0 - f_2) + y_0(f_3 - f_1) + y_2(f_1 - f_3) + y_3(f_2 - f_0)]$$
$$= \frac{1}{2}[(f_2 - f_0)(y_3 - y_1) + (f_1 - f_3)(y_2 - y_0)].$$

Therefore, as a result, we obtain the formula expressing the derivative at $P_{3/2}$ through the coordinates of nodes of the contour 0123 and the values of the function at these nodes:

$$\frac{\partial f}{\partial x} = \frac{(f_2 - f_0)(y_3 - y_1) + (f_1 - f_3)(y_2 - y_0)}{2S}. \tag{6.22}$$

Similarly,

$$\frac{\partial f}{\partial y} = \frac{(f_2 - f_0)(x_1 - x_3) + (f_1 - f_3)(x_0 - x_2)}{2S}. \tag{6.23}$$

Let us present the second method for calculating the partial derivatives. For this purpose, we express these partial derivatives through the directional derivatives and solve system (6.18) and (6.19) with respect to $\partial f/\partial x$ and $\partial f/\partial y$:

$$\frac{\partial f}{\partial l_1} = \frac{\partial f}{\partial x}\frac{\partial x}{\partial l_1} + \frac{\partial f}{\partial y}\frac{\partial y}{\partial l_1},$$

$$\frac{\partial f}{\partial l_2} = \frac{\partial f}{\partial x}\frac{\partial x}{\partial l_2} + \frac{\partial f}{\partial y}\frac{\partial y}{\partial l_2}$$

or

$$\frac{\partial f}{\partial l_1} = \frac{\partial f}{\partial x}\cos\alpha_1 + \frac{\partial f}{\partial y}\sin\alpha_1,$$

$$\frac{\partial f}{\partial l_2} = \frac{\partial f}{\partial x}\cos\alpha_2 + \frac{\partial f}{\partial y}\sin\alpha_2,$$

whence

$$\frac{\partial f}{\partial x} = \frac{\dfrac{\partial f}{\partial l_1}\sin\alpha_2 - \dfrac{\partial f}{\partial l_2}\sin\alpha_1}{\sin\alpha_2\cos\alpha_1 - \sin\alpha_1\cos\alpha_2}, \tag{6.24}$$

$$\frac{\partial f}{\partial y} = \frac{\dfrac{\partial f}{\partial l_1}\cos\alpha_2 - \dfrac{\partial f}{\partial l_2}\cos\alpha_1}{\sin\alpha_1\cos\alpha_2 - \sin\alpha_2\cos\alpha_1}. \tag{6.25}$$

Using formulas (6.8), (6.9), and (6.10), we obtain

$$\frac{\partial f}{\partial x} = -\frac{\dfrac{\partial f}{\partial l_1}\sin\alpha_2 - \dfrac{\partial f}{\partial l_2}\sin\alpha_1}{\sin\alpha_2\cos\alpha_1 - \sin\alpha_1\cos\alpha_2}, \tag{6.26}$$

$$\frac{\partial f}{\partial y} = \frac{\dfrac{\partial f}{\partial l_1}\cos\alpha_2 - \dfrac{\partial f}{\partial l_2}\cos\alpha_1}{\sin\alpha_1\cos\alpha_2 - \sin\alpha_2\cos\alpha_1}. \tag{6.27}$$

Transforming the latter relation, we obtain

$$\frac{\partial f}{\partial x} = \frac{\dfrac{\partial f}{\partial l_1} n_{2x} - \dfrac{\partial f}{\partial l_2} n_{1x}}{-n_{2x} n_{1y} + n_{1x} n_{2y}},$$
(6.28)

$$\frac{\partial f}{\partial y} = -\frac{\dfrac{\partial f}{\partial l_1} n_{2y} + \dfrac{\partial f}{\partial l_2} n_{1y}}{-n_{1x} n_{2y} + n_{2x} n_{1y}}.$$
(6.29)

Applying Eq. (6.20), we have

$$\frac{\partial f}{\partial x} = \frac{\dfrac{f_1 - f_3}{L_{13}} \dfrac{y_0 - y_2}{L_{20}} - \dfrac{f_0 - f_2}{L_{20}} \dfrac{y_1 - y_3}{L_{13}}}{-\dfrac{y_0 - y_2}{L_{20}} \dfrac{x_3 - x_1}{L_{13}} + \dfrac{y_1 - y_3}{L_{13}} \dfrac{x_2 - x_0}{L_{20}}}$$

$$= \frac{-(f_1 - f_3)(y_2 - y_0) + (f_2 - f_0)(y_1 - y_3)}{-(y_2 - y_0)(x_1 - x_3) + (y_1 - y_3)(x_2 - x_0)}$$

$$= \frac{(f_2 - f_0)(y_1 - y_3) - (f_1 - f_3)(y_2 - y_0)}{-2S}$$

$$= \frac{(f_2 - f_0)(y_3 - y_1) + (f_1 - f_3)(y_2 - y_0)}{2S},$$

which coincides with Eq. (6.22). The expression for $\partial f / \partial y$ is written similarly. Therefore, both methods for calculating the partial derivatives lead to the same result.

Let us express the components of the vector W through the gas-dynamic quantities ρ, u, and p and the partial derivatives calculated above, substitute these expression in approximation (6.5), and write the finite-difference algorithm for computing the flow on the nonstructured grid.

6.4 Finite-Difference Schemes for Two-Dimensional Flows

Let us present a finite-difference scheme for computing supersonic two-dimensional plane flows.

The equations for calculating the density, the components of the velocity, and the energy and also the expression for recalculating the pressure have the following form:

$$\widehat{\rho}_i = \rho_i - \frac{\Delta t}{S} \sum_k \left(j^x_{m,k+1/2} n^x_{k+1/2} + j^y_{m,k+1/2} n^y_{k+1/2} \right) L_k,$$

$$\widehat{u^x}_i = \frac{1}{\widehat{\rho}_i} \left(\rho_i u^x_i + \frac{\Delta t}{S} \sum_k \left(\left[\Pi^{xx}_{k+1/2} - j^x_{m,k+1/2} u^x_{k+1/2} - p \right] n^x_{k+1/2} \right. \right.$$

$$\left. \left. + \left[\Pi^{xy}_{k+1/2} - j^x_{m,k+1/2} u^y_{k+1/2} \right] n^y_{k+1/2} \right) L_k \right),$$

$$\widehat{u^y}_i = \frac{1}{\widehat{\rho}_i} \left(\rho_i u^y_i + \frac{\Delta t}{S} \sum_k \left(\left[\Pi^{yx}_{k+1/2} - j^y_{m,k+1/2} u^x_{k+1/2} \right] n^x_{k+1/2} \right. \right.$$

$$\left. \left. + \left[\Pi^{yy}_{k+1/2} - j^y_{m,k+1/2} u^y_{k+1/2} - p \right] n^y_{k+1/2} \right) L_k \right),$$

$$\widehat{E}_i = E_i + \frac{\Delta t}{S} \sum_k \left(\left[\Pi^{xx}_{k+1/2} u^x_{k+1/2} + \Pi^{xy}_{k+1/2} u^y_{k+1/2} \right. \right.$$

$$\left. - q^x_{k+1/2} - \frac{E_{k+1/2} + p_{k+1/2}}{\rho_{k+1/2}} j^x_{m,k+1/2} \right] n^x_{k+1/2}$$

$$+ \left[\Pi^{yx}_{k+1/2} u^x_{k+1/2} + \Pi^{yy}_{k+1/2} u^y_{k+1/2} \right.$$

$$\left. \left. - q^y_{k+1/2} - \frac{E_{k+1/2} + p_{k+1/2}}{\rho_{k+1/2}} j^y_{m,k+1/2} \right] n^y_{k+1/2} \right) L_k,$$

$$\widehat{p}_i = (\gamma - 1) \left(\widehat{E}_i - \widehat{\rho}_i \frac{(\widehat{u^x}_i)^2 + (\widehat{u^y}_i)^2}{2} \right),$$

where the subscript i is the number of the grid node and the subscript k is the number of node of the contour L. The relaxation parameter and the viscosity coefficient are calculated as follows:

$$\tau_k = \alpha \frac{\sum_{k=0}^{K} L_k}{K} \left(\gamma \frac{p_k}{\rho_k} \right)^{-1/2} + \frac{\text{Ma}}{\text{Re}} \left(\frac{p_k \gamma}{\rho_k} \right)^{\omega} \frac{1}{p_k \, \text{Sc}}, \tag{6.30}$$

$$\mu_k = p_k \, \text{Sc} \, \tau_k, \quad \mu_{k+1/2} = \frac{\mu_k + \mu_{k+1}}{2}, \quad \tau_{k+1/2} = \frac{\tau_k + \tau_{k+1}}{2},$$

where K is the number of nodes in the contour L, Ma is the Mach number, Re is the Reynolds number, Sc is the Schmidt number, γ is the specific heat ratio, and α is a numerical coefficient from $[0, 1]$.

The gas-dynamic parameters at half-integer points are equal to

$$\rho_{k+1/2} = \frac{\rho_k + \rho_{k+1}}{2}, \quad p_{k+1/2} = \frac{p_k + p_{k+1}}{2}, \quad u_{k+1/2} = \frac{u_k + u_{k+1}}{2},$$

$$E_{k+1/2} = \rho_{k+1/2} \frac{(u^x_{k+1/2})^2 + (u^y_{k+1/2})^2}{2} + \frac{1}{\gamma - 1} p_{k+1/2}.$$

The fluxes entering the system of finite-difference equations are approximated at the half-integer points as follows:

$$j^x_{m,k+1/2} = \rho(u^x_{k+1/2} - w^x_{k+1/2}), \quad j^y_{m,k+1/2} = \rho(u^y_{k+1/2} - w^y_{k+1/2}),$$

$$w^x_{k+1/2} = \frac{\tau_{k+1/2}}{\rho_{k+1/2}}\left[\left(\frac{\partial}{\partial x}\rho(u^x)^2\right)_{k+1/2} + \left(\frac{\partial}{\partial y}\rho u^x u^y\right)_{k+1/2} + \left(\frac{\partial p}{\partial x}\right)_{k+1/2}\right],$$

$$w^y_{k+1/2} = \frac{\tau_{k+1/2}}{\rho_{k+1/2}}\left[\left(\frac{\partial}{\partial x}\rho u^x u^y\right)_{k+1/2} + \left(\frac{\partial}{\partial y}\rho(u^y)^2\right)_{k+1/2} + \left(\frac{\partial p}{\partial y}\right)_{k+1/2}\right],$$

$$\Pi^{xx}_{ns\,k+1/2} = \mu_{k+1/2}\left[\frac{4}{3}\left(\frac{\partial u^x}{\partial x}\right)_{k+1/2} - \frac{2}{3}\left(\frac{\partial u^y}{\partial y}\right)_{k+1/2}\right],$$

$$\Pi^{xy}_{ns\,k+1/2} = \Pi^{yx}_{ns\,k+1/2} = \mu_{k+1/2}\left[\left(\frac{\partial u^x}{\partial y}\right)_{k+1/2} + \left(\frac{\partial u^y}{\partial x}\right)_{k+1/2}\right],$$

$$\Pi^{yy}_{ns\,k+1/2} = \mu_{k+1/2}\left[\frac{4}{3}\left(\frac{\partial u^y}{\partial y}\right)_{k+1/2} - \frac{2}{3}\left(\frac{\partial u^x}{\partial x}\right)_{k+1/2}\right],$$

$$\Pi^{xx}_{k+1/2} = \Pi^{xx}_{ns\,k+1/2} + u^x_{k+1/2}w^{x*}_{k+1/2} + R^*_{k+1/2},$$

$$\Pi^{xy}_{k+1/2} = \Pi^{xy}_{ns\,k+1/2} + u^x_{k+1/2}w^{y*}_{k+1/2},$$

$$\Pi^{yx}_{k+1/2} = \Pi^{yx}_{ns\,k+1/2} + u^y_{k+1/2}w^{x*}_{k+1/2},$$

$$\Pi^{yy}_{k+1/2} = \Pi^{yy}_{ns\,k+1/2} + u^y_{k+1/2}w^{y*}_{k+1/2} + R^*_{k+1/2},$$

where the auxiliary quantities $w^{x*}_{k+1/2}$, $w^{y*}_{k+1/2}$, and $R^*_{k+1/2}$ are calculated by the formulas

$$w^{x*}_{k+1/2} = \tau_{k+1/2}\left[\rho_{k+1/2}\left(u^x_{k+1/2}\left(\frac{\partial u^x}{\partial x}\right)_{k+1/2}\right.\right.$$
$$\left.\left. + u^y_{k+1/2}\left(\frac{\partial u^x}{\partial y}\right)_{k+1/2}\right) + \left(\frac{\partial p}{\partial x}\right)_{k+1/2}\right],$$

$$w^{y*}_{k+1/2} = \tau_{k+1/2}\left[\rho_{k+1/2}\left(u^x_{k+1/2}\left(\frac{\partial u^y}{\partial x}\right)_{k+1/2}\right.\right.$$
$$\left.\left. + u^y_{k+1/2}\left(\frac{\partial u^y}{\partial y}\right)_{k+1/2}\right) + \left(\frac{\partial p}{\partial y}\right)_{k+1/2}\right],$$

$$R^*_{k+1/2} = \tau_{k+1/2} \left[u^x_{k+1/2} \left(\frac{\partial p}{\partial x} \right)_{k+1/2} + u^y_{k+1/2} \left(\frac{\partial p}{\partial y} \right)_{k+1/2} \right.$$

$$\left. + \gamma p_{k+1/2} \left(\left(\frac{\partial u^x}{\partial x} \right)_{k+1/2} + \left(\frac{\partial u^y}{\partial y} \right)_{k+1/2} \right) \right],$$

$$q^x_{ns\,k+1/2} = -\frac{\mu_{k+1/2}}{\mathrm{Pr}} \frac{\gamma}{\gamma - 1} \left(\frac{\partial}{\partial x} \frac{p}{\rho} \right)_{k+1/2}.$$

The components of the heat flux are calculated as follows:

$$q^y_{ns\,k+1/2} = -\frac{\mu_{k+1/2}}{\mathrm{Pr}} \frac{\gamma}{\gamma - 1} \left(\frac{\partial}{\partial y} \frac{p}{\rho} \right)_{k+1/2},$$

$$q^x_{k+1/2} = q^x_{ns} - u^x_{k+1/2} R^q_{k+1/2}, \qquad q^y_{k+1/2} = q^y_{ns} - u^y_{k+1/2} R^q_{k+1/2},$$

where the following auxiliary quantity is introduced:

$$R^q_{k+1/2} = \tau_{k+1/2}\, \rho_{k+1/2}$$

$$\times \left[\frac{1}{\gamma - 1} \left(u^x_{k+1/2} \left(\frac{\partial}{\partial x} \frac{p}{\rho} \right)_{k+1/2} + u^y_{k+1/2} \left(\frac{\partial}{\partial y} \frac{p}{\rho} \right)_{k+1/2} \right) \right.$$

$$\left. + p_{k+1/2} \left(u^x_{k+1/2} \left(\frac{\partial}{\partial x} \frac{1}{\rho} \right)_{k+1/2} + u^y_{k+1/2} \left(\frac{\partial}{\partial y} \frac{1}{\rho} \right)_{k+1/2} \right) \right].$$

The finite-difference approximation of the QGD equations presented here is constructed for two-dimensional flows by using these equations in the form (5.1), (5.2), (5.3), and (5.4). The difference approximation of the equations in the cylindrical geometry is constructed similarly by using these equations in the form (5.15), (5.16), (5.17), and (5.18):

$$\widehat{\rho}_i = \rho_i - \frac{\Delta t}{S} \sum_k (j^r_{m,k+1/2}\, n^r_{k+1/2} + j^z_{m,k+1/2}\, n^z_{k+1/2}) L_k,$$

$$\widehat{u^r}_i = \frac{1}{\widehat{\rho}_i} \left(\rho_i u^r_i + \frac{\Delta t}{S} \sum_k ([\Pi^{rr}_{k+1/2} - j^r_{m,k+1/2} u^r_{k+1/2} - p] n^r_{k+1/2} \right.$$

$$\left. + [\Pi^{rz}_{k+1/2} - j^r_{m,k+1/2} u^z_{k+1/2}] n^z_{k+1/2}) L_k - \frac{\Pi^{\phi\phi}_i}{r_i} \right),$$

$$
\widehat{u^z}_i = \frac{1}{\hat{\rho}_i}\left(\rho_i\, u_i^z + \frac{\Delta t}{S}\sum_k ([\Pi_{k+1/2}^{zr} - j_{m,k+1/2}^z u_{k+1/2}^r]n_{k+1/2}^r\right.
$$

$$
\left. + [\Pi_{k+1/2}^{zz} - j_{m,k+1/2}^z u_{k+1/2}^z - p]n_{k+1/2}^z)L_k\right),
$$

$$
\widehat{E}_i = E_i + \frac{\Delta t}{S}\sum_k \left(\left[\Pi_{k+1/2}^{rr}u_{k+1/2}^r + \Pi_{k+1/2}^{rz}u_{k+1/2}^z - q_{k+1/2}^r\right.\right.
$$

$$
\left. - \frac{E_{k+1/2} + p_{k+1/2}}{\rho_{k+1/2}}j_{m,k+1/2}^r\right]n_{k+1/2}^r + \left[\Pi_{k+1/2}^{zr}u_{k+1/2}^r\right.
$$

$$
\left.\left. + \Pi_{k+1/2}^{zz}u_{k+1/2}^z - q_{k+1/2}^z - \frac{E_{k+1/2} + p_{k+1/2}}{\rho_{k+1/2}}j_{m,k+1/2}^z\right]n_{k+1/2}^z\right)L_k,
$$

$$
\widehat{p}_i = (\gamma - 1)\left(\widehat{E}_i - \hat{\rho}_i\frac{(\widehat{u^r}_i)^2 + (\widehat{u^z}_i)^2}{2}\right),
$$

where the subscript i is the number of the grid node and k is the number of the node of the contour L.

The relaxation parameter and the viscosity coefficient are calculated by the formulas (6.30). The quantities at half-integer points are calculated as the arithmetical means at the nearest nodes:

$$
\mu_{k+1/2} = \frac{\mu_k + \mu_{k+1}}{2}, \qquad \tau_{k+1/2} = \frac{\tau_k + \tau_{k+1}}{2},
$$

$$
\rho_{k+1/2} = \frac{\rho_k + \rho_{k+1}}{2}, \qquad p_{k+1/2} = \frac{p_k + p_{k+1}}{2}, \qquad u_{k+1/2} = \frac{u_k + u_{k+1}}{2},
$$

$$
E_{k+1/2} = \rho_{k+1/2}\frac{(u_{k+1/2}^r)^2 + (u_{k+1/2}^z)^2}{2} + \frac{1}{\gamma - 1}p_{k+1/2}.
$$

The fluxes entering the equations of the main system are calculated by the formulas

$$
w_{k+1/2}^r = \frac{\tau_{k+1/2}}{\rho_{k+1/2}}\left[\left(\frac{\partial}{\partial r}\rho(u^r)^2\right)_{k+1/2} + \left(\frac{\rho(u^r)^2}{r}\right)_{k+1/2}\right.
$$

$$
\left. + \left(\frac{\partial}{\partial z}\rho u^r u^z\right)_{k+1/2} + \left(\frac{\partial p}{\partial r}\right)_{k+1/2}\right],
$$

$$
w_{k+1/2}^z = \frac{\tau_{k+1/2}}{\rho_{k+1/2}}\left[\left(\frac{\partial}{\partial r}\rho u^r u^z\right)_{k+1/2} + \left(\frac{\rho u^r u^z}{r}\right)_{k+1/2}\right.
$$

$$
\left. + \left(\frac{\partial}{\partial z}\rho(u^z)^2\right)_{k+1/2} + \left(\frac{\partial p}{\partial z}\right)_{k+1/2}\right],
$$

$$j^r_{m,k+1/2} = \rho(u^r_{k+1/2} - w^r_{k+1/2}), \quad j^z_{m,k+1/2} = \rho(u^z_{k+1/2} - w^z_{k+1/2}).$$

The components of the viscous shear-stress tensor have the form

$$\Pi^{rr}_{ns\,k+1/2} = \mu_{k+1/2}\left[\frac{4}{3}\left(\frac{\partial u^r}{\partial r}\right)_{k+1/2} - \frac{2}{3}\left(\frac{u^r}{r}\right)_{k+1/2} - \frac{2}{3}\left(\frac{\partial u^z}{\partial z}\right)_{k+1/2}\right],$$

$$\Pi^{rz}_{ns\,k+1/2} = \mu_{k+1/2}\left[\left(\frac{\partial u^r}{\partial z}\right)_{k+1/2} + \left(\frac{\partial u^z}{\partial r}\right)_{k+1/2}\right],$$

$$\Pi^{zr}_{ns\,k+1/2} = \mu_{k+1/2}\left[\left(\frac{\partial u^z}{\partial r}\right)_{k+1/2} + \left(\frac{\partial u^r}{\partial z}\right)_{k+1/2}\right],$$

$$\Pi^{zz}_{ns\,k+1/2} = \mu_{k+1/2}\left[\frac{4}{3}\left(\frac{\partial u^z}{\partial z}\right)_{k+1/2} - \frac{2}{3}\left(\frac{u^r}{r}\right)_{k+1/2} - \frac{2}{3}\left(\frac{\partial u^r}{\partial r}\right)_{k+1/2}\right],$$

$$\Pi^{\phi\phi}_{ns\,i} = \mu_i\left[2\left(\frac{u^r}{r}\right)_i - \frac{2}{3}\left(\frac{\partial u^r}{\partial r}\right)_i - \frac{2}{3}\left(\frac{\partial u^z}{\partial z}\right)_i\right],$$

$$\Pi^{rr}_{k+1/2} = \Pi^{rr}_{ns\,k+1/2} + u^r_{k+1/2}w^{*r}_{k+1/2} + R^*,$$

$$\Pi^{rz}_{k+1/2} = \Pi^{rz}_{ns\,k+1/2} + u^r_{k+1/2}w^{*z}_{k+1/2},$$

$$\Pi^{zr}_{k+1/2} = \Pi^{zr}_{ns\,k+1/2} + u^z_{k+1/2}w^{*r}_{k+1/2},$$

$$\Pi^{zz}_{k+1/2} = \Pi^{zz}_{ns\,k+1/2} + u^z_{k+1/2}w^{*z}_{k+1/2} + R^*,$$

$$\Pi^{\phi\phi}_i = \Pi^{\phi\phi}_{ns\,i} + R^*.$$

Here we have introduced the following auxiliary quantities:

$$w^{*r}_{k+1/2} = \tau_{k+1/2}\left(\rho_{k+1/2}u^z_{k+1/2}\left(\frac{\partial u^r}{\partial z}\right)_{k+1/2}\right.$$

$$\left. + \rho_{k+1/2}u^r_{k+1/2}\left(\frac{\partial u^r}{\partial r}\right)_{k+1/2} + \left(\frac{\partial p}{\partial r}\right)_{k+1/2}\right),$$

$$w^{*z}_{k+1/2} = \tau_{k+1/2}\left(\rho_{k+1/2}u^z_{k+1/2}\left(\frac{\partial u^z}{\partial z}\right)_{k+1/2}\right.$$

$$\left. + \rho_{k+1/2}u^r_{k+1/2}\left(\frac{\partial u^z}{\partial r}\right)_{k+1/2} + \left(\frac{\partial p}{\partial z}\right)_{k+1/2}\right),$$

$$R^* = \tau_{k+1/2}\left[u^z_{k+1/2}\left(\frac{\partial p}{\partial z}\right)_{k+1/2} + u^r_{k+1/2}\left(\frac{\partial p}{\partial r}\right)_{k+1/2}\right.$$

$$\left. + \gamma p_{k+1/2}\left(\left(\frac{\partial u^r}{\partial r}\right)_{k+1/2} + \frac{u^r_{k+1/2}}{r} + \left(\frac{\partial u^z}{\partial z}\right)_{k+1/2}\right)\right].$$

The radial and axial components of the heat flux are calculated by the formulas

$$q^r_{k+1/2} = q^r_{ns} - u^r_{k+1/2} R^q_{k+1/2}, \qquad q^z_{k+1/2} = q^z_{ns} - u^z_{k+1/2} R^q_{k+1/2},$$

where

$$R^q_{k+1/2} = \tau_{k+1/2}\, \rho_{k+1/2} \times \left[\frac{1}{\gamma - 1} \left(u^r_{k+1/2} \left(\frac{\partial}{\partial r} \frac{p}{\rho} \right)_{k+1/2} + u^z_{k+1/2} \left(\frac{\partial}{\partial z} \frac{p}{\rho} \right)_{k+1/2} \right) \right.$$
$$\left. + p_{k+1/2} \left(u^r_{k+1/2} \left(\frac{\partial}{\partial r} \frac{1}{\rho} \right)_{k+1/2} + u^z_{k+1/2} \left(\frac{\partial}{\partial z} \frac{1}{\rho} \right)_{k+1/2} \right) \right].$$

6.5 Approximation of Boundary Conditions

To close the finite-difference form of the QGD system, we construct an approximation of boundary conditions.

Let the spatial grid be chosen in such a way that its nodes lie on the boundary of the computational domain (see Fig. 6.4).

In Fig. 6.4, B_j is a boundary node and M_i are interior nodes of the grid. Let us consider an approximation of the boundary conditions for a certain gas-dynamic quantity $f = (u, \rho, p, T)$ at the node B_j. For other boundary nodes, the approximation is performed similarly.

We approximate the Dirichlet boundary condition

$$f \big|_\Gamma = f_0$$

as follows:

$$f(B_j) = f_0,$$

where Γ is the boundary of the domain.

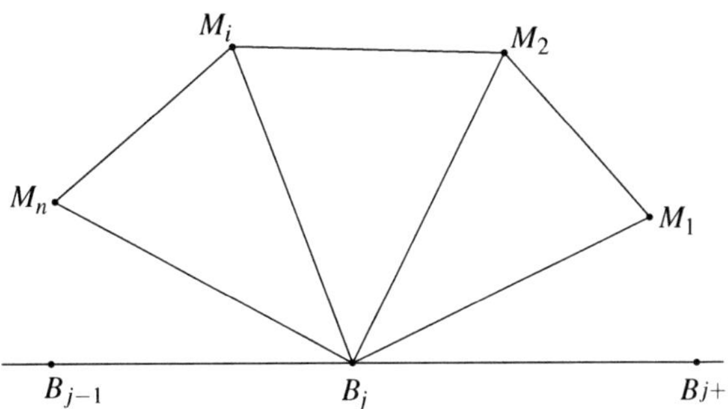

Fig. 6.4 Approximation of boundary conditions

We approximate the boundary conditions of the second and third kinds using interior nodes of the grid.

As a result of computation according to the finite-difference scheme (6.5), at a certain step in time, let the values of the quantity f be found at the interior nodes of the grid. Let us find the values on the boundary through the mean-weighted values of f at the interior nodes whose triangles contain this boundary node as a vertex. For the boundary node B_j, these are the interior nodes $M_1, M_2, \ldots, M_n$.

We write the Neumann boundary condition

$$\left. \frac{\partial f}{\partial n} \right|_\Gamma = q_0$$

in the form

$$f(B_j) = \frac{1}{N} \sum_{i=1}^{N} f(M_i) + \frac{q_0}{N} \sum_{i=1}^{N} L(B_j, M_i),$$

where $L(B_j, M_i)$ is the projection of the segment $B_j M_i$ on the inward normal to the boundary.

We approximate the boundary condition of the third kind

$$\left[\frac{\partial f}{\partial n} + \beta f \right]\Bigg|_\Gamma = q_0$$

as follows:

$$f(B_j) = \frac{\sum_{i=1}^{N} f(M_i) + q_0 \sum_{i=1}^{N} L(B_j, M_i)}{N + \beta \sum_{i=1}^{N} L(B_j, M_i)}.$$

6.6 Computation of the Flow in a Neighborhood of a Cylinder

As an example of using the algorithm written above, we present the numerical simulation of the problem on the transverse streamline flow of a circular cylinder by the flow of a subsonic viscous gas. This problem is the object of studies of many authors (see, e.g., [23, 38, 50, 138, 165, 209]). The interest to it is because of the existence of various flow regimes in the wake of the cylinder corresponding to the same boundary conditions of the problem. The problem considered is a known test for verifying the efficiency of numerical algorithms (see [115]).

The flow structure in the wake of viscous incompressible flows is determined by the Reynolds number

$$\mathrm{Re} = \frac{D \rho_0 u_0}{\mu_0}, \tag{6.31}$$

where D is the diameter of the cylinder, u_0 is the velocity of the incident flow, and μ_0 and ρ_0 are the dynamical viscosity and the density of the fluid in the incident flow, respectively. It was found experimentally that for $1 < \mathrm{Re} < \mathrm{Re}_1$ ($\mathrm{Re}_1 \approx 40$), the regime of the viscous flow is stable, and it is a system of two symmetric vortex structures after the cylinder; for $\mathrm{Re}_1 < \mathrm{Re} < \mathrm{Re}_2$ ($\mathrm{Re}_2 \approx 150$), after the cylinder, regular periodic derangements arise, which compose the so-called Karman street in the wake of the cylinder. The number Re_1 is the critical Reynolds number. To characterize the frequency of the arising oscillations, one introduces the dimensionless quantity, the Struchal number

$$\mathrm{Sh} = \frac{D}{T u_0}, \tag{6.32}$$

where T is the period of oscillation.

Rayleigh[1] suggested an empirical dependence of the frequency of oscillations on the Reynolds number in the interval $(\mathrm{Re}_1, \mathrm{Re}_2)$ in the form

$$\mathrm{Sh} = a \left(1 - \frac{b}{\mathrm{Re}} \right). \tag{6.33}$$

According to studies of Rayleigh, the numerical coefficients are $a = 0.195$ and $b = 16.3$. More later improved data (see [38, 165]) define these coefficients as $a = 0.212$ and $b = 21.2$.

We consider the problem on streamline flow of the cylinder in the two-dimensional setting in the Cartesian coordinate system. The computation is performed on a nonstructured grid whose example is presented in Fig. 6.5.

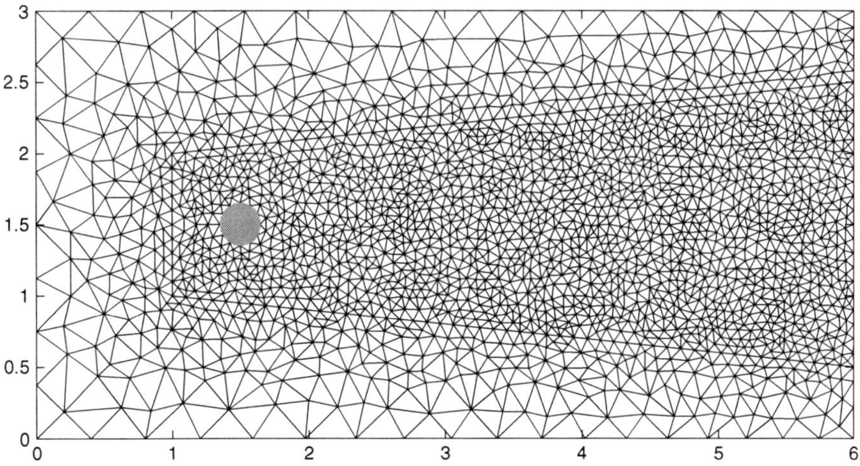

Fig. 6.5 Computation domain and distribution of nodes of the grid

---

[1] L. Rayleigh, "Aeolian tones," *Philos. Mag.*, **29**, 433 (1915).

The reduction of the problem to the dimensionless form and the choice of boundary conditions are performed in accordance with the procedure described in Sect. 5.10, where the diameter of the cylinder D is chosen as a characteristic linear size.

The initial conditions are the uniform gas flow with the density $\rho_0 = 1$, the pressure $p_0 = 1/(\gamma\,\mathrm{Ma}^2)$, the Mach number $\mathrm{Ma} = 0.1$, the speed $u_{0x} = 1$ along the axis x, and the speed $u_{0y} = 0$ along the axis y. The conditions on the upper and lower boundaries of the computational domain have the form

$$\frac{\partial \rho}{\partial n} = 0, \quad \frac{\partial p}{\partial n} = 0, \quad u_x = 1, \quad u_y = 0,$$

on the left boundary, they are

$$\rho = 1, \quad \frac{\partial p}{\partial n} = -0.1, \quad u_x = 1, \quad u_y = 0,$$

and on the right boundary, they are

$$\frac{\partial \rho}{\partial n} = 0, \quad p = \frac{1}{\gamma\,\mathrm{Ma}^2}, \quad \frac{\partial u_x}{\partial n} = 0, \quad \frac{\partial u_y}{\partial n} = 0.$$

On the lateral boundary of the cylinder, we set

$$\frac{\partial \rho}{\partial n} = 0, \quad \frac{\partial p}{\partial n} = 0, \quad u_x = 0, \quad u_y = 0.$$

The computation is performed for the Reynolds numbers corresponding to two flow regimes, namely, for $\mathrm{Re} = 20$, which corresponds to the stationary case, and for $\mathrm{Re} = 90$ and 100, which correspond to the self-oscillatory regime. The effects related to the compressibility of the medium are of order $O(\mathrm{Ma}^2)$, and we can assume that they are unessential in this problem.

In Fig. 6.6, we present the distribution of the density and streamlines for the stationary regime, $\mathrm{Re} = 20$. The computation was carried out on the grid with 2018 nodes and 3840 triangular elements.

The results of computations for the oscillatory regime for $\mathrm{Re} = 90$ are carried out in the dimensional form. The streamline flow of the cylinder of radius $D = 0.3$ m by air being in normal conditions with speed $u_0 = 35.31$ m/s was considered. The grid consisting of 2191 nodes and 4307 triangles was used (Fig. 6.5).

In Fig. 6.7, we present the dependence of the speed u_x on time t for the case of the self-oscillatory regime for $\mathrm{Re} = 90$ at the points with coordinates $(3, 1.5)$.

According to Fig. 6.7, the period of oscillations is equal to $T = 0.059$ s, which corresponds to $\mathrm{Sh} = 0.147$. The period of oscillations and the dimensionless frequency calculated by the Rayleigh formula (6.33) for the incompressible flow with coefficients $a = 0.212$ and $b = 21.2$ are equal to $T = 0.0524$ s and $\mathrm{Sh} = 0.162$. The latter values are sufficiently close to those obtained in numerical computing. If

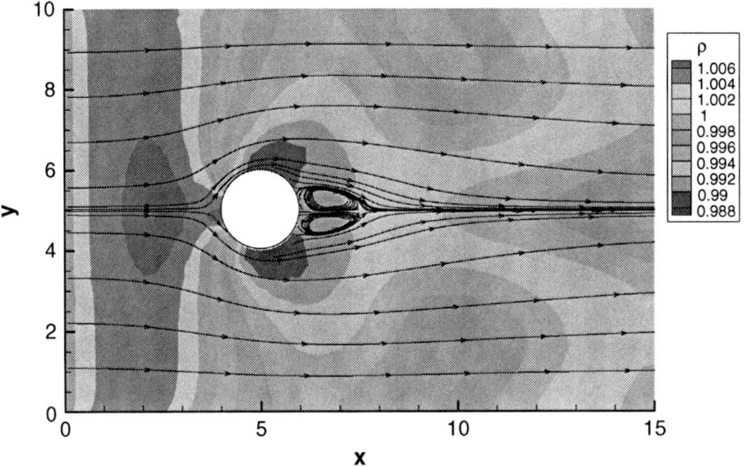

Fig. 6.6 A fragment of the distribution of the density ρ and streamlines for Re $= 20$

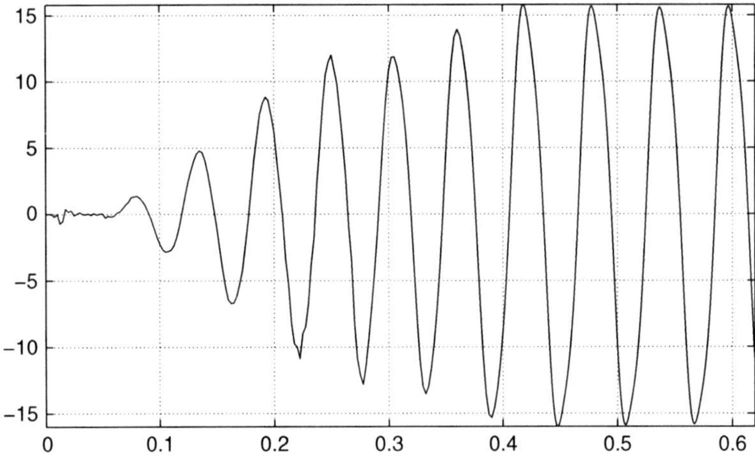

Fig. 6.7 Time dependence of the velocity component u_y for Re $= 90$

we thicken the grid in a neighborhood of the boundary of the cylinder (3472 nodes and 6839 triangles), then the accuracy of computations increases.

The streamlines and the isosurfaces of the square u^2 of the speed at the instant of time $t = 0.5$ s are presented in Fig. 6.8 for the whole computation domain (upper figure) and its fragment (lower figure). In the upper figure, we additionally show the grid structure, and on the lower figure, we show the streamlines.

This computation shows that the QGD algorithm adequately describes the peculiarities and frequency characteristics of nonstationary flows of a viscous compressible gas.

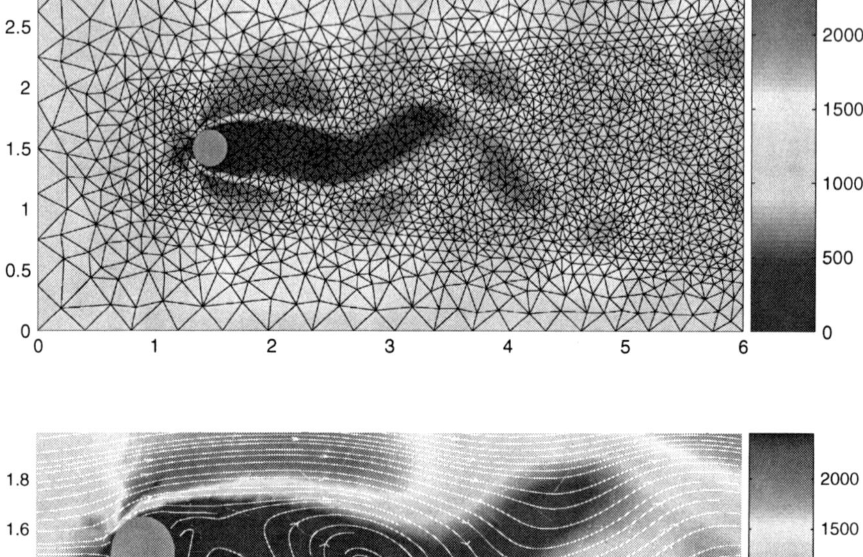

Fig. 6.8 Streamlines and the grid in the self-oscillation process for the Reynolds number Re $= 90$ and $t = 0.5$ s

In conclusion, we call attention to the fact that in all problems of numerical modelling of gas-dynamic flows considered above, one- or two-dimensional, stationary or not, by using rectangular or nonstructured grids, the same QGD algorithm was in fact applied. A unique adjustment parameter was the parameter τ whose variation from one computation to another is minimal. This universality of the algorithm is an incontestable merit of the QGD approach. Another merit of the QGD algorithms is the simplicity of their construction and numerical implementation, which is of especial importance in constructing numerical algorithms by using complicated three-dimensional spatial grids and an implementation of parallel computing.

Chapter 7
Quasi-hydrodynamic Equations and Flows of Viscous Incompressible Fluids

This chapter is devoted to the study of the quasi-hydrodynamic system. This system was proposed and developed in the works of Yu. V. Sheretov (see, e.g., [180–182, 184]). In particular, Sheretov carried out a detailed study of this system, obtained it in the case of flows of viscous incompressible fluids in the Oberbeck–Boussinesq approximation, and constructed a number of exact solutions, which were compared with the corresponding solutions of the Navier–Stokes equations.

In this chapter, we describe algorithms for simulation of flows of viscous incompressible fluids based on the quasi-hydrodynamic equations and present examples in which these equations are used for modelling two- and three-dimensional nonstationary flows. In the presentation, we use the results of [63–66, 73, 74, 111, 174].

7.1 Quasi-hydrodynamic System

The quasi-hydrodynamic Sheretov system (QHD) was written in Chap. 1 in the following form:

$$\frac{\partial \rho}{\partial t} + \operatorname{div} j_m = 0, \tag{7.1}$$

$$\frac{\partial (\rho u)}{\partial t} + \operatorname{div}(j_m \otimes u) + \nabla p = \rho F + \operatorname{div} \Pi, \tag{7.2}$$

$$\frac{\partial}{\partial t}\left[\rho\left(\frac{u^2}{2} + \varepsilon\right)\right] + \operatorname{div}\left[j_m\left(\frac{u^2}{2} + \varepsilon + \frac{p}{\rho}\right)\right] + \operatorname{div} q = (j_m \cdot F) + \operatorname{div}(\Pi \cdot u). \tag{7.3}$$

The closing relations for this system have the form

$$j_m = \rho(u - w), \tag{7.4}$$

$$\Pi = \Pi_{\mathrm{NS}} + \rho u \otimes w, \tag{7.5}$$

$$q = -\varkappa \nabla T, \tag{7.6}$$

$$w = \frac{\tau}{\rho}\left[\rho(u \cdot \nabla)u + \nabla p - \rho F\right], \tag{7.7}$$

T.G. Elizarova, *Quasi-Gas Dynamic Equations*, Computational Fluid and Solid Mechanics, DOI 10.1007/978-3-642-00292-2_7,

where

$$\Pi_{\mathrm{NS}} = \mu\left[(\nabla \otimes u) + (\nabla \otimes u)^T - \frac{2}{3}I \operatorname{div} u\right].$$

The quantity j_m is interpreted as the spatial-time vector of the mass flux density and ρu as the spatial-time momentum of the unit fluid volume.

Substituting expressions (7.4), (7.5), and (7.6) in Eqs. (7.1), (7.2), and (7.3), we obtain the following form of the quasi-hydrodynamic system:

$$\frac{\partial \rho}{\partial t} + \operatorname{div}(\rho u) = \operatorname{div}(\rho w), \tag{7.8}$$

$$\frac{\partial(\rho u)}{\partial t} + \operatorname{div}(\rho u \otimes u) + \nabla p = \rho F + \operatorname{div} \Pi_{\mathrm{NS}} + \operatorname{div}\left[(\rho w \otimes u) + (\rho u \otimes w)\right], \tag{7.9}$$

$$\frac{\partial}{\partial t}\left[\rho\left(\frac{u^2}{2} + \varepsilon\right)\right] + \operatorname{div}\left[\rho u\left(\frac{u^2}{2} + \varepsilon\right) + pu\right] + \operatorname{div} q = \rho F \cdot (u - w)$$
$$+ \operatorname{div}(\Pi_{\mathrm{NS}} \cdot u) + \operatorname{div}\left[\rho w\left(\frac{u^2}{2} + \varepsilon\right) + pw + \rho u(w \cdot u)\right]. \tag{7.10}$$

This system is complemented by the state equations (1.16). The entropy balance equation has the form (1.67).

System (7.8), (7.9), and (7.10) was obtained in [180] and studied in [181, 182, 184] in detail. This system was constructed on the basis of conservation laws by the technique presented in Sect. 3.6. Moreover, in expressions (3.61), it is necessary to set $\rho^{\star} = \rho$ and $p^{\star} = p$.

For many hydrodynamic flows, we can neglect variations of the fluid density. Assuming that ρ is constant and taking its variation into account only in the summand with the buoyancy force, we write system (7.8), (7.9), and (7.10) in the Oberbeck–Boussinesq approximation (see, e.g., [134]):

$$\operatorname{div} u = \operatorname{div} w, \tag{7.11}$$

$$\frac{\partial u}{\partial t} + \operatorname{div}(u \otimes u) + \frac{1}{\rho}\nabla p = \frac{1}{\rho}\operatorname{div} \Pi_{\mathrm{NS}} + \operatorname{div}[(w \otimes u) + (u \otimes w)] - \beta g T, \tag{7.12}$$

$$\frac{\partial T}{\partial t} + \operatorname{div}(uT) = \operatorname{div}(wT) + \chi \Delta T, \tag{7.13}$$

where $\rho = \operatorname{const} > 0$ is the mean value of the density, $u = u(x, t)$ is the hydrodynamic velocity, $p = p(x, t)$ is the pressure measured from the hydrostatic level, and $T = T(x, t)$ is the deviation of the temperature from its mean value T_0. Here we assume that $F = -\beta g T$, g is the gravitational acceleration, and β is the thermal expansion coefficient.

The Navier–Stokes viscous stress tensor and the vector w are calculated as follows:

$$\Pi_{NS} = \mu\left[(\nabla \otimes u) + (\nabla \otimes u)^T\right], \tag{7.14}$$

$$w = \tau\left[(u \cdot \nabla)u + \frac{1}{\rho}\nabla p + \beta g T\right]. \tag{7.15}$$

In Eqs. (7.17), (7.18), and (7.19) and also in expressions (7.14) and (7.15), the dynamical viscosity coefficient μ, the kinematic viscosity coefficient $v = \mu/\rho$, the thermal diffusivity $\chi = \varkappa/\rho$, the thermal expansion coefficient of the fluid β, and the parameter τ are assumed to be positive constants. The parameter τ can be calculated by the formula

$$\tau = \frac{\mu}{\rho c^2} = \frac{v}{c^2}, \tag{7.16}$$

where c is the sound speed in the fluid under the temperature T_0.

As $\tau \to 0$, the QHD system passes to the classical Navier–Stokes system in the Oberbeck–Boussinesq approximation (see [134]).

In [182], a number of exact solutions of the QHD equations corresponding to a number of the known solutions of the classical Navier–Stokes system were constructed, for example, the Archimedean law, the Couette and Poiseuille flows, the solutions of nonstationary Stokes and Rayleigh problems, and the solution of the problem on the flow in planar vertical and horizontal layers.

The QHD system in the Oberbeck–Boussinesq approximation can be reduced to the equivalent nondivergence form

$$\operatorname{div}(u - w) = 0, \tag{7.17}$$

$$\frac{\partial u}{\partial t} + (u - w) \cdot \nabla u + \frac{1}{\rho}\nabla p = \frac{1}{\rho}\operatorname{div}\Pi - \beta g T, \tag{7.18}$$

$$\frac{\partial T}{\partial t} + (u - w) \cdot \nabla T = \chi \Delta T. \tag{7.19}$$

The QHD system (7.17), (7.18), and (7.19) is dissipative. The equation describing the change in time of the kinetic energy with nonnegative dissipative function in the right-hand side holds for it. In the case where exterior forces vanish, taking the inner product of Eq. (7.18) and ρu and transforming the relation obtained taking Eq. (7.17) into account, we obtain the kinetic energy balance equation in the form

$$\frac{\partial}{\partial t}\left(\rho \frac{u^2}{2}\right) + \operatorname{div}\left[(u - w)\left(\rho \frac{u^2}{2} + p\right) - \rho u(w \cdot u) - (\Pi_{NS} \cdot u)\right] = -\Phi_{QHD},$$

where Φ_{QHD} is the dissipative function (1.67). This equation implies the theorem on the nondecrease of the total kinetic energy of the system in the form

$$\frac{dE}{dt} = -\int_{V_0} \Phi_{\mathrm{QHD}} dx,$$

where

$$E(t) = \int_{V_0} \frac{\rho u^2}{2} dx$$

is the total kinetic energy of the fluid in the volume V_0. A similar law on the change of the kinetic energy also holds for the Navier–Stokes system.

In studying flows in closed domains, one uses traditional boundary conditions for the QHD system conventional for the Navier–Stokes theory and complemented by the mass non-leakage condition in the form $(j_m \cdot n) = 0$, where n is the field of outward unit normal to the surface.

The QHD equations for the viscous incompressible fluid flow in the absence of exterior forces for the case of planar ($k = 0$) and axially symmetric ($k = 1$) flows can be written in the following form:

$$\frac{1}{r^k}\frac{\partial(r^k u_r)}{\partial r} + \frac{\partial u_z}{\partial z} = \frac{1}{r^k}\frac{\partial(r^k w_r)}{\partial r} + \frac{\partial w_z}{\partial z},$$

$$\frac{\partial u_r}{\partial t} + \frac{1}{r^k}\frac{\partial(r^k u_r^2)}{\partial r} + \frac{\partial(u_z u_r)}{\partial z} + \frac{1}{\rho}\frac{\partial p}{\partial r}$$

$$= \frac{2}{r^k}\frac{\partial}{\partial r}\left(r^k \nu \frac{\partial u_r}{\partial r}\right) + \frac{\partial}{\partial z}\left[\nu\left(\frac{\partial u_z}{\partial r} + \frac{\partial u_r}{\partial z}\right)\right] - k\frac{2\nu u_r}{r^2}$$

$$+ \frac{2}{r^k}\frac{\partial(r^k u_r w_r)}{\partial r} + \frac{\partial(u_r w_z)}{\partial z} + \frac{\partial(u_z w_r)}{\partial z},$$

$$\frac{\partial u_z}{\partial t} + \frac{1}{r^k}\frac{\partial(r^k u_r u_z)}{\partial r} + \frac{\partial(u_z^2)}{\partial z} + \frac{1}{\rho}\frac{\partial p}{\partial z}$$

$$= \frac{1}{r^k}\frac{\partial}{\partial r}\left[r^k \nu\left(\frac{\partial u_z}{\partial r} + \frac{\partial u_r}{\partial z}\right)\right] + 2\frac{\partial}{\partial z}\left(\nu \frac{\partial u_z}{\partial z}\right)$$

$$+ \frac{1}{r^k}\frac{\partial(r^k u_z w_r)}{\partial r} + 2\frac{\partial(u_z w_z)}{\partial z} + \frac{1}{r^k}\frac{\partial(r^k u_r w_z)}{\partial r},$$

$$\frac{\partial T}{\partial t} + \frac{1}{r^k}\frac{\partial}{\partial r}(r^k u_r T) + \frac{\partial}{\partial z}(u_z T) = \frac{1}{r^k}\frac{\partial}{\partial r}\left(r^k \chi \frac{\partial T}{\partial r}\right) + \frac{\partial}{\partial z}\chi\frac{\partial T}{\partial z}$$

$$+ \frac{1}{r^k}\frac{\partial}{\partial r}(r^k w_r T) + \frac{\partial}{\partial z}(w_z T),$$

where the corrections to the velocity have the form

$$w_r = \tau\left(u_r \frac{\partial u_r}{\partial r} + u_z \frac{\partial u_r}{\partial z} + \frac{1}{\rho}\frac{\partial p}{\partial r}\right), \quad w_z = \tau\left(u_r \frac{\partial u_z}{\partial r} + u_z \frac{\partial u_z}{\partial z} + \frac{1}{\rho}\frac{\partial p}{\partial z}\right).$$

7.2 Computational Algorithm

The QHD equations for describing viscous incompressible flows allow one to construct effective numerical algorithms for solving hydrodynamic equations in the variables velocity–pressure. These new algorithms have a number of advantages as compared with the traditional numerical methods constructed on the basis of the Navier–Stokes equations.

The QHD algorithms for the simulation of viscous incompressible flows have the following peculiarities that differ from the algorithms based on the Navier–Stokes system:

1. Additional dissipative terms in the system of the QHD equations serve as an effective regularizer of the numerical algorithm. This allows one to use the central difference approximations for all spatial derivatives, including convective terms. As a rule, in algorithms based on the Navier–Stokes system, more complicated approximations of the convective summands are used.
2. In the framework of the QHD approach, the Poisson equation for the pressure directly follows from the continuity equation. In the Navier–Stokes model, the equation for the pressure is a consequence of the equations of motion and the continuity equation, which complicates the procedure of the numerical simulation of the system.
3. In contrast to the Navier–Stokes system, in the QHD model, the boundary conditions for the pressure necessary for the solution of the Poisson equation are written in an explicit way. These conditions directly follow from the given conditions for the vector of the mass flux density j_m on the boundary.

The latter fact simplifies the numerical solution of the Poisson equation. In particular, this gives a possibility of finding the values of the pressure and the components of the velocity in the same nodes of the finite-difference scheme, which allow us to avoid the introduction of the so-called scattered spatial grids. The use of such grids is especially difficult for the computation on nonstructured or three-dimensional meshes.

Let us present an example of the finite-difference algorithm based on the QHD equations. For this purpose, we write system (7.11), (7.12), and (7.13) in the Oberbeck-Boussinesq approximation in the dimensionless form for the case of planar nonstationary flows in the case where $g = (0, g)$ (i.e., the gravitational force acts in the direction of the y-axis):

$$\frac{\partial u}{\partial x} + \frac{\partial v}{\partial y} = \tau \frac{\partial}{\partial x}\left(u\frac{\partial u}{\partial x} + v\frac{\partial u}{\partial y} + \frac{\partial p}{\partial x}\right) + \tau \frac{\partial}{\partial y}\left(u\frac{\partial v}{\partial x} + v\frac{\partial v}{\partial y} + \frac{\partial p}{\partial y} - \mathrm{Gr}\,T\right),$$

$$(7.20)$$

$$\frac{\partial u}{\partial t}+\frac{\partial}{\partial x}(u^2)+\frac{\partial}{\partial y}(uv)+\frac{\partial p}{\partial x}$$

$$=\frac{2}{\mathrm{Re}}\frac{\partial}{\partial x}\left(\frac{\partial u}{\partial x}\right)+\frac{1}{\mathrm{Re}}\frac{\partial}{\partial y}\left(\frac{\partial u}{\partial y}\right)+\frac{1}{\mathrm{Re}}\frac{\partial}{\partial y}\left(\frac{\partial v}{\partial x}\right)$$

$$+2\tau\frac{\partial}{\partial x}u\left(u\frac{\partial u}{\partial x}+v\frac{\partial u}{\partial y}+\frac{\partial p}{\partial x}\right)+\tau\frac{\partial}{\partial y}v\left(u\frac{\partial u}{\partial x}+v\frac{\partial u}{\partial y}+\frac{\partial p}{\partial x}\right)$$

$$+\tau\frac{\partial}{\partial y}u\left(u\frac{\partial v}{\partial x}+v\frac{\partial v}{\partial y}+\frac{\partial p}{\partial y}-\mathrm{Gr}\,T\right),\tag{7.21}$$

$$\frac{\partial v}{\partial t}+\frac{\partial}{\partial y}(v^2)+\frac{\partial}{\partial x}(uv)+\frac{\partial p}{\partial y}=\frac{2}{\mathrm{Re}}\frac{\partial}{\partial y}\left(\frac{\partial v}{\partial y}\right)+\frac{1}{\mathrm{Re}}\frac{\partial}{\partial x}\left(\frac{\partial v}{\partial x}\right)$$

$$+\frac{1}{\mathrm{Re}}\frac{\partial}{\partial x}\left(\frac{\partial u}{\partial y}\right)+2\tau\frac{\partial}{\partial y}v\left(u\frac{\partial v}{\partial x}+v\frac{\partial v}{\partial y}+\frac{\partial p}{\partial y}-\mathrm{Gr}\,T\right)$$

$$+\tau\frac{\partial}{\partial x}u\left(u\frac{\partial v}{\partial x}+v\frac{\partial v}{\partial y}+\frac{\partial p}{\partial y}-\mathrm{Gr}\,T\right)$$

$$+\tau\frac{\partial}{\partial x}v\left(u\frac{\partial u}{\partial x}+v\frac{\partial u}{\partial y}+\frac{\partial p}{\partial x}\right)+\mathrm{Gr}\,T,\tag{7.22}$$

$$\frac{\partial T}{\partial t}+\frac{\partial}{\partial x}(uT)+\frac{\partial}{\partial y}(vT)$$

$$=\mathrm{Pr}^{-1}\left(\frac{\partial^2 T}{\partial x^2}+\frac{\partial^2 T}{\partial y^2}\right)+\tau\frac{\partial}{\partial x}\left[T\left(u\frac{\partial u}{\partial x}+v\frac{\partial u}{\partial y}+\frac{\partial p}{\partial x}\right)\right]$$

$$+\tau\frac{\partial}{\partial y}\left[T\left(u\frac{\partial v}{\partial x}+v\frac{\partial v}{\partial y}+\frac{\partial p}{\partial y}-\mathrm{Gr}\,T\right)\right].\tag{7.23}$$

Here, u and v are the velocity components; Re, Pr, and Gr are the Reynolds number, the Prandtl number, and the Grashof number, respectively, and τ is the relaxation parameter (7.16) also written in the dimensionless form. The method for reducing system (7.11), (7.12), and (7.13) to the dimensionless form and the conditions on the boundaries are determined by a concrete problem and are described below.

For numerical solution of system (7.20), (7.21), (7.22), and (7.23), we use the finite-volume method. The system is approximated in the same way as the system of the quasi-gas-dynamic equations for describing flows of a viscous compressible gas (Chaps. 5 and 6).

All hydrodynamic quantities (the components of the velocity, the pressure, and the temperature) are referred to nodes of the computational grid. The values of quantities at half-integer nodes are found as half-sums of their values at the nearest integer nodes. The mixed derivatives are approximated by using the values of quantities at the centers of cells, which are calculated as the arithmetical mean of these quantities at the adjacent nodes. The spatial derivatives in system (7.20), (7.21), (7.22), and (7.23) are approximated by central differences.

The boundary of the computational domain lies at half-integer nodes of the grid. The approximation of the boundary conditions for the velocity and the temperature is performed by calculating the corresponding derivatives with second order of accuracy and is ensured by introducing additional fictitious layers of nodes along the exterior boundaries of the computational domain.

The derivatives in time are approximated by the forward differences with the first-order accuracy. The velocity and temperature fields at the next time step are found by using the explicit scheme from the finite-difference analogs of Eqs. (7.21), (7.22), and (7.23). The stability of the numerical algorithm is determined by the value of the parameter τ. Its value is referred either to the step of the spatial grid [174, 175] or to the values of dimensionless parameters of the problem, the Reynolds, Grashof, or Marangoni numbers. In the first case, the quantity τ is calculated as follows:

$$\tau = \alpha\sqrt{h_i^2 + h_j^2}, \tag{7.24}$$

where α is the numerical coefficient chosen in the computation process for ensuring the stability of computations. In the second case, for example, this parameter is chosen in the form $\tau = \alpha/\mathrm{Re}$, where $0 < \alpha < 1$. As numerical simulations show, the finite-difference algorithm is conditionally stable. The time step Δt is related with the parameter τ by the formula $\Delta t = \beta\tau$, where β is a numerical coefficient of order 1.

A flow is assumed to be stationary if the following condition holds:

$$\varepsilon_u = \max_{ij}\left|\frac{u_{ij}^{n+1} - u_{ij}^{n}}{\Delta t}\right| \le 0.001, \tag{7.25}$$

where n is the number of step in time.

At each step in time, the pressure field is found by the velocity and temperature fields by solving the continuity equation of the QHD system, for example, in the form of the Poisson equation

$$\frac{\partial^2 p}{\partial x^2} + \frac{\partial^2 p}{\partial y^2}$$
$$= \frac{1}{\tau}\left(\frac{\partial u}{\partial x} + \frac{\partial v}{\partial y}\right) - \frac{\partial}{\partial x}\left(u\frac{\partial u}{\partial x} + v\frac{\partial u}{\partial y}\right) - \frac{\partial}{\partial y}\left(u\frac{\partial v}{\partial x} + v\frac{\partial v}{\partial y} - \mathrm{Gr}\,T\right), \tag{7.26}$$

which is an equivalent representation of Eq. (7.20) for $\tau = $ const. Note that the right-hand side of Eq. (7.26) differs from the right-hand side of the Poisson equation in the Navier–Stokes model only by the term with the coefficient $1/\tau$.

The finite-difference approximation of this equation is constructed in the same way as that for the equations of motion. The boundary conditions for the pressure follow from the conditions for the velocity and the mass flux density on the boundary. The conditions for the pressure are approximated with the second order of accuracy by introducing the layer of fictitious nodes along the boundary. As a rule, the computational expenditures in solving Eq. (7.26) determine the effectiveness of the algorithm as a whole.

To solve Eq. (7.26), one can use exact as well as iterative methods for solution. In particular, in [175], the method of steepest gradient decent was applied. To solve problems of heat convection, one uses the preconditioned generalized method of conjugate gradients, where the preconditioner is constructed by using a pointwise incomplete decomposition of the matrix of the linear system of equations $Ax = B$ (see [151]). Also, the sweep-like methods [191] and the Cholesky conjugate gradient method [74] were used. The terminating condition for the iteration described below has the form

$$\varepsilon_p = \left[\sum_{ij} \left(p_{\bar{x}x,ij} + p_{\bar{y}y,ij} + f_{ij} \right)^2 h^2 \right]^{1/2} \le 10^{-5}, \qquad (7.27)$$

where f_{ij} is the right-hand side of the finite-difference analog of Eq. (7.26).

It is convenient to perform the analysis and visualization of the results of simulation of two-dimensional flows by using the stream function. The stream function ψ for the QHD equations is constructed on the basis of the vector field $u - w$ for which the solenoidal condition $\mathrm{div}(u - w) = 0$ holds:

$$u - w_1 = \frac{\partial \psi}{\partial y}, \quad v - w_2 = -\frac{\partial \psi}{\partial x}, \qquad (7.28)$$

where $w = (w_1, w_2)^T$. For small values of the parameter τ, the vectors u and $u - w$ are close.

In the next sections, we present examples of numerical simulation of the problems on the flows in a channel and a cavity and also the problems of heat convection carried out by using the QHD algorithm.

An example of the adaptation of the numerical algorithm for simulations on non-structured grids can be found in [92].

7.3 Backward-Facing Step Flows in a Channel

For laminar flows, the size of the detach zone after the backward-facing step is a sensitive characteristic, which substantially depends on the velocity of the flow and the geometry of the problem. The dependence of the size of the detach zone on the

Reynolds number and on the relative height of the step is presented, for example, in [196]. The length of the detach zone grows almost linearly as the Reynolds number grows. Laminar flows after a backward-facing step admit successful numerical modelling, and for nonlarge Reynolds numbers, the computation results of various authors in the two-dimensional setting correspond to each other and to experimental data (see, e.g., [112, 196]). This allows one to use this problem as a test for the approbation of new numerical algorithms.

The results presented below are based on [65, 66].

7.3.1 Formulation of the Problem

Consider a planar two-dimensional flow of isothermal fluid in a channel of height H and length L. In the input section of the channel, there is a narrowing, the size of which is determined by the height h of the step. The scheme of the computational domain and the flow formed is presented in Fig. 7.1.

This flow is described by system (7.11) and (7.12) in the absence of the buoyancy force, i.e., where $\beta g T = 0$.

Let us reduce the QHD system to the dimensionless form using the relations

$$x = \tilde{x}h, \quad y = \tilde{y}h, \quad u = \tilde{u}U_0, \quad v = \tilde{v}U_0,$$

$$p = \tilde{p}\rho U_0^2, \quad t = \frac{\tilde{t}h}{U_0}, \quad \mathrm{Re} = \frac{U_0 h}{\nu},$$

where

$$U_0 = \frac{1}{H - h} \int_h^H u_0(y)\,dy$$

is the velocity of the fluid in the channel averaged in the entrance section, ν is the kinematic viscosity coefficient, and $u_0(y)$ is a given velocity profile in the input section. In the dimensionless form, the system obtained coincides with system (7.20), (7.21), and (7.22) for $\mathrm{Gr} = 0$. We complement the system by the following boundary conditions:

$$y = 0, \quad 0 < x < L, \quad u = v = 0, \quad \frac{\partial p}{\partial y} = 0$$

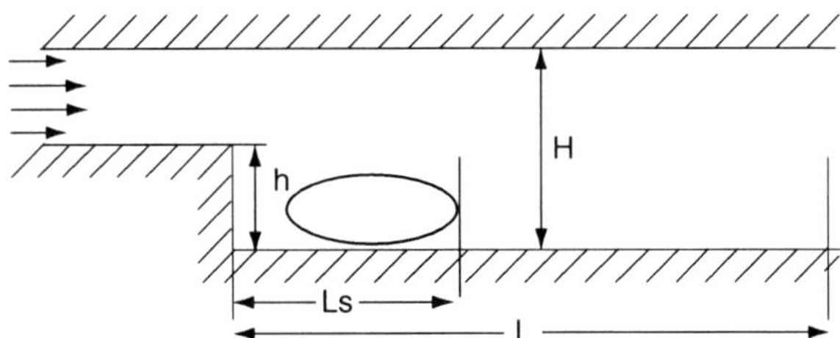

Fig. 7.1 Scheme of the computational domain

on the lower rigid wall;

$$y = H, \quad 0 < x < L, \quad u = v = 0, \quad \frac{\partial p}{\partial y} = 0$$

on the upper rigid wall;

$$x = 0, \quad 0 < y < h, \quad u = v = 0, \quad \frac{\partial p}{\partial x} = 0$$

on the left rigid wall;

$$x = 0, \quad h < y < H, \quad u = u_0(y), \quad v = 0, \quad \frac{\partial p}{\partial x} = \text{const},$$

the inflow on the left boundary; and

$$x = L, \quad 0 < y < H, \quad \frac{\partial u}{\partial x} = \frac{\partial v}{\partial x} = 0, \quad p = 0$$

on the right boundary.

The boundary conditions for the pressure on the rigid walls follow from the adhesion conditions for the velocity components, $u = 0$ and $v = 0$, and from the non-leakage conditions for the mass flow density (7.4). The relations $w_x = 0$ and $w_y = 0$, which hold on the walls, are a consequence of the later condition.

The gradient of the pressure in the input of the channel can be calculated as follows. Assign the velocity profile in the input of the channel in the form of the Poiseuille parabola (see [134, 144]):

$$u_0(y) = \frac{\text{Re}}{2} \frac{\partial p}{\partial x} (H - y)(h - y). \tag{7.29}$$

The rate of flow in the input section is calculated by the formula

$$J = \int_h^H [u(0, y) - w_x(0, y)] dy = -\frac{\text{Re}}{12} (H - h)^3 \frac{\partial p}{\partial x} - \tau (H - h) \frac{\partial p}{\partial x}. \tag{7.30}$$

From Eq. (7.30), we find the value of the pressure gradient in the input:

$$\frac{\partial p}{\partial x} = -\frac{12J}{\text{Re}(H - h)^3} \left[1 + \frac{12\tau}{\text{Re}(H - h)^2} \right]^{-1}. \tag{7.31}$$

As the initial condition, we take the equilibrium state $u = v = 0$. The gradient of the pressure at the initial instant of time is assumed to be constant in the whole flow field.

According to Eq. (7.16), for laminar flows, the dimensionless smoothing parameter τ is calculated as follows:

$$\tau = \frac{\mathrm{Ma}}{\mathrm{Re}_s} = \frac{\mathrm{Ma}^2}{\mathrm{Re}}, \quad \text{where} \quad \mathrm{Ma} = \frac{U_0}{c}, \quad \mathrm{Re}_s = \frac{ch}{\nu}, \tag{7.32}$$

Ma is the Mach number, Re is the Reynolds number, and Re_s is the Reynolds number calculated by the sound speed. For example, for air under a normal temperature, $c = 3.4 \times 10^4$ cm/s and $\nu = 0.15\,\mathrm{cm}^2$/s. Let $h = 10$ cm, then $\mathrm{Re}_s = 2 \times 10^6$. For incompressible or weakly compressible fluids, $\mathrm{Ma} \ll 1$, and the smoothing parameter turns out to be small.

To ensure the stability of computations, we choose the value of the parameter τ in the computation process and consider it as a small quantity, $0 < \tau < 1/\mathrm{Re}$. Since the Mach number in fluids is very small, to ensure the stability of computations, we must use artificial regularizers with overstated values of τ. The substitution of the true value of τ leads to a sharp diminishing of the step in time. In this problem, the parameter is chosen in the form $\tau = \alpha/\mathrm{Re}$, where α is a numerical coefficient.

7.3.2 Results of Numerical Simulation

The flow problem after the backward-facing step is solved for the Reynolds numbers $\mathrm{Re} = 100, 200, 300$, and 400; $h/H = 1/2$. The velocity profile in the input section is the Poiseuille parabola (7.29). The dimensionless rate of flow J is assumed to be equal to unity, which corresponds to the choice of the pressure gradient in the input of the channel in the form

$$\frac{\partial p}{\partial x} = -\frac{12}{\mathrm{Re}}\left[1 + \frac{12\tau}{\mathrm{Re}}\right]^{-1}.$$

For small τ and large Re, we can assume that

$$\frac{\partial p}{\partial x} = -\frac{12}{\mathrm{Re}}.$$

The calculated length of the detach zone after the backward-facing step is compared with the data of [112, 117, 196] and also with the graphs from [14].

The results obtained are systematized in Table 7.1. Here, L is the dimensionless length of the computational domain; N_x and N_y are the numbers of computational points in the directions of the x- and y-axes, respectively; L_s is the length of the detach zone; and N_t is the number of steps in time up to the convergence. The computations are performed on the spatial grid with equal steps $h_x = h_y = 0.025$. It is known that the use of equal h_x and h_y improves the accuracy of description of the detach flow.

In the computations listed in Table 7.1, the value of the smoothing parameter (7.32) is $\tau = 0.5/\mathrm{Re}$. The step of integration Δt with respect to time is the same for all variants and is equal to 10^{-4}.

Table 7.1 Comparison of QHD computations with the computations of [14, 112, 117, 196] and with the experimental data [14]

Re	100	200	300	400
L	7.5	5.0	7.5	10
$N_x \times N_y$	300×40	200×40	300×40	400×40
τ	0.005	0.0025	0.00166	0.00125
N_t	19800	~ 20000	~ 60000	~ 110000
L_s/h (QHD computation)	5.0	8.2	10.1	14.8
L_s/h (computation of [196])	4.43	7.5	10.0	–
L_s/h (computation of [117])	5	8.3	10.5	12
L_s/h (computation of [112])	5	8.5	12	–
L_s/h (computation of [14])	5.0	8.3	8.4	7.8
L_s/h (experiment of [14])	5.0	8.5	11.3	14.2

The computation was terminated if

$$
\delta p = \max \left| \frac{p^{n+1} - p^n}{\Delta t} \right| < 10^{-3},
$$

where n is the number of integration steps in time.

In all variants, the flow attains the stationary regime. The length L_s of the detach zone is found by the position of the zero streamline and is presented with accuracy of order 0.2. The comparison of the QHD computations with those based on the Navier–Stokes system and also with the experiment data of [14] shows a good correspondence of the length of the detach zone and the general picture of flow. In computations, we observe an almost linear growth of values of L_s as the number Re grows. We see that for Re $= 400$, the data of the QHD computation are in a good agreement with the experimental data of [14]. Very close results were obtained in a numerical computation of the gas flow after the backward-facing step for small Mach and Reynolds numbers. These results are presented in Appendix D.

For Re $= 100$ and 200, the process of stabilization of the flow is the birth and the subsequent growth of one vortex formation after the backward-facing step. For Re $= 300$ and 400, the stabilization process has oscillatory character and is accompanied by the birth and detach of vortex formations, but, in contrast to the flow regimes considered below, for large Reynolds numbers Re, this oscillatory process decays and leads to the formation of a single stationary vortex after the backward-facing step. Isolines of the stream function ψ constructed in accordance with Eq. (7.28) that illustrate the process of stabilization of the flow in time for the variants Re $= 100$ and 400 are presented in Figs. 7.2 and 7.3. The isolines are equidistant.

For a further growth of the Reynolds number, it is impossible to obtain a stationary solution, and the flow becomes turbulent (see [86, 87]).

For the variant Re $= 100$, we studied the influence of the regularization parameter τ and the convergence of the numerical solution on the grid. We have obtained that the length of the detach zone and the general structure of the flow are practically independent of the value of the regularization parameter τ as well as of the value of

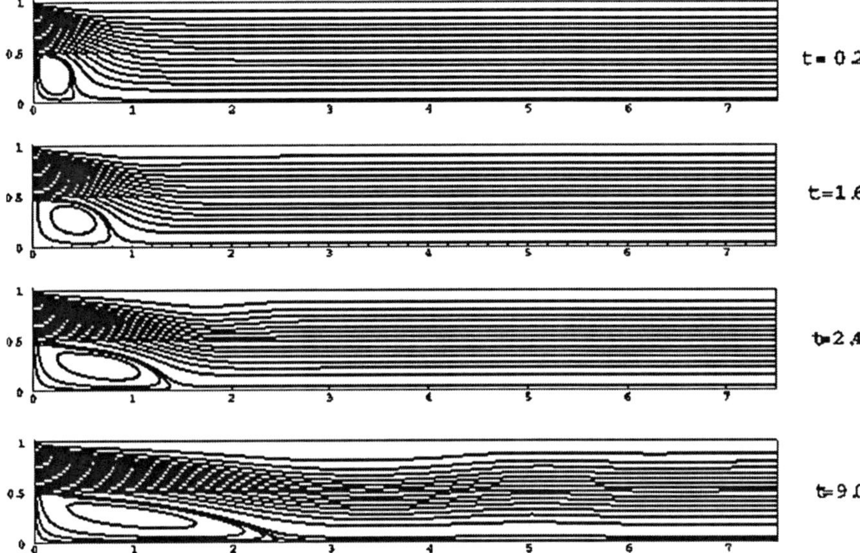

Fig. 7.2 Isolines of the stream functions for Re $= 100$ in the process of ascertainment of the flow

step in the space. The increase of τ leads to the smoothing of the flow structure and allows us to enlarge the step of integration of the system in time. The decrease of the spatial step allowed us to resolve the flow structure in more detail. For a constant flow rate, the solution is weakly sensitive to the choice of the pressure gradient in the input section.

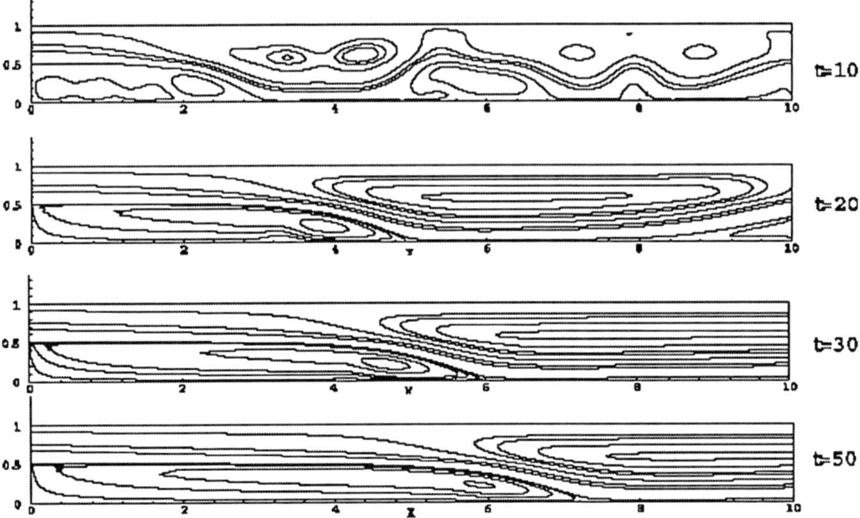

Fig. 7.3 Isolines of the stream function for Re $= 400$ in the process of ascertainment of the flow

These results are consistent with theoretical estimates according to which for the stationary flows, the additional QHD terms are small and the solution of the QHD equations is close to the solution of the Navier–Stokes equations. Moreover, the additional terms play the role of the regularizers that allow us to use a sufficiently simple, stable, and accurate numerical algorithm.

7.4 Heat Convection in a Square Cavity

We consider the flow of a heat-conducting, viscous, incompressible fluid in a square cavity with two vertical isothermal walls. The flow arises due to the difference of temperatures of the walls. The horizontal walls are adiabatic. The flow is described by system (7.20), (7.21), (7.22), and (7.23), where Re $= 1$ with the following boundary conditions:

$$y = 0, \quad 0 < x < 1, \quad u = 0, \quad v = 0, \quad \frac{\partial p}{\partial y} = \text{Gr } T, \quad \frac{\partial T}{\partial y} = 0$$

on the lower wall;

$$y = 1, \quad 0 < x < 1, \quad u = 0, \quad v = 0, \quad \frac{\partial p}{\partial y} = \text{Gr } T, \quad \frac{\partial T}{\partial y} = 0$$

on the upper wall;

$$x = 0, \quad 0 < y < 1, \quad u = 0, \quad v = 0, \quad \frac{\partial p}{\partial x} = 0, \quad T = 1$$

on the left lateral wall; and

$$x = 1, \quad 0 < y < 1, \quad u = 0, \quad v = 0, \quad \frac{\partial p}{\partial x} = 0, \quad T = 0$$

on the right lateral wall.

As in the previous problem, the boundary conditions for the pressure follow from the non-leakage condition for the mass flux on the boundary, which leads to the conditions $w_x = 0$ and $w_y = 0$ in Eq. (7.15).

As the initial condition, we use the unperturbed fields of velocity and temperature. System (7.20), (7.21), (7.22), and (7.23) is reduced to the dimensionless form by using the relations

$$x = \tilde{x} H, \quad y = \tilde{y} H, \quad u = \tilde{u} \frac{v}{H}, \quad v = \tilde{v} \frac{v}{H},$$

$$t = \tilde{t} \frac{H^2}{v}, \quad p = \tilde{p} \rho \left(\frac{v}{H} \right)^2, \quad T = \tilde{T} \Delta T, \tag{7.33}$$

where $\Delta T = T_1 - T_2$ is the difference of temperatures between the left and right walls of the domain considered and H is the size of the cavity. For the chosen reduction to the dimensionless form, the Grashof number is $\mathrm{Gr} = \beta g \Delta T H^3 / \nu^2$.

The dimensionless parameter τ has the form

$$\tau = \frac{\nu^2}{H^2 c^2} = \mathrm{Ma}_s^2,$$

where

$$\mathrm{Ma}_s = \frac{U}{c}, \quad U = \frac{\nu}{H}.$$

For the problem considered, the value of the smoothing parameter τ is very small and, therefore, in performing the calculations, we choose τ from the interval $0 < \tau < 1$.

The heat-convection problem in a closed cavity heating from the left is solved for moderate Grashof numbers $\mathrm{Gr} = 10^2$, 10^4, and 10^5, the Prandtl number $\mathrm{Pr} = 1$, and the parameter τ in the interval 10^{-5}–10^{-2}. Uniform grids 21×21, 41×41, and 81×81 were used.

The results of simulation are presented in Tables 7.2 and 7.3, where they are compared with the data of [47, 203]. In [203], the mass-heat-transfer problem in the Oberbeck-Boussinesq approximation is calculated on the basis of the Navier–Stokes system in the variables "stream function–velocity curl" on detailed grids, and the results are compared with the experiment data of [47]. The results of these works for moderate Grashof numbers can be considered as standard.

In the tables, we use the following notation:

$|\psi|_{\mathrm{mid}}$ is the absolute value of the stream function at the center of the cavity,

$|\psi|_{\mathrm{max}}$ is the maximum of the module of the stream function in the computational domain,

u_{max} is the maximum of the horizontal velocity in the middle vertical section,

v_{max} is the maximum of the vertical velocity in the middle vertical section,

Nu_0 is the mean Nusselt number on the left face of the cavity,

Table 7.2 Results of numerical simulation and their comparison with the computational data of [203] and the experimental data [47], $\mathrm{Gr} = 10^4$, $\mathrm{Pr} = 1$

τ	grid	$|\psi|_{\mathrm{max}}$	u_{max}	v_{max}	Nu_0	$\mathrm{Nu}_{\mathrm{max}}$	$\mathrm{Nu}_{\mathrm{min}}$
10^{-4}	21×21	5.044	15.938	19.513	2.306	3.939	0.579
	[203]	5.277	16.144	19.363	2.253	3.615	0.591
10^{-4}	41×41	5.099	16.005	19.663	2.281	3.708	0.591
	[203]	5.125	16.262	19.602	2.249	3.563	0.586
10^{-4}	81×81	5.113	16.070	19.663	2.275	3.649	0.581
	[203]	5.086	16.219	19.648	2.247	3.541	0.585
	[47]	5.071	16.178	19.617	2.238	3.528	0.586
10^{-3}	21×21	5.195	15.587	18.565	2.431	4.318	0.529
10^{-2}	21×21	6.723	13.182	11.542	3.850	9.268	0.352

Table 7.3 Results of numerical simulation and their comparison with the computational data of [203] and the experimental data [47], $Gr = 10^5$, $Pr = 1$

τ	grid	$|\psi|_{mid}$	$|\psi|_{max}$	u_{max}	v_{max}	Nu_0	Nu_{max}	Nu_{min}
10^{-5}	21×21	9.264	9.666	32.33	67.70	4.86	9.777	0.204
	[203]	10.259	10.86	40.30	65.07	4.53	8.123	0.762
10^{-5}	41×41	9.502	9.909	33.24	70.91	4.68	8.733	0.729
	[203]	9.388	9.918	36.63	68.11	4.55	7.968	0.730
	[47]	9.111	9.612	34.73	68.59	4.51	7.717	0.729
10^{-4}	21×21	9.358	9.754	32.20	66.71	4.91	9.931	-0.033
10^{-3}	21×21	10.400	10.74	33.12	57.86	5.73	12.091	-0.611

Nu_{max} is the maximum value of the Nusselt number on this face,

Nu_{min} is the minimum value of the Nusselt number on this face.

To calculate the dimensionless heat flux on the left lateral face (the Nusselt number Nu), we use the expression

$$Nu(y) = -\frac{\partial T(0, y)}{\partial x},$$

and the mean Nusselt number is calculated as follows:

$$Nu_0 = \int_0^1 Nu(y)dy.$$

For computing the local heat fluxes, we use the right finite-difference derivatives, and for the mean Nusselt number, we use the trapezium quadrature formula.

It follows from the tables that for small τ, even on the grid 21×21, the solution coincides well with the standard solution, i.e., small values of τ correspond to the convection regime in the test problem [203]. When the step of the spatial net is refined, the accuracy of solution increases, which is determined by the possibility of resolving the boundary layers.

In Fig. 7.4, we present the isolines of the stream function, the isotherms, and the isobars for the Grashof number $Gr = 10^4$ and the Prandtl number $Pr = 1$ (grid

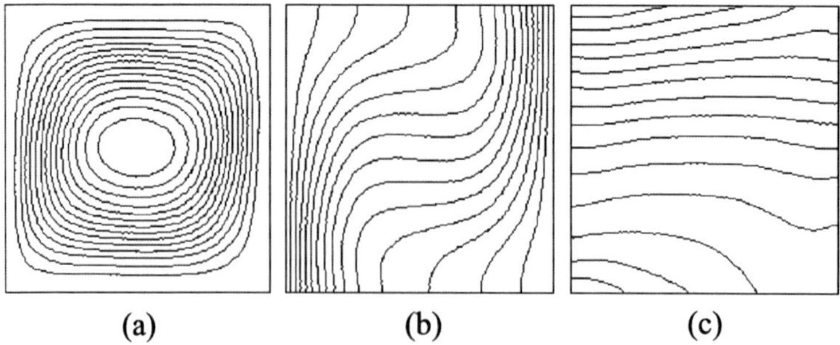

(a)   (b)   (c)

Fig. 7.4 Streamlines (**a**), isotherms (**b**), and isobars (**c**) for $Gr = 10^4$, $Pr = 1$, and $\tau = 10^{-4}$

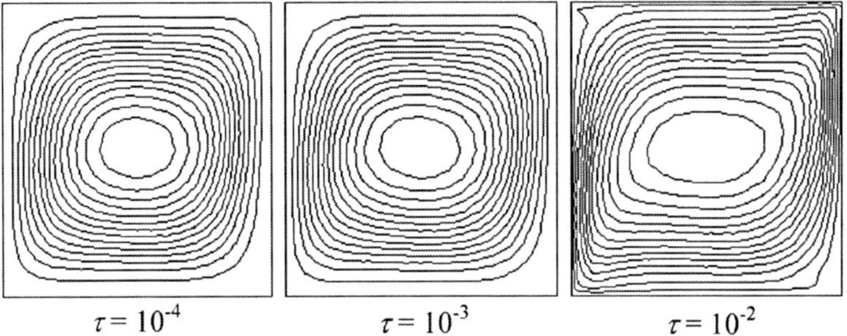

Fig. 7.5 Streamlines for different values of the parameter τ and for Gr $= 10^4$ and Pr $= 1$

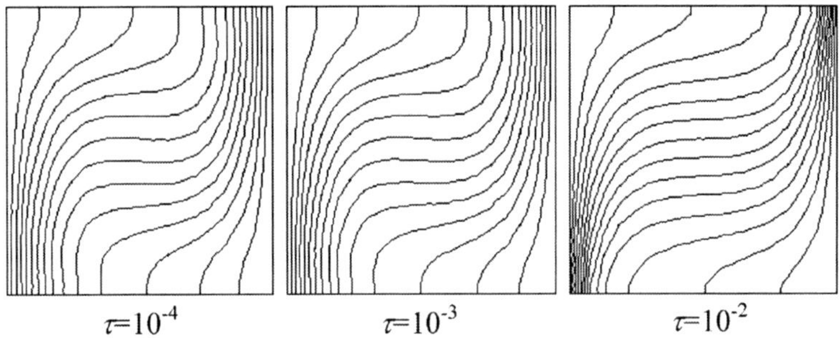

Fig. 7.6 Isotherms for different values of the parameter τ and for Gr $= 10^4$ and Pr $= 1$

41×41). It is clearly seen from the presented figures and Tables 7.2 and 7.3 that there is a good correspondence between the streamlines and isotherms and the results of [203]. The increase of τ leads to the strengthening of the nonlinear interaction in the medium and, as a consequence, to the distortion of isolines of the stream function and the isotherms (Figs. 7.5 and 7.6).

When τ decreases, to ensure the stability of computation, it is necessary to diminish the step in time.

It follows from the results presented that the numerical solutions of the QHD system coincide well with the standard solutions of the Navier–Stokes system for moderate Grashof numbers Gr and for $\tau \leq$ Gr^{-1}. The coincidence is improved when the spatial grid condenses.

7.5 Heat Convection for Low Prandtl Numbers

Let us consider an example of mathematically modelling the heat gravitational convection of an incompressible fluid in a rectangular cavity of height H and length L for small Prandtl numbers Pr (see [63]). This problem is the known test suggested in 1987 for analyzing numerical methods for computing convective flows in melts. The practical necessity of such computations is related to the fact that periodic oscillations of temperature in metallic melts (fluids with small Prandtl number) lead to serious difficulties in growing semiconducting crystals (see [20]).

The flow is described by system (7.20), (7.21), (7.22), and (7.23) with $\mathrm{Re} = 1$, which is reduced to the dimensionless form by using the relations

$$x = \tilde{x}H, \quad y = \tilde{y}H, \quad u = \tilde{u}\frac{\nu}{H}, \quad v = \tilde{v}\frac{\nu}{H},$$
$$t = \tilde{t}\frac{H^2}{\nu}, \quad p = \tilde{p}\rho\left(\frac{\nu}{H}\right)^2, \quad T = \tilde{T}\frac{\Delta T}{A}, \tag{7.34}$$

where $A = L/H$ and $\Delta T = T_1 - T_2$ is the difference of temperatures between the left and right walls. For the chosen reduction to the dimensionless form, we have $\mathrm{Gr} = \beta g \Delta T H^4/L\nu^2$, $\mathrm{Re} = 1$, $\mathrm{Pr} = \nu/\chi$, and $\tau = \mathrm{Ma}^2$, where $\mathrm{Ma} = \nu/(Hc)$ is the Mach number.

The cavity has the rigid lower wall, and its upper boundary can be either rigid (R-R case) or free (R-F case).

We add to system (7.20), (7.21), (7.22), and (7.23) the following boundary conditions:

$$u = 0, \quad v = 0, \quad \frac{\partial p}{\partial y} = \mathrm{Gr}\, T, \quad T(x) = A - x$$

on the lower wall,

$$u = 0, \quad v = 0, \quad \frac{\partial p}{\partial y} = \mathrm{Gr}\, T, \quad T(x) = A - x$$

on the upper boundary for the R-R case or

$$\frac{\partial u}{\partial y} = 0, \quad v = 0, \quad \frac{\partial p}{\partial y} = \mathrm{Gr}\, T, \quad T(x) = A - x$$

for the R-F case;

$$u = 0, \quad v = 0, \quad \frac{\partial p}{\partial x} = 0, \quad T = A$$

on the left lateral wall, and

$$u = 0, \quad v = 0, \quad \frac{\partial p}{\partial x} = 0, \quad T = 0$$

on the right lateral wall.

We can connect the dimensionless parameter τ with the Grashof number by the relation

$$\tau = \frac{1}{\mathrm{Gr}} \frac{\beta g \Delta T H^2}{L c^2} = \alpha / \mathrm{Gr}, \tag{7.35}$$

where α is very small. So, for example, for the heat convection of air for $\Delta T = 100^\circ$C in the square cavity of height $H = 1$ m, we have $\alpha \sim 3 \times 10^{-5}$. In performing computations, the parameter α must be chosen from the interval $0 < \alpha \leq 1$.

In most computations, α was chosen to be equal to unity. As numerical experiments show, the decrease of the parameter α leads to an insignificant improvement of the computational results, but, at the same time, this leads to a deterioration of stability of the numerical algorithm, which requires the diminishing of the step Δt of integration in time.

The problem on the heat convection in the rectangular cavity ($A = 4$) heating from the left is solved for moderate Grashof numbers and for small Prandtl number $\mathrm{Pr} = 0.015$ on the uniform spatial grid 22×82 with step $h = 1/20$. In all variants, the step in time is chosen to be equal to 10^{-6}.

For the R-R case and for $\mathrm{Gr} = 4 \times 10^4$, as the initial conditions, we use the velocity and temperature fields obtained in the computation with $\mathrm{Gr} = 3 \times 10^4$. In all other computations, as the initial conditions, we use the unperturbed velocity and temperature fields.

7.5.1 Results of Computations for the R-R Case

The computations are carried out for the moderate Grashof numbers $\mathrm{Gr} = 2 \times 10^4$, 3×10^4, and 4×10^4.

For $\mathrm{Gr} = 2 \times 10^4$, we obtain a stationary regime of flow (see Fig. 7.7). The streamlines form a single vortex expanded in length. Qualitatively, as well as quantitatively, the results obtained are in good correspondence with the data from [20,21], where the computations were carried out on very detailed spatial grids (in [21], this was done on the grid 81×321). In Table 7.4, the characteristic values of the stream function and the components of the velocity are given together in comparison with the analogous quantities from [20] and, moreover, they turn out to be very close. Since the reduction of the velocity to the dimensionless form in [20] was different from the present work, namely, $u = \tilde{u} v \sqrt{\mathrm{Gr}}/H$ and $v = \tilde{v} v \sqrt{\mathrm{Gr}}/H$, it follows that

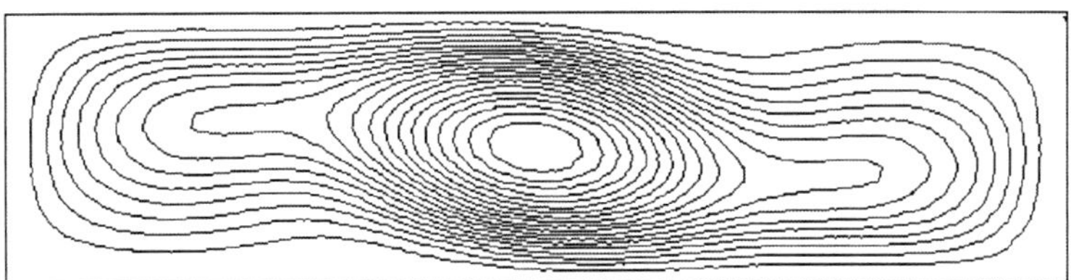

Fig. 7.7 Streamlines for $\mathrm{Gr} = 2 \times 10^4$

Table 7.4 Characteristic (normalized) values of the streamline function and the components of the velocity

	V^*	U^*	ψ^*
QHD algorithm	0.448	0.672	0.409
[20]	0.452–0.482	0.669–0.704	0.406–0.409

for comparison with the results of [20], in Table 7.4, we present the values of the following quantities:

$$\Psi^* = \max_{xy} \frac{|\Psi|}{\sqrt{Gr}}, \quad U^* = \max_{y} \frac{|u(y)|}{\sqrt{Gr}} \quad \text{for} \quad x = \frac{3A}{4},$$

$$V^* = \max_{x} \frac{|v(x)|}{\sqrt{Gr}} \quad \text{for} \quad y = \frac{1}{2}.$$

For $Gr = 3 \times 10^4$, we obtain the stationary regime of flow (Figs. 7.8 and 7.9) whose stabilization process has the form of damped oscillations. In Fig. 7.8, we carried out the comparison of evolutional curves of the horizontal component of the velocity at the center of the domain obtained in computations for $\alpha = 1$ (continuous line) and $\alpha = 0.1$ (dotted line). The curves obtained practically coincide, which testifies a weak dependence of solution on the regularization parameter in its range chosen. In most works, in this case, the oscillatory flow regime is obtained, but, however, as is shown in [24, 29], a stationary regime is possible. The stationary regime is the main vortex located near the center and two additional vortices in the

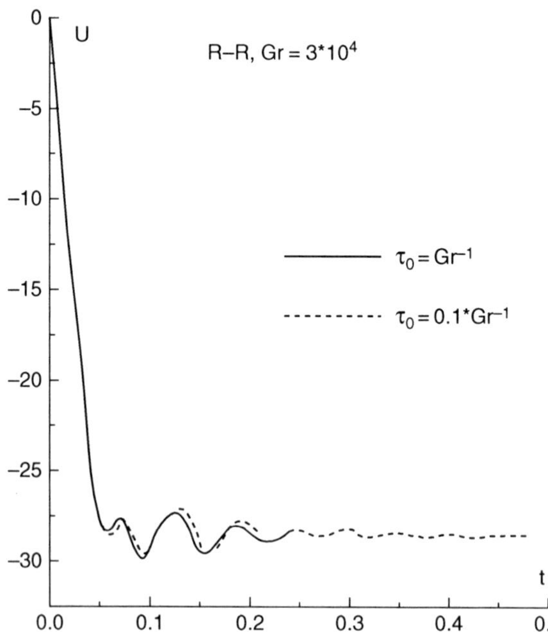

Fig. 7.8 Evolutional curves of the horizontal component of the velocity at the center of the domain for $Gr = 3 \times 10^4$

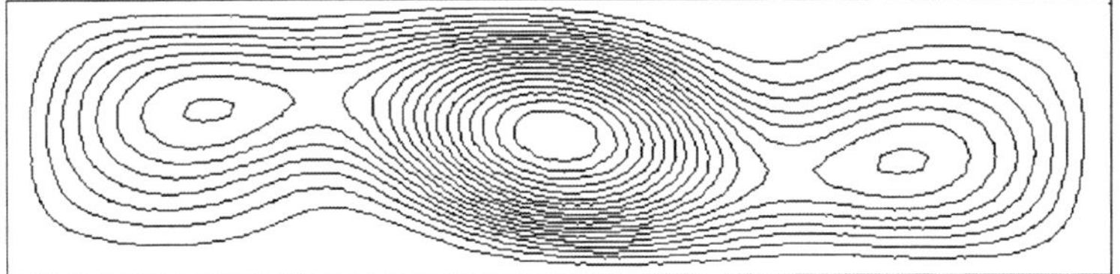

Fig. 7.9 Streamlines for Gr $= 3 \times 10^4$

left and right parts of the cavity. In [29], the stationary solution is obtained for this case and, moreover, the comparison of the results presented in this work with the results of computations of the author shows a good correspondence. In particular, we compare the profiles of the horizontal velocity along the vertical section lying on a distance $x = 3A/4$ from the left wall, the distributions of the vertical velocity in the section $y = 1/2$, and also the streamlines.

For Gr $= 4 \times 10^4$, we obtain the oscillatory flow regime (Figs. 7.10 and 7.11) whose period of oscillations can be estimated as $T_{\text{vib}} = 0.047$ (see Fig. 7.10), which corresponds to the frequency $f_1 = 1/T_{\text{vib}} = 21.28$. This result is very close to that presented in [20], where the results of computations of many authors are summarized. The frequency of oscillations obtained by different authors for this case is from 21.186 upto 22.35. The flow in this problem has a structure similar to the

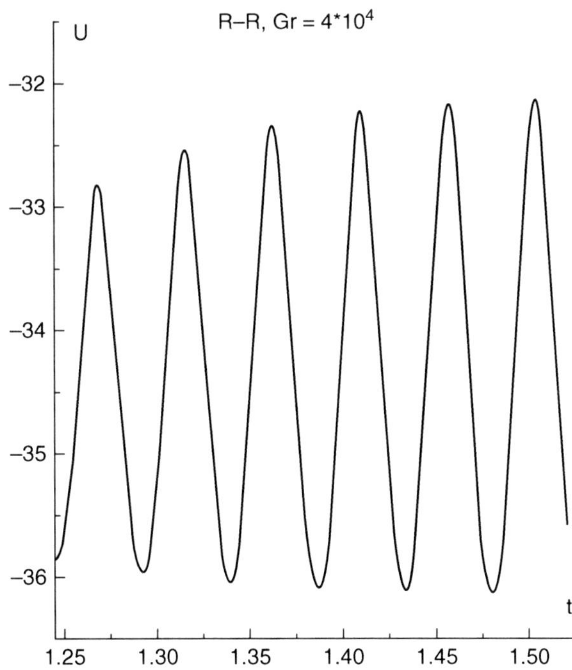

Fig. 7.10 Evolutional curves of the horizontal component of the velocity at the center of the domain for Gr $= 4 \times 10^4$

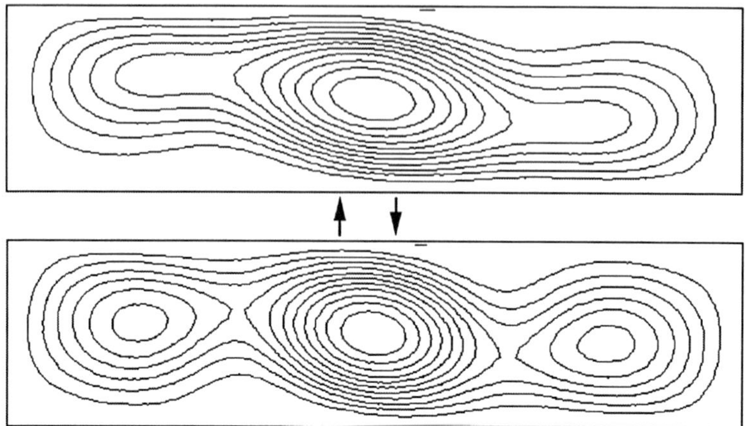

Fig. 7.11 Streamlines for $Gr = 4 \times 10^4$

previous variant, and, moreover, the oscillation process is the change of intensity of the vortex formations and is in good correspondence with the results of [29].

7.5.2 Results of Computations for the R-F Case

In contrast to the previous case, in these variants, on the upper wall, we pose the sliding condition for the tangential component of velocity, which simplifies the process of oscillation birth.

For $Gr = 10^4$, we obtain a stationary flow regime (Fig. 7.12) in which an asymmetric structure consisting of two cells is formed. The larger of them lies near the cold wall and the other, lesser in value, lies between the center of cavity and the hot left wall. The figures presented are in a close correspondence to analogous figures presented in [153].

For $Gr = 2 \times 10^4$, we obtain the oscillatory flow regime (Figs. 7.13 and 7.14) whose period of oscillations can be estimated as $T_{vib} = 0.0646$ (see Fig. 7.13), which corresponds to the frequency $f_1 = 15.84$. In the survey [20], the frequency of oscillations considered by different authors varies from 15.580 upto 17.2. The flow in this problem is a vortex displaced to the right (cold) wall of the cavity. In the

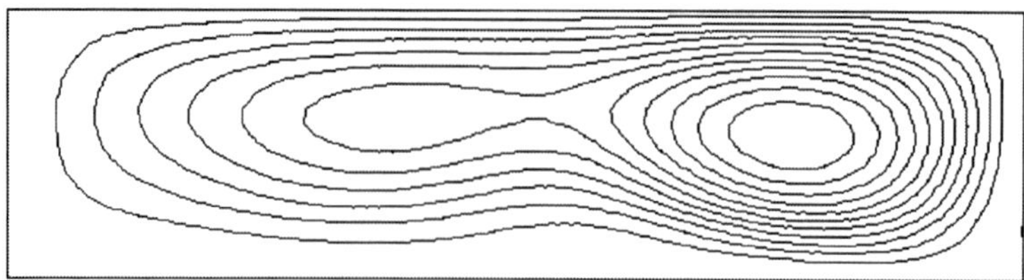

Fig. 7.12 Streamlines for $Gr = 10^4$

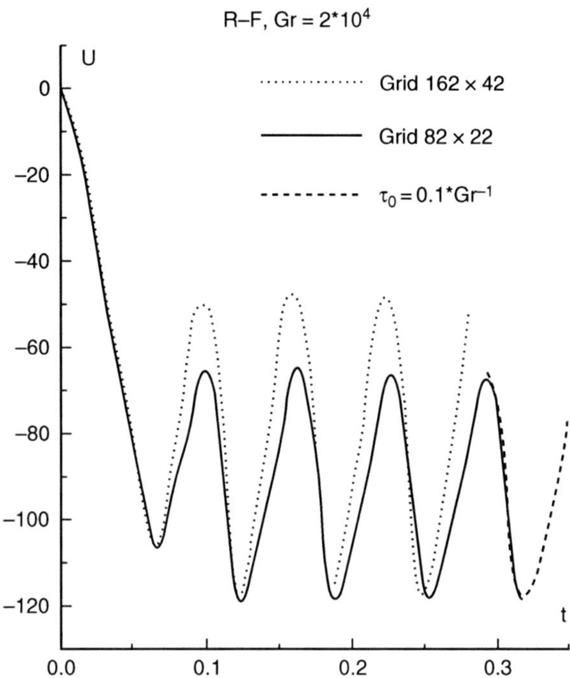

Fig. 7.13 Evolutional curves of the horizontal component of the velocity at the center of the domain for $Gr = 2 \times 10^4$

left part of the cavity, the second, weaker in intensity, vortex is formed (Fig. 7.14). The oscillation process is the change of intensities of these formations. In this case, we explicitly see the time evolution of the secondary vortices whose picture is in good correspondence with the analogous figures presented in [153].

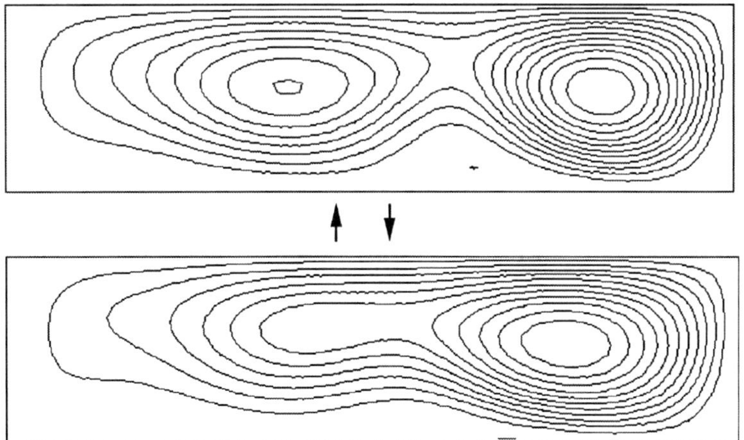

Fig. 7.14 Streamlines for $Gr = 2 \times 10^4$

To reveal the influence of the parameter α, we calculate this variant with $\alpha = 1$ (continuous line in Fig. 7.13) and $\alpha = 0.1$ (dotted line), and to verify the accuracy, we recompute the variant with $\alpha = 1$ on the doubled grid 42×162 (dotted line). The comparison of the time evolution of the velocity for these variants is presented in Fig. 7.13. It is seen that the decrease of α changes neither the period nor the amplitude of oscillations; the corresponding curves are practically not distinguished. The condensation of the grid has no influence on the frequency of oscillation but slightly changes its amplitude.

Therefore, the results of computing the viscous, incompressible fluid flow problems for small Prandtl numbers on the basis of the QHD system completely coincide with analogous computations based on the Navier–Stokes system, in which more complicated computational algorithms and denser spatial grids were used. Moreover, along with the stationary flow regime, we also succeeded in computing the secondary regimes, i.e., the nonstationary regimes. It is shown that the period of oscillations is independent of the grid step and the regularization parameter.

7.6 Marangoni Convection in the Zero Gravity

We consider below the problem on the thermocapillary convection of a viscous incompressible fluid under the absence of gravitation (see [152, 166]). A large number of studies are devoted to this problem, since in many problems of cosmic technology (directed crystallization and crucibleless zone melting), the surface of the fluid (melt) is free and the thermocapillary effect leads to the appearance of convective motion in the melt.

The convective flow is stipulated only by the surface tension forces, which can be written in the form of the following equilibrium condition of surface tension forces and the viscous friction forces on the upper free boundary:

$$\Pi_{yx} = \frac{\partial \sigma}{\partial T} \frac{\partial T}{\partial x},$$

where $\sigma = \sigma(T)$ is the surface tension coefficient of the fluid.

According to Eqs. (7.5), (7.14), and (7.15), in the two-dimensional case, we have

$$
\Pi = \mu
\begin{pmatrix}
2\dfrac{\partial u}{\partial x} & \dfrac{\partial u}{\partial y} + \dfrac{\partial v}{\partial x} \\[2ex]
\dfrac{\partial u}{\partial y} + \dfrac{\partial v}{\partial x} & 2\dfrac{\partial v}{\partial y}
\end{pmatrix}
$$
$$
+\tau
\begin{pmatrix}
u\left(\rho\left(u\dfrac{\partial u}{\partial x} + v\dfrac{\partial u}{\partial y}\right) + \dfrac{\partial p}{\partial x}\right) & u\left(\rho\left(u\dfrac{\partial v}{\partial x} + v\dfrac{\partial v}{\partial y}\right) + \dfrac{\partial p}{\partial y}\right) \\[3ex]
v\left(\rho\left(u\dfrac{\partial u}{\partial x} + v\dfrac{\partial u}{\partial y}\right) + \dfrac{\partial p}{\partial x}\right) & v\left(\rho\left(u\dfrac{\partial v}{\partial x} + v\dfrac{\partial v}{\partial y}\right) + \dfrac{\partial p}{\partial y}\right)
\end{pmatrix}.
$$

Since $\partial p/\partial y = 0$ and $v = 0$ on the upper boundary, it follows that $\partial v/\partial x = 0$, and the tensor Π simplifies:

$$\Pi = \mu \begin{pmatrix} 2\dfrac{\partial u}{\partial x} & \dfrac{\partial u}{\partial y} \\[2mm] \dfrac{\partial u}{\partial y} & 2\dfrac{\partial v}{\partial y} \end{pmatrix} + \tau \begin{pmatrix} u\left(\rho u\dfrac{\partial u}{\partial x} + \dfrac{\partial p}{\partial x}\right) & 0 \\[2mm] 0 & 0 \end{pmatrix}.$$

Finally, we obtain the condition on the upper boundary in the dimensional form:

$$\Pi_{yx} = \mu \frac{\partial u}{\partial y} = \frac{\partial \sigma}{\partial T} \frac{\partial T}{\partial x},$$

which coincides with the traditional boundary condition in the thermocapillary convection problem for the Navier–Stokes equations. Note that $\partial \sigma/\partial T < 0$ for most fluids.

We consider a flow in a rectangular cavity of height H and length L. The flow is described by system (7.11), (7.12), and (7.13) in which there are no exterior forces ($g = 0$). The system in the two-dimensional case is written in the form (7.20), (7.21), (7.22), and (7.23) with Gr $= 0$. The passage to the dimensionless form is performed by the formulas

$$x = \tilde{x}H, \quad y = \tilde{y}H, \quad u = \tilde{u}\frac{\nu}{H}, \quad v = \tilde{v}\frac{\nu}{H}, \quad t = \tilde{t}\frac{H^2}{\nu},$$

$$p = \tilde{p}\rho\left(\frac{\nu}{H}\right)^2, \quad T = \tilde{T}\frac{\Delta T}{A}.$$

For the chosen reduction to the dimensionless form, the Marangoni number is

$$\mathrm{Ma} = -\frac{\partial \sigma}{\partial T} \frac{\Delta T}{A} \frac{H}{\mu\chi},$$

and also

$$\mathrm{Re} = 1, \quad \mathrm{Pr} = \frac{\nu}{\chi}, \quad \tau = M^2,$$

where $M = \nu/(Hc)$ is the Mach number, which is small for problems considered.

System (7.20), (7.21), (7.22), and (7.23) is closed by the following boundary conditions:

$$u = 0, \quad v = 0, \quad \frac{\partial p}{\partial y} = 0, \quad \frac{\partial T}{\partial y} = 0$$

on the lower wall;

$$\frac{\partial u}{\partial y} = -\frac{\mathrm{Ma} \cdot A}{\mathrm{Pr}} \frac{\partial T}{\partial x}, \quad v = 0, \quad \frac{\partial p}{\partial y} = 0, \quad \frac{\partial T}{\partial y} = 0$$

on the upper wall;

$$u = 0, \quad v = 0, \quad \frac{\partial p}{\partial x} = 0, \quad T = 1$$

on the left lateral wall; and

$$u = 0, \quad v = 0, \quad \frac{\partial p}{\partial x} = 0, \quad T = 0$$

on the right lateral wall.

As in the heat-convection problem, the dimensionless parameter τ can be connected with the dimensionless parameters of the problem by the relation

$$\tau = \alpha \frac{\mathrm{Pr}}{\mathrm{Ma} \cdot A},$$

where

$$\alpha = \left| \frac{\partial \sigma}{\partial T} \right| \Delta T \cdot \frac{1}{H c^2}.$$

The quantity α turns out to be very small. So, for example, for silicon at $\Delta T = 1000°\mathrm{C}$ and a cavity of height $H = 1\,\mathrm{cm}$, we have $\alpha \sim 10^{-9}$. In numerical simulations, we need to choose the value of this parameter in the interval $0 < \alpha \leq 1$. In simulation of this problem, we choose $\alpha = 1$ and, respectively, the terms with τ are considered as regularizers.

The Marangoni convection problem in the rectangular cavity ($A = 4$) heating from the left is solved for $\mathrm{Ma} = 5, 10, 20, 100$, and 400 and $\mathrm{Pr} = 0.015$ on the uniform 27×102 spatial grid. The step in time is 5×10^{-6}. As the initial conditions, the unperturbed velocity and temperature fields are used.

For $\mathrm{Ma} = 400$, the results of computations are compared with the data of [152], and for $\mathrm{Ma} \leq 400$, with the data of [166].

In Figs. 7.15, 7.16, and 7.17, we present the streamlines for $\mathrm{Ma} = 5$, $\mathrm{Ma} = 100$, and $\mathrm{Ma} = 400$. Since the Marangoni numbers are positive, it follows that along the

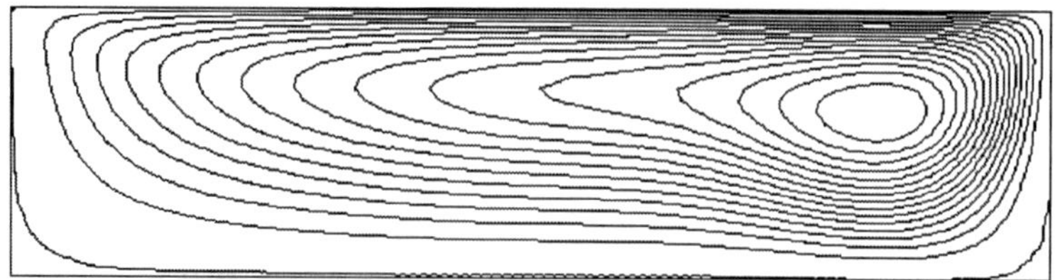

Fig. 7.15 Streamlines for $\mathrm{Ma} = 5$

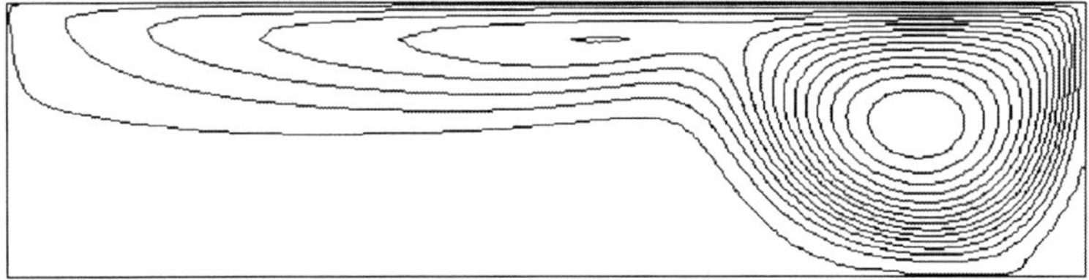

Fig. 7.16 Streamlines for Ma $= 100$

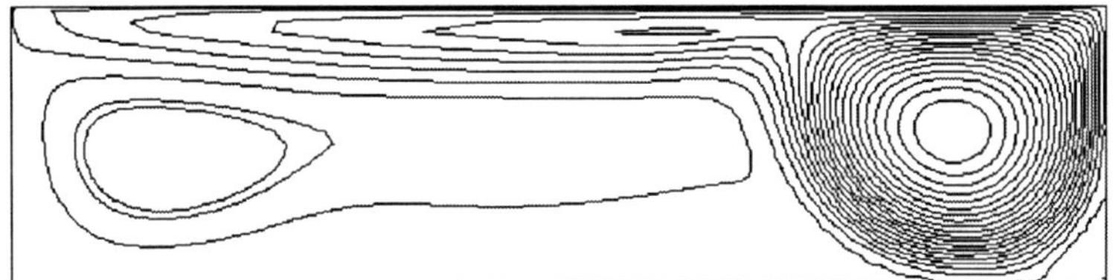

Fig. 7.17 Streamlines for Ma $= 400$

upper boundary, the fluid moves from the hot wall to the cold wall. A vortex flow arises whose center is displaced to the right wall. When Ma increases, the velocity of motion also increases and isotherms start to distort. For Ma $= 100$, additional vortices arise; they rotate in the same direction as the main vortex. In [166], analogous graphs for the same Marangoni numbers are presented, and, moreover, we see their good correspondence to the results of the present work. For Ma $= 400$, in the left lower part of the domain, a vertex rotating in the opposite direction arises. An analogous picture is described in [152]. The profiles of the horizontal velocity in the section $x = A/2$ are compared with the corresponding profiles of [166] for different Maragoni numbers. In this case, to compare the results, we calculate the velocity as

$$U(y) = \frac{\mathrm{Pr}}{\mathrm{Ma}} u(y),$$

which corresponds to the reduction to the dimensionless form used in [166]. The graphs obtained practically coincide with analogous curves of [166].

Therefore, the results obtained for Ma ≤ 100 are in good correspondence with the data of [166], and for Ma $= 400$, with the results of [152].

7.7 Flows in a Cubic Cavity with a Movable Lid

The fluid flow problem in a cavity with a movable lid is a known and sufficiently complicated test for estimating the efficiency of numerical methods. The first problem on which the capacity for work of the QHD algorithm for computing viscous incompressible fluids is the simulation of the two-dimensional fluid flow in a cavity [111]. Here, in accordance with [73, 74], we present the results of numerical simulation of the three-dimensional flow of a viscous, incompressible, isothermal fluid in a cubic cavity with a movable lid.

For nonlarge Reynolds numbers Re < 1000, the flow is laminar and stationary and is practically a single vortex with center near the center of the domain. The flow in the symmetry plane of the cavity is practically two-dimensional and is well described by the existing two-dimensional models. In this case, the velocity distributions obtained in computations by different methodologies are in good correspondence with each other and with the data of the natural experiments [126]. When the Reynolds number grows, the structure of the flow rapidly complicates, is stratified, becomes nonstationary, and then turbulent.

The computation of spatial flows is a laborious computational problem; to realize it, power computational complexes are necessary. The computational algorithm described is implemented for a multi-processor cluster-type computation system with distributed memory MBC1000M in which the data-exchange interface is organized according to the MPI (message passing interface) principle (see [95]).

We consider a flow of an isothermal fluid in a cubic cavity with edge H. The upper lid of the cavity moves with a constant speed U_0. The scheme of the computational domain and the coordinate system used are shown in Fig. 7.18.

We introduce the dimensionless quantities using the relations

$$x = \tilde{x}H, \quad y = \tilde{y}H, \quad z = \tilde{z}H, \quad u_x = \tilde{u}_x U_0, \quad u_y = \tilde{u}_y U_0, \quad u_z = \tilde{u}_z U_0,$$

$$p = \tilde{p}\rho U_0^2, \quad t = \frac{\tilde{t}H}{U_0}, \quad \mathrm{Re} = \frac{U_0 H}{\nu}.$$

After the reduction to the dimensionless form, the QHD system for the planar isothermal flows has the form

$$\frac{\partial u_x}{\partial x} + \frac{\partial u_y}{\partial y} + \frac{\partial u_z}{\partial z} = \frac{\partial w_x}{\partial x} + \frac{\partial w_y}{\partial y} + \frac{\partial w_z}{\partial z}, \tag{7.36}$$

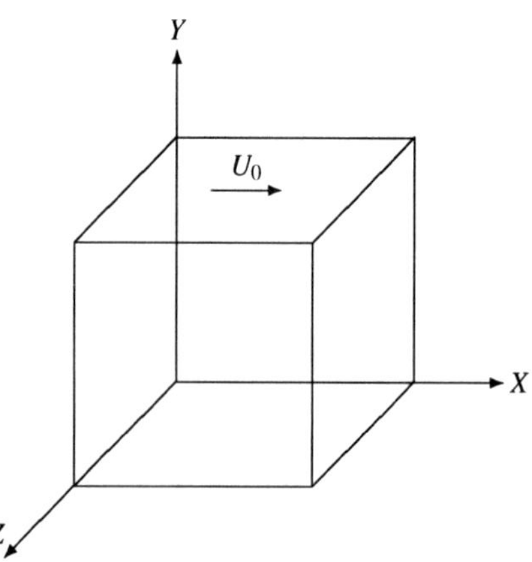

Fig. 7.18 Scheme of the computational domain and the coordinate system

$$\frac{\partial u_x}{\partial t} + \frac{\partial (u_x^2)}{\partial x} + \frac{\partial (u_x u_y)}{\partial y} + \frac{\partial (u_x u_z)}{\partial z} + \frac{\partial p}{\partial x}$$

$$= \frac{2}{Re} \frac{\partial^2 u_x}{\partial x^2} + \frac{1}{Re} \frac{\partial}{\partial y} \left(\frac{\partial u_x}{\partial y} + \frac{\partial u_y}{\partial x} \right) + \frac{1}{Re} \frac{\partial}{\partial z} \left(\frac{\partial u_x}{\partial z} + \frac{\partial u_z}{\partial y} \right)$$

$$+ 2 \frac{\partial (u_x w_x)}{\partial x} + \frac{\partial (u_y w_x)}{\partial y} + \frac{\partial (u_x w_y)}{\partial y} + \frac{\partial (u_z w_x)}{\partial z} + \frac{\partial (u_x w_z)}{\partial z}, \qquad (7.37)$$

$$\frac{\partial u_y}{\partial t} + \frac{\partial (u_x u_y)}{\partial x} + \frac{\partial (u_y^2)}{\partial y} + \frac{\partial (u_z u_y)}{\partial z} + \frac{\partial p}{\partial y}$$

$$= \frac{1}{Re} \frac{\partial}{\partial x} \left(\frac{\partial u_x}{\partial y} + \frac{\partial u_y}{\partial x} \right) + \frac{2}{Re} \frac{\partial^2 u_y}{\partial y^2} + \frac{1}{Re} \frac{\partial}{\partial z} \left(\frac{\partial u_y}{\partial z} + \frac{\partial u_z}{\partial y} \right)$$

$$+ \frac{\partial (u_x w_y)}{\partial x} + \frac{\partial (u_y w_x)}{\partial x} + 2 \frac{\partial (u_y w_y)}{\partial y} + \frac{\partial (u_z w_y)}{\partial z} + \frac{\partial (u_y w_z)}{\partial z}, \qquad (7.38)$$

$$\frac{\partial u_z}{\partial t} + \frac{\partial (u_x u_z)}{\partial x} + \frac{\partial (u_y u_z)}{\partial y} + \frac{\partial (u_z^2)}{\partial z} + \frac{\partial p}{\partial z}$$

$$= \frac{1}{Re} \frac{\partial}{\partial x} \left(\frac{\partial u_x}{\partial z} + \frac{\partial u_z}{\partial x} \right) + \frac{1}{Re} \frac{\partial}{\partial y} \left(\frac{\partial u_y}{\partial z} + \frac{\partial u_z}{\partial y} \right) + \frac{2}{Re} \frac{\partial^2 u_z}{\partial z^2}$$

$$+ \frac{\partial (u_x w_z)}{\partial x} + \frac{\partial (u_z w_x)}{\partial x} + \frac{\partial (u_y w_z)}{\partial y} + \frac{\partial (u_z w_y)}{\partial y} + 2 \frac{\partial (u_z w_z)}{\partial z}, \qquad (7.39)$$

where

$$w_x = \tau \left(u_x \frac{\partial u_x}{\partial x} + u_y \frac{\partial u_x}{\partial y} + u_z \frac{\partial u_x}{\partial z} + \frac{\partial p}{\partial x} \right),$$

$$w_y = \tau \left(u_x \frac{\partial u_y}{\partial x} + u_y \frac{\partial u_y}{\partial y} + u_z \frac{\partial u_y}{\partial z} + \frac{\partial p}{\partial y} \right), \qquad (7.40)$$

$$w_z = \tau \left(u_x \frac{\partial u_z}{\partial x} + u_y \frac{\partial u_z}{\partial y} + u_z \frac{\partial u_z}{\partial z} + \frac{\partial p}{\partial z} \right).$$

The unknown quantities are the components $u_x = u_x(x, y, z, t)$, $u_y = u_y(x, y, z, t)$, $u_z = u_z(x, y, z, t)$ of the velocity vector and the pressure $p = p(x, y, t)$.

The pressure field is found by the already known velocity field by solving the Poisson equation

$$\frac{\partial^2 p}{\partial x^2} + \frac{\partial^2 p}{\partial y^2} + \frac{\partial^2 p}{\partial z^2} = \frac{1}{\tau} \left(\frac{\partial u_x}{\partial x} + \frac{\partial u_y}{\partial y} + \frac{\partial u_z}{\partial z} \right)$$

$$- \frac{\partial}{\partial x} \left(u_x \frac{\partial u_x}{\partial x} + u_y \frac{\partial u_x}{\partial y} + u_z \frac{\partial u_x}{\partial z} \right)$$

$$- \frac{\partial}{\partial y} \left(u_x \frac{\partial u_y}{\partial x} + u_y \frac{\partial u_y}{\partial y} + u_z \frac{\partial u_y}{\partial z} \right)$$

$$- \frac{\partial}{\partial z} \left(u_x \frac{\partial u_z}{\partial x} + u_y \frac{\partial u_z}{\partial y} + u_z \frac{\partial u_z}{\partial z} \right), \qquad (7.41)$$

which is an equivalent representation of Eq. (7.36) for $\tau = $ const.

We complement system (7.37), (7.38), (7.39), (7.40), and (7.41) by boundary conditions. On the fixed rigid surface, we use the adhesion conditions $u = 0$ for the velocity. On the surface $y = 1$, we set the conditions $u_x = U_0$, $u_y = 0$, and $u_z = 0$. The boundary conditions for the pressure follow from the non-leakage conditions and have the form

$$\frac{\partial p}{\partial n} = 0, \tag{7.42}$$

where n is the normal to the surface. For example, on the face $x = 0$, we have $\partial p / \partial x = 0$, and on the edge $x = 0$, $y = 0$, condition (7.42) leads to the relations

$$\frac{\partial p}{\partial x} = 0, \quad \frac{\partial p}{\partial y} = 0.$$

At the vertex $x = 0$, $y = 0$, $z = 0$, we have

$$\frac{\partial p}{\partial x} = 0, \quad \frac{\partial p}{\partial y} = 0, \quad \frac{\partial p}{\partial z} = 0.$$

As the initial condition, we choose the equilibrium state $u_x = u_y = u_z = 0$. At the initial instant of time, the pressure is assumed to be constant in the whole fluid field: $p = 0$. To remove the lack of uniqueness in calculating the pressure, its value in the calculation at the cube vertex is assumed to be constant.

To solve the three-dimensional finite-difference equation $Ay = f$ for the pressure, we use the conjugated gradient method with preconditioning of the modified incomplete Cholesky decomposition without filling.

In numerical simulation, the value of the parameter τ was chosen from the accuracy and stability conditions of the algorithm in the form $\tau = 1/\text{Re}$. As the practice of two-dimensional computations of flows in a cavity [111] and flows after a backward-facing step (Sect. 7.3) shows, such a choice of the regularization parameters ensures a sufficient stability and accuracy of the computational algorithm.

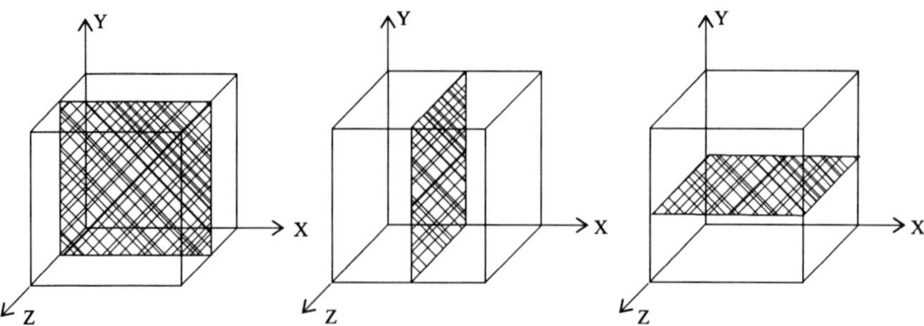

Fig. 7.19 Scheme of sections of the computational domain

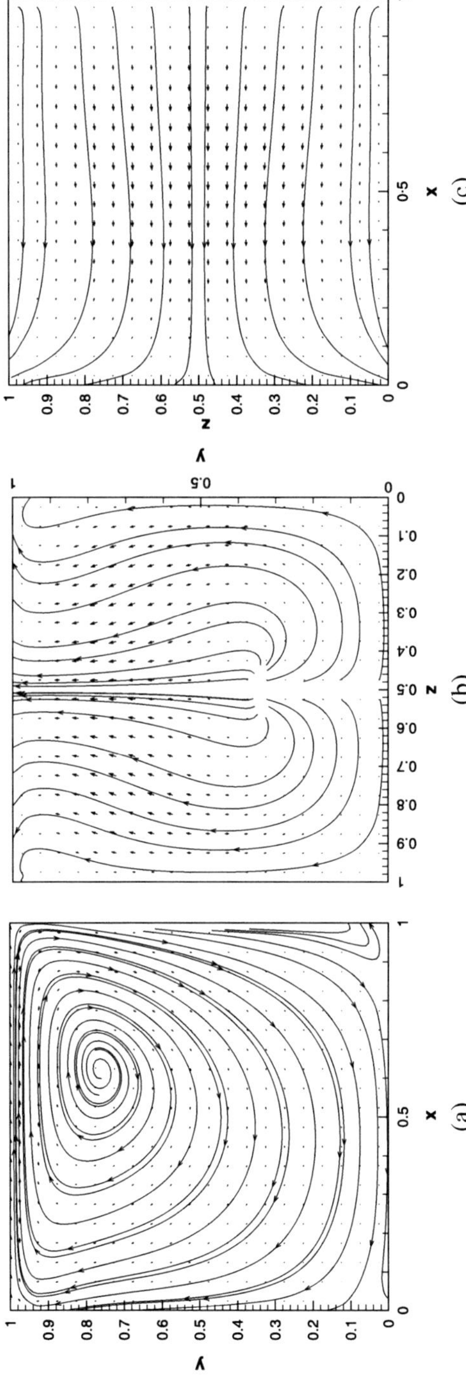

Fig. 7.20 Re $= 100$; streamlines for u_x and u_y in the plane $z = 0.5$ (**a**); u_y and u_z in the plane $x = 0.5$ (**b**); u_x and u_z in the plane $y = 0.5$ (**c**)

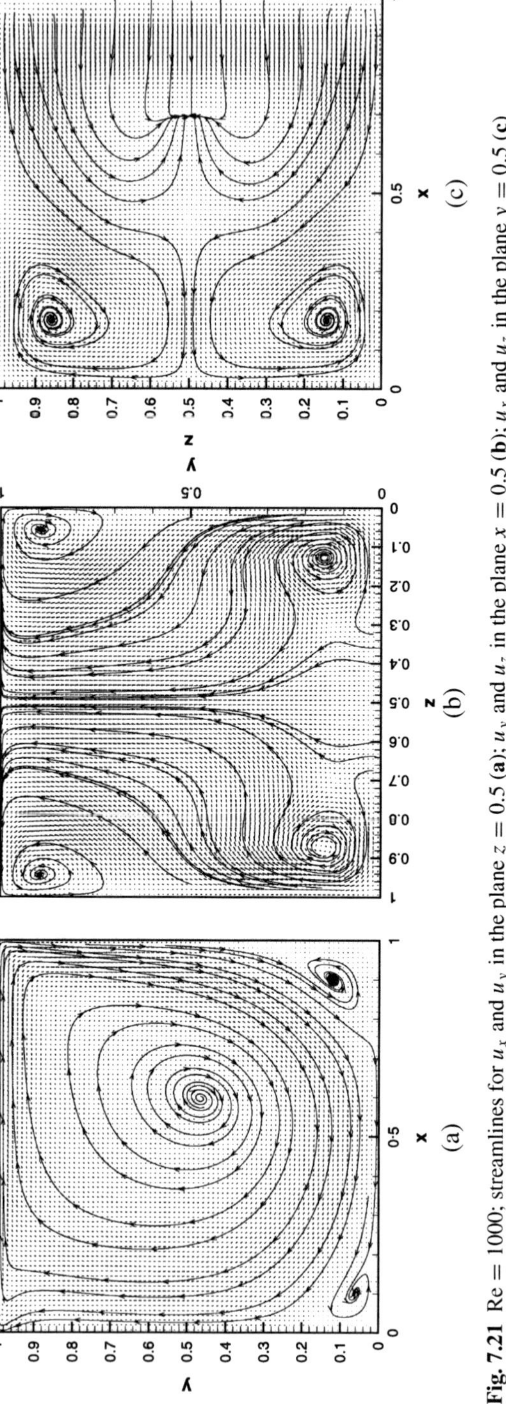

Fig. 7.21 Re = 1000; streamlines for u_x and u_y in the plane $z = 0.5$ (**a**); u_y and u_z in the plane $x = 0.5$ (**b**); u_x and u_z in the plane $y = 0.5$ (**c**)

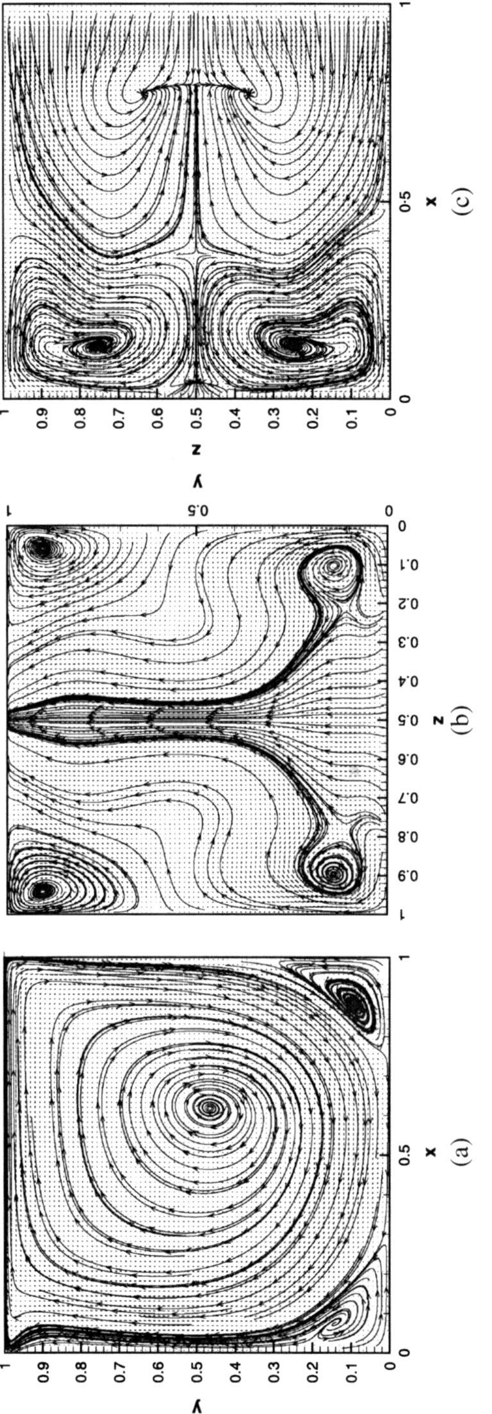

Fig. 7.22 Re = 2000; streamlines for u_x and u_y in the plane $z = 0.5$ (**a**); u_y and u_z in the plane $x = 0.5$ (**b**); u_x and u_z in the plane $y = 0.5$ (**c**)

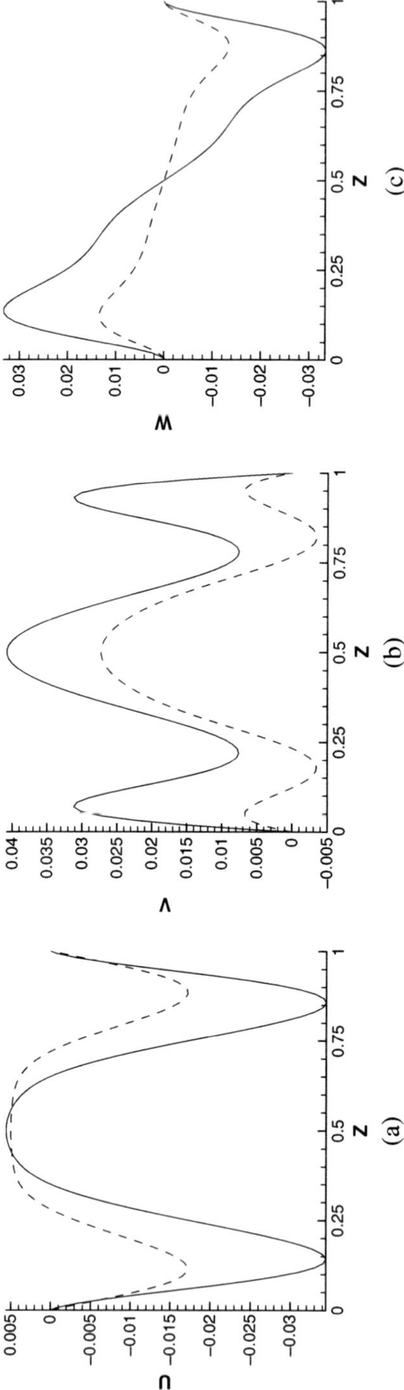

Fig. 7.23 *Continuous line* Re $= 1000$, *dotted line* Re $= 2000$; $u_x(0.5, 0.5, z)$ (**a**); $u_y(0.5, 0.5, z)$ (**b**); $u_z(0.5, 0.5, z)$ (**c**)

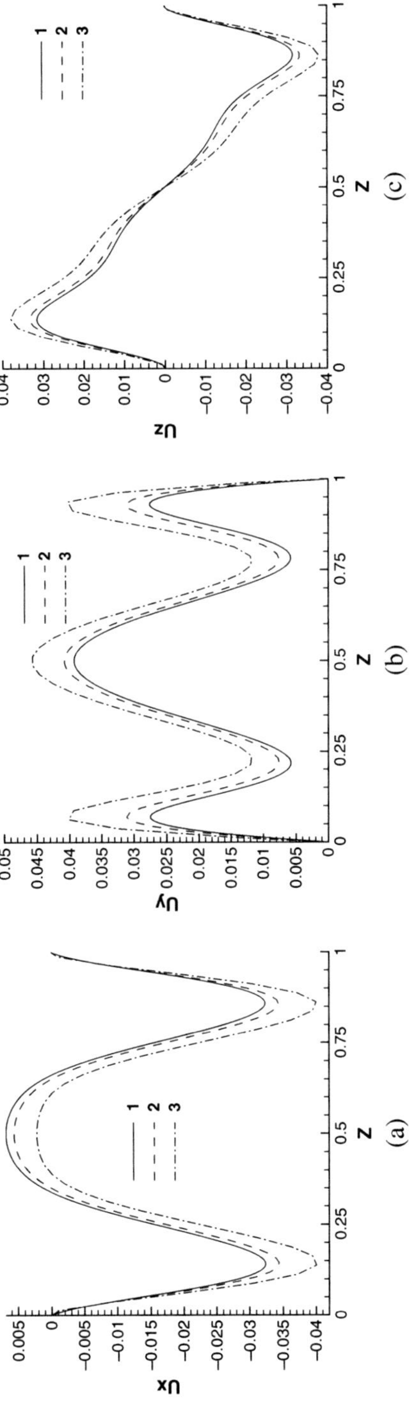

Fig. 7.24 $\mathrm{Re} = 1000$; *line 1* for $N_h = 162$, *line 2* for $N_h = 82$, *line 3* for $N_h = 42$: $u_x(0.5, 0.5, z)$ (**a**); $u_y(0.5, 0.5, z)$ (**b**); $u_z(0.5, 0.5, z)$ (**c**)

The problems were solved for the Reynolds numbers Re = 100, 1000, and 2000 on uniform spatial grids with the same number of nodes in all three directions N_h. For Re = 100, we use grids with number of nodes equal to N_h = 22 and 42; for Re = 1000, the number of computational nodes of the grid in the space is equal to N_h = 42, 82, and 162. For Re = 2000, N_h = 162. The steps in time are varied in the limits $\Delta t = 0.02$–0.001.

The schemes of three central sections of the cavity with coordinates $z = 0.5$, $x = 0.5$, and $y = 0.5$ are presented in Fig. 7.19. In Figs. 7.20, 7.21 and 7.22, we show the streamlines in these sections. In this problem, the section $z = 0.5$ is a symmetry plane. We show the pictures for Re = 100, the grid $22 \times 22 \times 22$, and Re = 1000 and Re = 2000, the grids $82 \times 82 \times 82$. In the last two figures, we see an essentially three-dimensional character of the flow, which is accompanied with the formation of sources and sinks in these planes. In the symmetry plane, the components of the velocity in the direction z are small, and the flow is close to two-dimensional (see [111]).

Figure 7.23 shows the one-dimensional distributions of the velocity components along the axis z for Re = 1000 and Re = 2000.

In Figs. 7.24 and 7.25, for Re = 1000, we present the one-dimensional distributions of the components of the velocity computed on a sequence of condensing grids. Lines 1 (continuous lines) correspond to computations with N_h = 162, lines 2 (dotted lines) to computations with N_h = 82, and lines 3 (prime-dotted lines) to computations with N_h = 42. The convergence of the numerical solution as the spatial grid condenses is clearly seen.

The results presented in this chapter justify the universality of the constructed QHD algorithm and demonstrate its efficiency for numerically modelling a wide class of viscous fluid flows, e.g., [67, 91].

All computations presented above were performed by using an explicit in time finite-difference scheme for which the value of the parameter τ was artificially

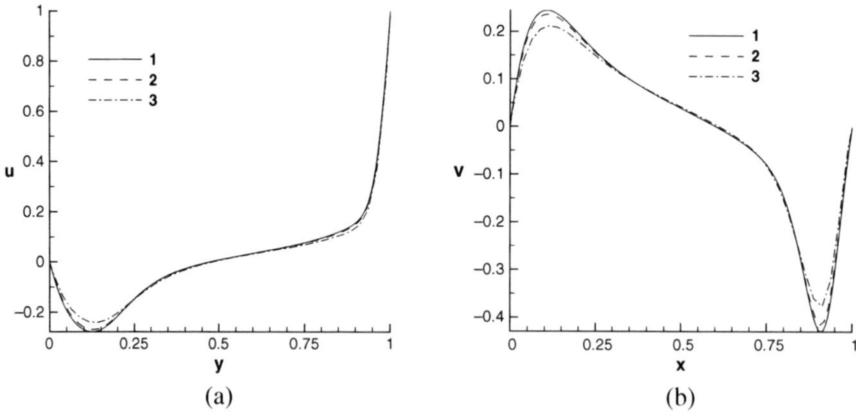

Fig. 7.25 Re = 1000; *line 1* for N_h = 162, *line 2* for N_h = 82, *line 3* for N_h = 42; $u_x(0.5, y, 0.5)$ (**a**); $u_y(x, 0.5, 0.5)$ (**b**)

overrated. However, the QHD equations admit other numerical realizations. For example, for solving the QHD system one can use analogs of the known algorithms (explicit and implicit) for the Navier–Stokes equation in the Oberbeck-Boussinesq approximation (see, e.g., [7, 163]). Application of finite-element methods, and, in particular, methods that suppress numerical oscillations by mixed interpolations on velocities and pressure, to the QHD equations also seems to be a promising approach (see [17–19, 32, 123, 124]).

Chapter 8
Quasi-gas-dynamic Equations for Nonequilibrium Gas Flows

In this chapter, we generalize the quasi-gas-dynamic equations to the case of gas flows with translational–rotational temperature nonequilibrium. Such a nonequilibrium is characteristic for moderately rarefied gases consisting of two-atom or poly-atom molecules [26–28, 122, 137, 168, 169, 212].

In the model constructed here, a molecule is considered as a rigid body having only translational and rotational degrees of freedom. Such a description is correct if the gas temperature is not very high (oscillatory degrees of freedom are not excited) and is not very small (rotational degrees of freedom can be considered classically). For most of the molecular gases, the second condition is satisfied for usual temperatures, since, in this case, the distance between the rotational levels $\hbar^2/2I^0$ is small as compared with kT, where $\hbar$ is the Plank constant and I^0 is the moment of inertia of the molecule (see [137, 168, 169, 212]).

The contents of this chapter are based on [88, 192].

8.1 Molecular Models and Distribution Functions

Assume that all molecules are the same and are rigid rotators. In this case, the total energy of a molecule coincides with its kinetic energy and can be written as follows:

$$E_m = \frac{m_0}{2}\xi^2 + E_{\text{rot}},$$

where E_{rot} is the rotational energy, m_0 is the mass of the molecule, and $\xi = u + c$ is the velocity of its center of masses, which is represented as the sum of the macroscopic velocity u and the thermal velocity c.

The rotational energy of a poly-atom molecule is given by

$$E_{\text{rot}}^{(3)} = \frac{1}{2}(I_1^0\omega_1^2 + I_2^0\omega_2^2 + I_3^0\omega_3^2), \tag{8.1}$$

where I_1^0, I_2^0, and I_3^0 are the principal moments of inertia of the molecule and ω_1, ω_2, and ω_3 are angular velocities about three principal axes (see [133, p. 128]).

T.G. Elizarova, *Quasi-Gas Dynamic Equations*, Computational Fluid and Solid Mechanics, DOI 10.1007/978-3-642-00292-2_8,
© Springer-Verlag Berlin Heidelberg 2009

In this case, the molecule has six degrees of freedom: three translational and three rotational (the number of inner degrees of freedom satisfies $\zeta = 3$). In what follows, when referring to such a system, we use the notation "3R."

In the case of linear (in particular, two-atom) molecules having two rotational degrees of freedom ($\zeta = 2$), the rotational energy is written as follows:

$$E_{\text{rot}}^{(2)} = \frac{I^0}{2}(\omega_1^2 + \omega_2^2), \tag{8.2}$$

where I^0 is the principal moment of inertia with respect to the axis perpendicular to the symmetry axis of the molecule. When referring to this model, we use the notation "2R."

The state of a gas consisting of rigid rotators can be described by the one-particle distribution function $f(t, R, \xi, \omega)$ depending on time t, the coordinate R, the velocity ξ of the center of masses, and the vector ω of angular velocities. We normalize this distribution function by the relation

$$d\rho = f\, d\xi\, d\omega,$$

where ρ is the mass density of the gas.

Consider the locally equilibrium two-temperature Maxwell–Boltzmann distribution function (see [212, p. 109])

$$f_{0r} = f_0 \cdot f_{\text{rot}},$$

which is the product of the Maxwell distribution function

$$f_0 = \rho \left(2\pi \frac{\mathcal{R}_\star}{\mathcal{M}} T_{\text{tr}} \right)^{-3/2} \exp\left(-\frac{c^2}{2(\mathcal{R}_\star/\mathcal{M})T_{\text{tr}}} \right)$$

in translational degrees of freedom and the Hinshelwood distribution function (see [28, p. 104])

$$f_{\text{rot}} = \mathcal{A} \exp\left(-\frac{E_{\text{rot}}}{kT_{\text{rot}}} \right)$$

in rotational degrees of freedom, where $\mathcal{A}$ is the normalizing factor. For a gas with two rotational degrees of freedom (2R), this distribution has the form

$$f_{\text{rot}}^{2R} = \left(2\pi \frac{\mathcal{R}_\star}{I} T_{\text{rot}} \right)^{-1} \exp\left(-\frac{\omega_1^2 + \omega_2^2}{2(\mathcal{R}_\star/I)T_{\text{rot}}} \right),$$

and for a gas with three rotational degrees of freedom (3R), it is

$$f_{\text{rot}}^{3R} = \frac{\sqrt{I_1 I_2 I_3}}{2\pi \mathcal{R}_\star T_{\text{rot}}} \times \exp\left(-\frac{\omega_1^2}{2(\mathcal{R}_\star/I_1)T_{\text{rot}}} - \frac{\omega_2^2}{2(\mathcal{R}_\star/I_2)T_{\text{rot}}} - \frac{\omega_3^2}{2(\mathcal{R}_\star/I_3)T_{\text{rot}}} \right).$$

Here, $\mathcal{R}_\star$ is the universal gas constant, $\mathcal{M}$ is the molar mass of the gas considered, $I = I^0 N_A$, $I_\alpha = I_\alpha^0 N_A$, $\alpha = 1, 2, 3$, N_A is the Avogadro constant, $k_B = \mathcal{R}_\star / N_A$ is the Boltzmann constant, $\mathcal{R}_\star / \mathcal{M} = \mathcal{R}$ is the gas constant, T_{tr} is the translational temperature, which is the same for all translational degrees of freedom of the molecule, and T_{rot} is the rotational temperature, which is the same for all rotational degrees of freedom of a molecule.

The normalizing factor $\mathcal{A}$ is chosen from the condition

$$\int f_{rot}\, d\omega = 1,$$

where the integration is performed over the whole space of angular velocities (see Sect. 8.2). Such a representation of the distribution function is correct in the case where the gas temperature is not very high and, therefore, the oscillatory degrees of freedom are not excited, but it is not very small, i.e., the rotational degrees of freedom of the molecule can be considered in the classical approximation (see [137, 168, 169, 212]).

8.2 Coordinate Systems and Certain Integrals

Let us introduce the Cartesian coordinate system (x, y, z) in the space, and let R be the spatial radius-vector. For a vector-valued function $a(R)$ of spatial coordinates, a_x, a_y, and a_z denote its x-, y-, and z-projections, respectively.

Consider also an arbitrary orthonormal curvilinear coordinate system (μ^1, μ^2, μ^3). For the same vector-valued function $a(R)$, let a_i be a covariant coordinate, i.e.,

$$a_i = a \cdot R_i, \quad R_i = \frac{\partial R}{\partial \mu^i}.$$

The contravariant coordinates have the form $a^i = g^{ij} a_j$, where $g^{ij} = R^i \cdot R^j$ is the metric tensor. We find R^i from the equations

$$R_i \cdot R^j = \delta_i^j, \quad \delta_i^j = \begin{cases} 1, & i = j, \\ 0, & i \neq j. \end{cases}$$

Also, the following relations hold:

$$a^i = a \cdot R^i, \quad a_i = g_{ij} a^j, \quad g_{ik} g^{jk} = \delta_i^j, \quad g_{ij} = R_i \cdot R_j.$$

In the Cartesian coordinates, a^i is expressed as

$$a^i = a \cdot R^i = a_x R_x^i + a_y R_y^i + a_z R_z^i, \tag{8.3}$$

where R_x^i, R_y^i, and R_z^i are the x-, y-, and z-projections of the vector R^i, respectively. The metric tensor can be expressed as follows:

$$g^{ij} = R^i \cdot R^j = R^i_x R^j_x + R^i_y R^j_y + R^i_z R^j_z. \tag{8.4}$$

Also, consider two velocity spaces, the space of linear velocities and the space of angular velocities. In the space of linear velocities, we denote the radius-vector by ξ. Relative to the spatial coordinates, ξ can be considered as a constant vector field ξ^i:

$$\nabla_j \xi^i = 0, \tag{8.5}$$

where ∇_j is the covariant derivative. In the space of linear velocities, we introduce the Cartesian coordinate system (ξ_x, ξ_y, ξ_z) such that the directions of its coordinate axes coincide with those of the Cartesian coordinate system in the space R.

Also, in the space of angular velocities, we introduce the Cartesian coordinate system $(\omega_1, \omega_2, \omega_3)$. The directions of coordinate axes of this system coincide with the principal axes of the molecule considered as a rigid body and are different for different molecules.

Below, we perform the integration in the space of linear velocities ξ and in the space of angular velocities ω using the corresponding Cartesian coordinate systems. The system of moment equations is constructed in the curvilinear coordinate system (μ^1, μ^2, μ^3).

Let us calculate some integrals useful for what follows. Unless otherwise specified, the limits of integration are assumed to be infinite:

$$\int \ldots = \int_{-\infty}^{\infty} \ldots .$$

In the subsequent calculations, we use Eqs. (8.3) and (8.4). We have

$$\int c^i f_0 \, dc = \int (c_x R^i_x + c_y R^i_y + c_z R^i_z) f_0 \, dc$$

$$= R^i_x \iiint c_x f_0 \, dc_x dc_y dc_z + \cdots = 0;$$

$$\int c_i f_0 \, dc = g_{ik} \int c^k f_0 \, dc = 0;$$

$$\int c^i c^j f_0 \, dc = \int (c_x R^i_x + c_y R^i_y + c_z R^i_z)(c_x R^j_x + c_y R^j_y + c_z R^j_z) f_0 \, dc$$

$$= R^i_x R^j_x \int c_x^2 f_0 \, dc + R^i_y R^j_y \int c_y^2 f_0 \, dc + R^i_z R^j_z \int c_z^2 f_0 \, dc$$

$$= (R^i \cdot R^j) p_{tr} = g^{ij} p_{tr};$$

$$\int c^i c^j c^k f_0 \, dc = 0;$$

$$\int c_x^4 f_0 \, dc = \int c_y^4 f_0 \, dc = \int c_z^4 f_0 \, dc = 3\frac{p_{tr}^2}{\rho};$$

$$\int c_x^2 c_y^2 f_0 \, dc = \int c_x^2 c_z^2 f_0 \, dc = \int c_y^2 c_z^2 f_0 \, dc = \frac{p_{tr}^2}{\rho};$$

$$\int c^i c^j c^2 f_0 \, dc = 5\frac{p_{tr}^2}{\rho} g^{ij}.$$

Here, $p_{tr} = \rho(\mathcal{R}_\star/\mathcal{M})T_{tr}$ is the translational pressure (see Sect. 8.3). In calculating this integrals, we use the relations

$$\int\limits_{-\infty}^{\infty} e^{-y^2} \, dy = \sqrt{\pi}, \quad \int\limits_{-\infty}^{\infty} y^2 e^{-y^2} \, dy = \frac{\sqrt{\pi}}{2}, \quad \int\limits_{-\infty}^{\infty} y^4 e^{-y^2} \, dy = \frac{3\sqrt{\pi}}{4}.$$

8.3 Construction of Moment Equations

The behavior of the distribution function f is described by the Boltzmann equation

$$\frac{\partial f}{\partial t} + (\xi \nabla)f = \mathcal{I}(f, f), \tag{8.6}$$

where $\mathcal{I}(f, f)$ is the collision integral. To construct the moment equations describing flows of a viscous gas, one traditionally uses the approximations of the function f in the form of the expansion in a small parameter near the equilibrium value with the subsequent averaging of the obtained kinetic equation with summator invariants (see [27, 122]).

To construct the equations that take into account the translational–rotational nonequilibrium (QGDR systems), we replace the distribution function f by its approximate value f^{QGDR}, which is the expansion in a small parameter in a neighborhood of the equilibrium distribution function f_{0r} (see Eq. (8.1)), having the form

$$f^{QGDR} = f_{0r} - \tau(\xi\nabla)f_{0r},$$

where τ is the Maxwell relaxation time,

$$\tau = \frac{\mu}{p_{tr}},$$

and $\mu \sim T_{tr}^{\omega}$ is the gas viscosity coefficient calculated on the basis of the translational particle temperature (see [168]) and ω is determined by the molecule interaction law.

The formal change $f \to f^{\text{QGDR}}$ in the convective summand of the Boltzmann equation (8.6) leads to the regularized kinetic equation

$$\frac{\partial f}{\partial t} + (\xi \nabla) f_{0r} - (\xi \nabla) \tau (\xi \nabla) f_{0r} = \mathcal{I}(f, f). \tag{8.7}$$

A regularized kinetic equation of an analogous form was used in Chap. 3 for constructing the QGD equations.

In contrast to the classical kinetic Kagan–Maximov equation for a gas with rotational degrees of freedom obtained under similar assumptions, in this equation, we do not take into account the molecular translational acceleration under the exterior force action and the precession of the rotational moment of the molecule in the exterior magnetic or electric field (see [212]).

The macroscopic QGDR equations are obtained by using the moment averaging of Eq. (8.7) in the velocity spaces. The procedure for obtaining these equations is similar to those presented in [56] and Chap. 2.

We assume that the following relations hold:

$$\int f \, d\xi \, d\omega = \int f_{0r} \, d\xi \, d\omega = \rho, \quad \int c^i f \, d\xi \, d\omega = 0,$$

$$p_{\text{tr}} = \frac{1}{3} \int c^2 f \, d\xi \, d\omega = \frac{1}{3} \int c^2 f_{0r} \, d\xi \, d\omega = \rho \frac{\mathcal{R}_\star}{\mathcal{M}} T_{\text{tr}}. \tag{8.8}$$

In the 2R case,

$$p_{\text{rot}} = \int \varepsilon_\omega^{2R} f \, d\xi \, d\omega = \int \varepsilon_\omega^{2R} f_{0r} \, d\xi \, d\omega = \rho \frac{\mathcal{R}_\star}{\mathcal{M}} T_{\text{rot}},$$

$$\varepsilon_\omega^{2R} = \frac{I}{2\mathcal{M}} (\omega_1^2 + \omega_2^2). \tag{8.9}$$

In the 3R case,

$$p_{\text{rot}} = \frac{2}{3} \int \varepsilon_\omega^{3R} f \, d\xi \, d\omega = \frac{2}{3} \int \varepsilon_\omega^{3R} f_{0r} \, d\xi \, d\omega = \rho \frac{\mathcal{R}_\star}{\mathcal{M}} T_{\text{rot}},$$

$$\varepsilon_\omega^{3R} = \frac{1}{2\mathcal{M}} (I_1 \omega_1^2 + I_2 \omega_2^2 + I_3 \omega_3^2). \tag{8.10}$$

Using Eqs. (8.1) and (8.2), the total energy of the gas mass unit, including the energy E_{tr} of the translational motion and the energy E_{rot} of the rotational motion, is written in the form

$$E = E_{\text{tr}} + E_{\text{rot}} = \int \left(\frac{\xi^2}{2} + \varepsilon_\omega \right) f \, d\xi \, d\omega = \int \left(\frac{\xi^2}{2} + \varepsilon_\omega \right) f_{0r} \, d\xi \, d\omega,$$

where

$$E_{\text{tr}} = \frac{1}{2}\int \xi^2 f\,d\xi\,d\omega = \frac{1}{2}\int \xi^2 f_{0r}\,d\xi\,d\omega = \frac{\rho u^2}{2} + \frac{3}{2}p_{\text{tr}},$$

$$E_{\text{rot}}^{2R} = \int \varepsilon_\omega^{2R} f\,d\xi\,d\omega = \int \varepsilon_\omega^{2R} f_{0r}\,d\xi\,d\omega = p_{\text{rot}},$$

$$E_{\text{rot}}^{3R} = \int \varepsilon_\omega^{3R} f\,d\xi\,d\omega = \int \varepsilon_\omega^{3R} f_{0r}\,d\xi\,d\omega = \frac{3}{2}p_{\text{rot}}. \tag{8.11}$$

The average pressure is

$$p_{\text{av}} = \frac{3p_{\text{tr}} + \zeta p_{\text{rot}}}{3 + \zeta}. \tag{8.12}$$

For a 2R gas,

$$\zeta = 2, \quad p_{\text{av}} = \frac{3p_{\text{tr}} + 2p_{\text{rot}}}{5},$$

and for a 3R gas,

$$\zeta = 3, \quad p_{\text{av}} = \frac{p_{\text{tr}} + p_{\text{rot}}}{2}.$$

The average temperature is connected with the mean pressure by the state equation

$$p_{\text{av}} = \rho \frac{\mathcal{R}_\star}{\mathcal{M}} T_{\text{av}}.$$

The expression for the total energy in both cases can be written through the mean pressure:

$$E = \frac{1}{2}\rho u^2 + \frac{p_{\text{av}}}{\gamma - 1},$$

where p_{av} is defined by Eq. (8.12). Here γ is the specific heat ratio. For the perfect gas (see [28]),

$$\gamma = \frac{5 + \zeta}{3 + \zeta}. \tag{8.13}$$

Using Eq. (8.5), we rewrite Eq. (8.7) in the index form:

$$\frac{\partial f}{\partial t} + \nabla_i \xi^i f_{0r} - \nabla_i \tau \nabla_j \xi^i \xi^j f_{0r} = \mathcal{I}(f, f). \tag{8.14}$$

Integrating Eq. (8.14) with weight 1 and using relations (8.8), (8.9), (8.10), and (8.11) and expressions from Sect. 3.3, we obtain

$$\int \frac{\partial f}{\partial t}\, d\xi\, d\omega = \frac{\partial}{\partial t} \int f\, d\xi\, d\omega = \frac{\partial}{\partial t}\rho,$$

$$\int \nabla_i \xi^i f_{0r}\, d\xi\, d\omega = \nabla_i \int (u^i + c^i) f_0\, dc \int f_{\mathrm{rot}}\, d\omega = \nabla_i \rho u^i,$$

$$\int \nabla_i \tau \nabla_j \xi^i \xi^j f_{0r}\, d\xi\, d\omega = \nabla_i \tau \nabla_j \int (u^i + c^i)(u^j + c^j) f_0\, dc$$

$$= \nabla_i \tau \nabla_j \left(\rho u^i u^j + \int c^i c^j f_0\, dc \right)$$

$$= \nabla_i \tau \nabla_j \left(\rho u^i u^j + g^{ij} p_{\mathrm{tr}} \right).$$

By the properties of the collision integral (the unity is a summator invariant for the Boltzmann collision integral), the integral of the right-hand side vanishes:

$$\int \mathcal{I}(f, f)\, d\xi\, d\omega = 0.$$

Therefore, we obtain the equation for the density:

$$\frac{\partial}{\partial t}\rho + \nabla_i \rho u^i = \nabla_i \tau \nabla_j \left(\rho u^i u^j + g^{ij} p_{\mathrm{tr}} \right).$$

In what follows, we use Eqs. (8.8), (8.9), (8.10), and (8.11) omitting the corresponding references.

To obtain the equation for the momentum, we integrate Eq. (8.14) with the weight ξ^k:

$$\int \frac{\partial f}{\partial t} \xi^k\, d\xi\, d\omega = \frac{\partial}{\partial t}\rho u^k,$$

$$\int \nabla_i \xi^i f_{0r} \xi^k\, d\xi\, d\omega = \nabla_i \left(\rho u^i u^k + g^{ik} p_{\mathrm{tr}} \right),$$

$$\int \nabla_i \tau \nabla_j \xi^i \xi^j f_{0r} \xi^k\, d\xi\, d\omega = \left(\rho u^i u^j u^k + p_{\mathrm{tr}} \left(u^k g^{ij} + u^j g^{ik} + u^i g^{jk} \right) \right).$$

Since ξ^k is a summator invariant, it follows that in this case

$$\int \mathcal{I}(f, f) \xi^k\, d\xi\, d\omega = 0.$$

Combining the obtained expressions, we obtain the equation for ρu^k:

$$\frac{\partial}{\partial t}\rho u^k + \nabla_i\left(\rho u^i u^k + g^{ik} p_{\mathrm{tr}}\right)$$
$$= \nabla_i \tau \nabla_j\left(\rho u^i u^j u^k + p_{\mathrm{tr}}\left(u^k g^{ij} + u^j g^{ik} + u^i g^{jk}\right)\right).$$

To obtain the equation for the translational energy E_{tr}, we average Eq. (8.14) with the weight $\xi^2/2$:

$$\int \frac{\partial f}{\partial t}\frac{\xi^2}{2}\, d\xi\, d\omega = \frac{\partial}{\partial t} E_{\mathrm{tr}},$$

$$\int \nabla_i \xi^i f_{0r}\frac{\xi^2}{2}\, d\xi\, d\omega = \nabla_i \frac{1}{2}\int (u^i + c^i) f_0 \xi^2\, dc = \nabla_i u^i (E_{\mathrm{tr}} + p_{\mathrm{tr}}),$$

$$\int \nabla_i \tau \nabla_j \xi^i \xi^j f_{0r}\frac{\xi^2}{2}\, d\xi\, d\omega$$

$$= \nabla_i \tau \nabla_j \left(u^i u^j E_{\mathrm{tr}} + 2u^i u^j p_{\mathrm{tr}} + \frac{1}{2}u_k u^k g^{ij} p_{\mathrm{tr}} + \frac{5}{2}\frac{p_{\mathrm{tr}}^2}{\rho}g^{ij}\right).$$

For the collision integral $\mathcal{I}(f, f)$, the quantity $\xi^2/2$ is no longer a summator invariant since the exchange of energy between the translational and rotational degrees of freedom is possible, and the latter integral does not vanish. We call it the exchange term and denote by

$$\int \mathcal{I}(f, f)\frac{\xi^2}{2}\, d\xi\, d\omega = S_{\mathrm{tr}}.$$

Combining the expression obtained and differentiating by parts the term with squared pressure $5p_{\mathrm{tr}}^2/(2\rho)$, we obtain

$$\frac{\partial}{\partial t} E_{\mathrm{tr}} + \nabla_i u^i (E_{\mathrm{tr}} + p_{\mathrm{tr}}) = \nabla_i \tau \nabla_j \left(u^i u^j E_{\mathrm{tr}} + 2u^i u^j p_{\mathrm{tr}} + \frac{1}{2}u_k u^k g^{ij} p_{\mathrm{tr}}\right)$$

$$+ \frac{5}{2}\nabla_i \tau \frac{p_{\mathrm{tr}}}{\rho}\nabla_j p_{\mathrm{tr}} g^{ij} + \frac{5}{2}\nabla_i \tau p_{\mathrm{tr}}\nabla_j \frac{p_{\mathrm{tr}}}{\rho}g^{ij} + S_{\mathrm{tr}}.$$

To obtain the equation for the rotational energy E_{rot}, it is necessary to average Eq. (8.14) with weight $\varepsilon_\omega^{2\mathrm{R}}$ in the case of 2R molecules and with weight $\varepsilon_\omega^{3\mathrm{R}}$ in the case of 3R molecules; these quantities are defined in Eqs. (8.9) and (8.10).

For the first variant of averaging, we obtain

$$\int \frac{\partial f}{\partial t}\varepsilon_\omega\, d\xi\, d\omega = \frac{\partial}{\partial t} E_{\mathrm{rot}},$$

$$\int \nabla_i \xi^i f_{0r}\varepsilon_\omega\, d\xi\, d\omega = \nabla_i \int \xi^i f_0\, dc \int \varepsilon_\omega f_{\mathrm{rot}}\, d\omega = \nabla_i u^i E_{\mathrm{rot}},$$

$$\int \nabla_i \tau \nabla_j \xi^i \xi^j f_{0r} \varepsilon_\omega \, d\xi \, d\omega$$

$$= \nabla_i \tau \nabla_j \left(\rho u^i u^j + g^{ij} p_{tr} \right) \frac{E_{rot}}{\rho} = \nabla_i \tau \nabla_j \left(E_{rot} u^i u^j + g^{ij} \frac{p_{tr} p_{rot}}{\rho} \right).$$

Denote the exchange term by

$$\int \mathcal{I}(f, f) \varepsilon_\omega \, d\xi \, d\omega = S_{rot}.$$

Combining the expressions obtained and differentiating by parts the term containing the product of pressures, we obtain the following equation for the rotational energy of 2R molecules:

$$\frac{\partial}{\partial t} E_{rot} + \nabla_i u^i E_{rot} = \nabla_i \tau \nabla_j u^i u^j E_{rot}$$

$$+ \nabla_i \tau \frac{p_{rot}}{\rho} \nabla_j p_{tr} g^{ij} + \nabla_i \tau p_{tr} \nabla_j \frac{p_{rot}}{\rho} g^{ij} + S_{rot}.$$

In a similar way, we construct the equation for the rotational energy of the 3R gas. In this case, only the right-hand side of the equation changes:

$$\frac{\partial}{\partial t} E_{rot} + \nabla_i u^i E_{rot} = \nabla_i \tau \nabla_j u^i u^j E_{rot}$$

$$+ \frac{3}{2} \nabla_i \tau \frac{p_{rot}}{\rho} \nabla_j p_{tr} g^{ij} + \frac{3}{2} \nabla_i \tau p_{tr} \nabla_j \frac{p_{rot}}{\rho} g^{ij} + S_{rot}.$$

The method for obtaining the moment equations presented here leads to expressions for the heat flux with the Prandtl number equal to unity. To generalize the equations to the case of an arbitrary Prandtl number, we need to multiply the next-to-the-last terms in the equations for the translational and rotational energies by Pr^{-1}. To find the Prandtl number, we can use the Eucken approximation (see [199]):

$$Pr = \frac{4\gamma}{9\gamma - 5},$$

where γ is defined by Eq. (8.13); in the case of 2R gas, $\gamma = 7/5$ and $Pr = 14/19$, and in the case of 3R gas, $\gamma = 4/3$ and $Pr = 16/21$.

8.4 Calculation of Exchange Terms

The exchange terms in the right-hand sides of the equations obtained for energies are the moments of the Boltzmann collision integral (8.6). To calculate them, we use the relaxation model of the collision integral

$$\mathcal{I}(f, f) = \frac{f_{0r}^0 - f}{\tau_{\text{rot}}},$$

where f_{0r}^0 is the distribution function f_{0r} for the equilibrium case, i.e., for the case where $T_{\text{tr}} = T_{\text{rot}} = T_{\text{av}}$ and, respectively, $p_{\text{tr}} = p_{\text{rot}} = p_{\text{av}}$; τ_{rot} is the mean time of the rotational relaxation, which is usually several times greater than the Maxwell relaxation time τ:

$$\tau_{\text{rot}} = Z_{\text{rot}} \tau_c,$$

where $1/Z_{\text{rot}}$ is the relative frequency of nonelastic collisions, $\tau_c = \tau/\Omega$ is the mean free path time for elastic collisions, and $\Omega = 30/(7 - 2\omega)(5 - 2\omega)$ (see [28, 168]). A method for calculating Z_{rot} and its influence on the flow structure in shock waves is discussed in Sect. 3.4.2 and Appendix B.

Substituting the expression for the collision integral in the formula for the calculation of the exchange term, in the 2R case, we obtain

$$
\begin{aligned}
S_{\text{tr}} &= \int \frac{1}{\tau_{\text{rot}}}(f_{0r}^0 - f)\frac{\xi^2}{2}\, d\xi\, d\omega = \frac{1}{2\tau_{\text{rot}}} \int (f_{0r}^0 - f)(u^2 + 2c + c^2)\, d\xi\, d\omega \\
&= \frac{u^2}{2\tau_{\text{rot}}} \int (f_{0r}^0 - f)\, d\xi\, d\omega + \frac{u}{\tau_{\text{rot}}} \int c(f_{0r}^0 - f)\, d\xi\, d\omega \\
&\quad + \frac{1}{2\tau_{\text{rot}}} \int c^2(f_{0r}^0 - f)(u^2 + 2c + c^2)\, d\xi\, d\omega \\
&= 0 + 0 + \frac{1}{2\tau_{\text{rot}}}(3p_{\text{av}} - 3p_{\text{tr}}) = \frac{3}{2\tau_{\text{rot}}}(p_{\text{av}} - p_{\text{tr}}).
\end{aligned}
$$

Similarly, for the second exchange term, we obtain

$$S_{\text{rot}} = \int \frac{1}{\tau_{\text{rot}}}(f_{0r}^0 - f)\varepsilon_\omega^{2R}\, d\xi\, d\omega = \frac{1}{\tau_{\text{rot}}}(p_{\text{av}} - p_{\text{rot}}).$$

Substituting the expressions for the mean pressure (8.12), in the 2R case, we obtain

$$S_{\text{tr}} = \frac{3}{5\tau_{\text{rot}}}(p_{\text{rot}} - p_{\text{tr}}), \quad S_{\text{rot}} = -S_{\text{tr}}.$$

Similarly, in the 3R case, we obtain

$$S_{\text{tr}} = \frac{3}{4\tau_{\text{rot}}}(p_{\text{rot}} - p_{\text{tr}}), \quad S_{\text{rot}} = -S_{\text{tr}}.$$

Note that in accordance with the energy conservation law,

$$S_{\text{tr}} + S_{\text{rot}} = 0.$$

8.5 QGDR Equations for a Gas with Two or Three Rotational Degrees of Freedom

Let us present the final form of the QGDR equations for two-atom (2R) and poly-atom (3R) molecules in the form invariant to the coordinate system:

$$\frac{\partial}{\partial t}\rho + \nabla_i \rho u^i = \nabla_i \tau (\nabla_j \rho u^i u^j + \nabla^i p_{\text{tr}}), \tag{8.15}$$

$$\frac{\partial}{\partial t}\rho u^k + \nabla_i \rho u^i u^k + \nabla^k p_{\text{tr}} = \nabla_i \tau \nabla_j \rho u^i u^j u^k + \nabla_i \tau (\nabla^i p_{\text{tr}} u^k + \nabla^k p_{\text{tr}} u^i)$$
$$+ \nabla^k \tau \nabla_i p_{\text{tr}} u^i, \tag{8.16}$$

$$\frac{\partial}{\partial t}E_{\text{tr}} + \nabla_i u^i (E_{\text{tr}} + p_{\text{tr}}) = \nabla_i \tau \left(\nabla_j (E_{\text{tr}} + 2p_{\text{tr}})u^i u^j + \frac{1}{2}\nabla^i u_k u^k p_{\text{tr}} \right)$$
$$+ \frac{5}{2}\nabla_i \tau \frac{p_{\text{tr}}}{\rho}\nabla^i p_{\text{tr}} + \text{Pr}^{-1}\frac{5}{2}\nabla_i \tau p_{\text{tr}}\nabla^i \frac{p_{\text{tr}}}{\rho} + S_{\text{tr}}. \tag{8.17}$$

The equation for the rotational energy and the exchange terms in the case of 2R gas have the form

$$\frac{\partial}{\partial t}E_{\text{rot}} + \nabla_i u^i E_{\text{rot}} = \nabla_i \tau \nabla_j u^i u^j E_{\text{rot}}$$
$$+ \nabla_i \tau \frac{p_{\text{rot}}}{\rho}\nabla^i p_{\text{tr}} + \text{Pr}^{-1}\nabla_i \tau p_{\text{tr}}\nabla^i \frac{p_{\text{rot}}}{\rho} + S_{\text{rot}}, \tag{8.18}$$

$$S_{\text{tr}} = \frac{3}{5\tau_{\text{rot}}}(p_{\text{rot}} - p_{\text{tr}}), \quad S_{\text{rot}} = -S_{\text{tr}}, \quad E_{\text{tr}} = \frac{\rho u^2}{2} + \frac{3p_{\text{tr}}}{2}, \quad E_{\text{rot}} = p_{\text{rot}},$$

$$\gamma = \frac{7}{5}, \quad \text{Pr} = \frac{14}{19}, \quad p_{\text{tr}} = \rho\frac{\mathcal{R}_\star}{\mathcal{M}}T_{\text{tr}}, \quad p_{\text{rot}} = \rho\frac{\mathcal{R}_\star}{\mathcal{M}}T_{\text{rot}}.$$

The equation for the rotational energy and the exchange terms in the case of 3R gas have the form

$$\frac{\partial}{\partial t}E_{\text{rot}} + \nabla_i u^i E_{\text{rot}} = \nabla_i \tau \nabla_j u^i u^j E_{\text{rot}}$$
$$+ \frac{3}{2}\nabla_i \tau \frac{p_{\text{rot}}}{\rho}\nabla^i p_{\text{tr}} + \text{Pr}^{-1}\frac{3}{2}\nabla_i \tau p_{\text{tr}}\nabla^i \frac{p_{\text{rot}}}{\rho} + S_{\text{rot}}, \tag{8.19}$$

$$S_{\text{tr}} = \frac{3}{4\tau_{\text{rot}}}(p_{\text{rot}} - p_{\text{tr}}), \quad S_{\text{rot}} = -S_{\text{tr}}, \quad E_{\text{tr}} = \frac{\rho u^2}{2} + \frac{3p_{\text{tr}}}{2}, \quad E_{\text{rot}} = \frac{3}{2}p_{\text{rot}},$$

$$\gamma = \frac{4}{3}, \quad \text{Pr} = \frac{16}{21}, \quad p_{\text{tr}} = \rho\frac{\mathcal{R}_\star}{\mathcal{M}}T_{\text{tr}}, \quad p_{\text{rot}} = \rho\frac{\mathcal{R}_\star}{\mathcal{M}}T_{\text{rot}}.$$

The equation for the rotational energy E_r for the 2R and 3R variants can be approximated as follows:

$$\frac{\partial}{\partial t} T_r + u^i \nabla_i T_r = \frac{T_{\mathrm{av}} - T_r}{\tau_{\mathrm{rot}}}.$$

Such a form is often used in studying the relaxation processes (see [28, p. 117]).

The QGDR equations can be written in the form of conservation laws (1.54), (1.55), and (1.56) under the absence of exterior forces. In this case, the equation of continuity (1.54) and the equation of momentum (1.55) preserve their form in which the pressure p is replaced by p_{tr} and τ is calculated as $\tau = \mu/p_{\mathrm{tr}}$, where the viscosity coefficient is found by the translation temperature: $\mu = \mu(T_{\mathrm{tr}})$. In turn, in the QGDR model, the energy equation (1.56) splits into two equations: the equations for energy related to the translational degrees of freedom E_{tr} and the rotational degrees of freedom E_{rot}. In what follows, we present the form of this splitting for a 2R gas:

$$\frac{\partial E_{\mathrm{tr}}}{\partial t} + \nabla_i \frac{J^i}{\rho}(E_{\mathrm{tr}} + p_{\mathrm{tr}}) + \nabla_i q^i_{\mathrm{tr}} = \nabla_i(\Pi^{ik} u^k) + S_{\mathrm{tr}}, \tag{8.20}$$

$$\frac{\partial E_{\mathrm{rot}}}{\partial t} + \nabla_i u^i E_{\mathrm{rot}} + \nabla_i q^i_{\mathrm{rot}} = \nabla_i \tau \nabla_j u^i u^j E_{\mathrm{rot}} + \nabla_i \tau \frac{p_{\mathrm{rot}}}{\rho} \nabla^i p_{\mathrm{tr}} + S_{\mathrm{rot}}. \tag{8.21}$$

The general heat flux q^i also splits into two components:

$$q^i_{\mathrm{tr}} = -\frac{1}{\mathrm{Pr}} \frac{5}{2} \mu \nabla^i \frac{p_{\mathrm{tr}}}{\rho} - \tau \rho u^i \left[u^j \nabla_j \varepsilon + p_{\mathrm{tr}} u_j \nabla^j \left(\frac{1}{\rho}\right) \right], \tag{8.22}$$

where $\varepsilon = p_{\mathrm{tr}}/(\rho(\gamma - 1))$. This part determines a heat flux stipulated by the gradient of the translational temperature T_{tr}. The second component

$$q^i_{\mathrm{rot}} = -\frac{1}{\mathrm{Pr}} \mu \nabla^i \frac{p_{\mathrm{rot}}}{\rho} \tag{8.23}$$

defines the heat flux stipulated by the gradient of the rotational temperature T_{rot}. Note that in the QGDR model, both vectors q_{tr} and q_{rot} of heat flux in Eqs. (8.22) and (8.23) are proportional to $\mu(T_{\mathrm{tr}})$.

The systems written here and complemented by boundary conditions represent a closed model for simulation of flows of a moderately rarefied gas with a possible nonequilibrium between the translational and rotational degrees of freedom.

In the case of heat equilibrium of the gas between the translational and rotational degrees of freedom, i.e., in the case where $T_{\mathrm{tr}} = T_{\mathrm{rot}} = T$ and $p_{\mathrm{tr}} = p_{\mathrm{rot}} = p$, the total energy of the volume unit is given by

$$E = E_{\mathrm{tr}} + E_{\mathrm{rot}} = \frac{\rho u^2}{2} + \frac{p}{\gamma - 1},$$

where γ is found by Eq. (8.13).

In this case, the form of the equations for density and momentum does not change:

$$\frac{\partial}{\partial t}\rho + \nabla_i \rho u^i = \nabla_i \tau (\nabla_j \rho u^i u^j + \nabla^i p),$$

$$\frac{\partial}{\partial t}\rho u^k + \nabla_i \rho u^i u^k + \nabla^k p = \nabla_i \tau \nabla_j \rho u^i u^j u^k$$
$$+ \nabla_i \tau (\nabla^i p u^k + \nabla^k p u^i) + \nabla^k \tau \nabla_i p u^i.$$

Adding the equations for E_{tr} and E_{rot}, we obtain the equation for the total energy E of the form (3.10):

$$\frac{\partial}{\partial t}E + \nabla_i u^i (E + p) = \nabla_i \tau \left(\nabla_j (E + 2p)u^i u^j + \frac{1}{2}\nabla^i u_k u^k p \right)$$
$$+ \frac{\gamma}{\gamma - 1}\nabla_i \tau \frac{p}{\rho}\nabla^i p + \mathrm{Pr}^{-1}\frac{\gamma}{\gamma - 1}\nabla_i \tau p \nabla^i \frac{p}{\rho}.$$

Note that the energy equation presented above can be immediately obtained by the averaging of the approximate equation (8.14) with the summator invariant $\xi^2/2 + \varepsilon_\omega$ and a generalization of the obtained equation to the case of $\mathrm{Pr} \neq 1$.

Therefore, in the equilibrium case, the QGDR system passes to the QGD system with the corresponding values of γ and Pr.

Now we present the form of the QGDR equations for spatially one-dimensional flows:

$$\frac{\partial \rho}{\partial t} + \frac{1}{r^\nu}\frac{\partial}{\partial r}r^\nu \rho u = \frac{1}{r^\nu}\frac{\partial}{\partial r}\tau\frac{\partial}{\partial r}r^\nu \rho u^2 + \frac{1}{r^\nu}\frac{\partial}{\partial r}\tau r^\nu \frac{\partial}{\partial r}p_{\mathrm{tr}},$$

$$\frac{\partial \rho u}{\partial t} + \frac{1}{r^\nu}\frac{\partial}{\partial r}r^\nu \rho u^2 + \frac{\partial}{\partial r}p_{\mathrm{tr}} = \frac{1}{r^\nu}\frac{\partial}{\partial r}\tau\frac{\partial}{\partial r}r^\nu \rho u^3$$
$$+ 2\frac{1}{r^\nu}\frac{\partial}{\partial r}\tau r^\nu \frac{\partial}{\partial r}p_{\mathrm{tr}}u - 2\nu\frac{\tau}{r^2}p_{\mathrm{tr}}u + \frac{\partial}{\partial r}\frac{\tau}{r^\nu}\frac{\partial}{\partial r}r^\nu p_{\mathrm{tr}}u,$$

$$\frac{\partial E_{\mathrm{tr}}}{\partial t} + \frac{1}{r^\nu}\frac{\partial}{\partial r}r^\nu u(E_{\mathrm{tr}} + p_{\mathrm{tr}}) = \frac{1}{r^\nu}\frac{\partial}{\partial r}\tau\frac{\partial}{\partial r}r^\nu(E_{\mathrm{tr}} + 2p_{\mathrm{tr}})u^2$$
$$+ \frac{1}{r^\nu}\frac{\partial}{\partial r}\tau r^\nu \frac{\partial}{\partial r}\frac{1}{2}u^2 p_{\mathrm{tr}} + \frac{5}{2}\frac{1}{r^\nu}\frac{\partial}{\partial r}\tau\frac{p_{\mathrm{tr}}}{\rho}r^\nu \frac{\partial}{\partial r}p_{\mathrm{tr}}$$
$$+ \mathrm{Pr}^{-1}\frac{5}{2}\frac{1}{r^\nu}\frac{\partial}{\partial r}\tau p_{\mathrm{tr}}r^\nu \frac{\partial}{\partial r}\frac{p_{\mathrm{tr}}}{\rho} + S_{\mathrm{tr}}.$$

In the 2R case,

$$\frac{\partial E_{\mathrm{rot}}}{\partial t} + \frac{1}{r^\nu}\frac{\partial}{\partial r}r^\nu u E_{\mathrm{rot}} = \frac{1}{r^\nu}\frac{\partial}{\partial r}\tau\frac{\partial}{\partial r}r^\nu u^2 E_{\mathrm{rot}}$$
$$+ \frac{1}{r^\nu}\frac{\partial}{\partial r}\tau\frac{p_{\mathrm{rot}}}{\rho}r^\nu \frac{\partial}{\partial r}p_{\mathrm{tr}} + \mathrm{Pr}^{-1}\frac{1}{r^\nu}\frac{\partial}{\partial r}\tau p_{\mathrm{tr}}r^\nu \frac{\partial}{\partial r}\frac{p_{\mathrm{rot}}}{\rho} + S_{\mathrm{rot}}.$$

In the 3R case,

$$\frac{\partial E_{\text{rot}}}{\partial t} + \frac{1}{r^{\nu}}\frac{\partial}{\partial r} r^{\nu} u E_{\text{rot}} = \frac{1}{r^{\nu}}\frac{\partial}{\partial r}\tau\frac{\partial}{\partial r} r^{\nu} u^2 E_{\text{rot}}$$

$$+ \frac{3}{2}\frac{1}{r^{\nu}}\frac{\partial}{\partial r}\tau\frac{p_{\text{rot}}}{\rho} r^{\nu}\frac{\partial}{\partial r} p_{\text{tr}} + \text{Pr}^{-1}\frac{3}{2}\frac{1}{r^{\nu}}\frac{\partial}{\partial r}\tau p_{\text{tr}} r^{\nu}\frac{\partial}{\partial r}\frac{p_{\text{rot}}}{\rho} + S_{\text{rot}}.$$

Here $\nu = 0$ corresponds to the planar case, $\nu = 1$ to the cylindrical symmetry, and $\nu = 2$ to the spherical symmetry.

8.6 Examples of Numerical Computations

In this section, we present the results obtained on the basis of the QGDR equations for the problem of the flow of nitrogen, which is considered as a 2R gas with the parameters $Z_{\text{rot}} = 5$ and 10 for $\omega = 0.75$, where ω is the exponent in the law of the dependence of the viscosity on the temperature. The results obtained are compared with the computations carried out in the framework of the DSMC approach. We consider the following two one-dimensional problems: the relaxation problem of the translational and rotational temperatures in the flow and the problem on the shock wave structure. The numerical method is an explicit in time finite-difference scheme of the second order without artificial regularizers, which is analogous to that described in Chap. 5 (see Sect. 5.7).

Along with the one-dimensional problems presented below, a number of gas flow problems in underexpanded jet problems were solved (see [109, 146]).

8.6.1 Spatial Relaxation Problem

Consider a stationary one-dimensional gas flow ($\text{Ma} = 3.571$) with a nonequilibrium in the input section for $x = 0$, which is characterized by the temperatures $T_{\text{tr}\,1} \neq T_{\text{rot}\,1}$, the density ρ_1, and the velocity u_1. The rotational and translational temperatures are evolved to the equilibrium state as x grows.

For the planar one-dimensional flow, the QGDR system has the form

$$\frac{\partial\rho}{\partial t} + \frac{\partial}{\partial x}\rho u = \frac{\partial}{\partial x}\tau\frac{\partial}{\partial x}\left(\rho u^2 + p_{\text{tr}}\right),$$

$$\frac{\partial\rho u}{\partial t} + \frac{\partial}{\partial x}\left(\rho u^2 + p_{\text{tr}}\right) = \frac{\partial}{\partial x}\tau\frac{\partial}{\partial x}\left(\rho u^3 + 3 p_{\text{tr}} u\right),$$

$$\frac{\partial E_{\text{tr}}}{\partial t} + \frac{\partial}{\partial x} u(E_{\text{tr}} + p_{\text{tr}}) = \frac{\partial}{\partial x}\tau\frac{\partial}{\partial x}(E_{\text{tr}} + 2 p_{\text{tr}})u^2 + \frac{\partial}{\partial x}\tau\frac{\partial}{\partial x}\frac{1}{2}u^2 p_{\text{tr}}$$

$$+ \frac{5}{2}\frac{\partial}{\partial x}\tau\frac{p_{\text{tr}}}{\rho}\frac{\partial}{\partial x} p_{\text{tr}} + \frac{5}{2\,\text{Pr}}\frac{\partial}{\partial x}\tau p_{\text{tr}}\frac{\partial}{\partial x}\frac{p_{\text{tr}}}{\rho} + S_{\text{tr}},$$

$$\frac{\partial E_{\rm rot}}{\partial t} + \frac{\partial}{\partial x} u E_{\rm rot} = \frac{\partial}{\partial x} \tau \frac{\partial}{\partial x} u^2 E_{\rm rot}$$

$$+ \frac{\partial}{\partial x} \tau \frac{p_{\rm rot}}{\rho} \frac{\partial}{\partial x} p_{\rm tr} + \frac{1}{\rm Pr} \frac{\partial}{\partial x} \tau p_{\rm tr} \frac{\partial}{\partial x} \frac{p_{\rm rot}}{\rho} + S_{\rm rot}.$$

The exchange terms are

$$S_{\rm tr} = \frac{3}{2\tau_{\rm rot}} (p_{\rm av} - p_{\rm tr}), \quad S_{\rm rot} = -S_{\rm tr}, \quad \tau_{\rm rot} = \tau \Omega^{-1}(\omega) Z_{\rm rot},$$

where $\tau/\Omega = \tau_c$ is the mean free path time between collisions, $E_{\rm tr} = \rho u^2 + 3 p_{\rm tr}/2$ is the translational energy, $E_{\rm rot} = p_{\rm rot}$ is the rotational energy, and $\tau = \mu/p_{\rm tr}$. The mean values of the pressure and temperature are

$$p_{\rm av} = \frac{3 p_{\rm tr} + 2 p_{\rm rot}}{5} = \rho \mathcal{R} T_{\rm av}.$$

In computations by the DSMC method, we use the model of collisions, which approximately corresponds to the value $Z_{\rm rot} = 5$. The existence of a gradient of temperature on the input boundary of the computational domain calls difficulties in stating the boundary conditions for the DSMC computations. The conditions of the left input boundary are realized as follows: molecules with equilibrium temperatures $T_{\rm tr}$ and $T_{\rm rot}$ are injected to the computational domain from the zone lying at a point $x < 0$. The values of the gas-dynamic parameters obtained in the DSMC computations at the point $x = 0$ are taken as the boundary conditions for the QGDR computations. On the right output boundary, for the QGDR equations, we pose the flow drift condition or the "soft boundary conditions." A kinetic analog of the same conditions is also posed in the DSMC algorithm. The step of the spatial grid is chosen to be equal to $h = 1$ or 0.5.

The results obtained according to both methods turn out to be close, which characterize the accuracy of the constructed QGDR model (Fig. 8.1). In Fig. 8.1, the coordinate x is normalized to the mean free path λ_1, where

$$\lambda_1 = \frac{\mu(T_1)}{\rho_1 \sqrt{2(\mathcal{R}_\star/\mathcal{M})T_{\rm tr\,1}}} \frac{2(7-\omega)(5-\omega)}{15\sqrt{\pi}};$$

here $T_1 = T_{\rm av}$ at the point $x = 0$.

8.6.2 Problem of the Shock Wave Structure

The problem on the stationary shock wave structure is solved for the following variants: the Mach number of the incident flow is Ma $= 1.71$ for $Z_{\rm rot} = 5$ and 10, which corresponds to the DSMC computations presented in [28, p. 298], and for the

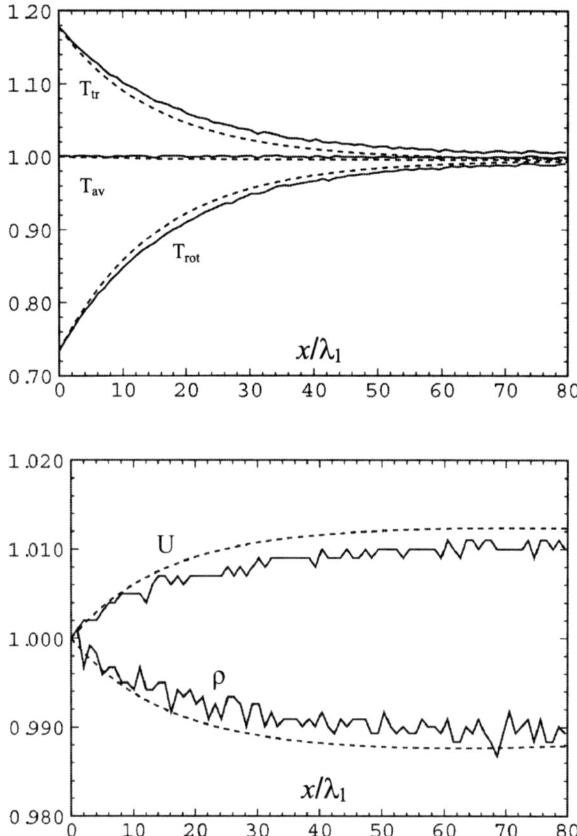

Fig. 8.1 Profiles of the temperature T_{tr} and T_{rot}, the density, and the velocity in the relaxation problem. The continuous line corresponds to the DSMC method and the dotted line to the QGDR model

Mach numbers Ma $= 7$ and 12.9, $Z_{rot} = 5$, which corresponds to DSMC computations in [129].

The profiles of density, velocity, and temperature obtained in the computations are presented in Figs. 8.2 and 8.3, where the value of the coordinate x is normalized to the mean free path in the incident flow. The ordinates are normalized in the usual way by using the Rankin–Hugoniot relations between the parameters before the shock wave (1) and after it (2), i.e.,

$$f_\rho = \frac{\rho - \rho^{(1)}}{\rho^{(2)} - \rho^{(1)}},$$

where f_ρ is the value of density in the figure and $\rho^{(1)}$ and $\rho^{(2)}$ are the values on the boundaries; the data for the temperature are analogous. For the velocity, we have

$$f_u = \frac{u - u^{(2)}}{u^{(1)} - u^{(2)}}.$$

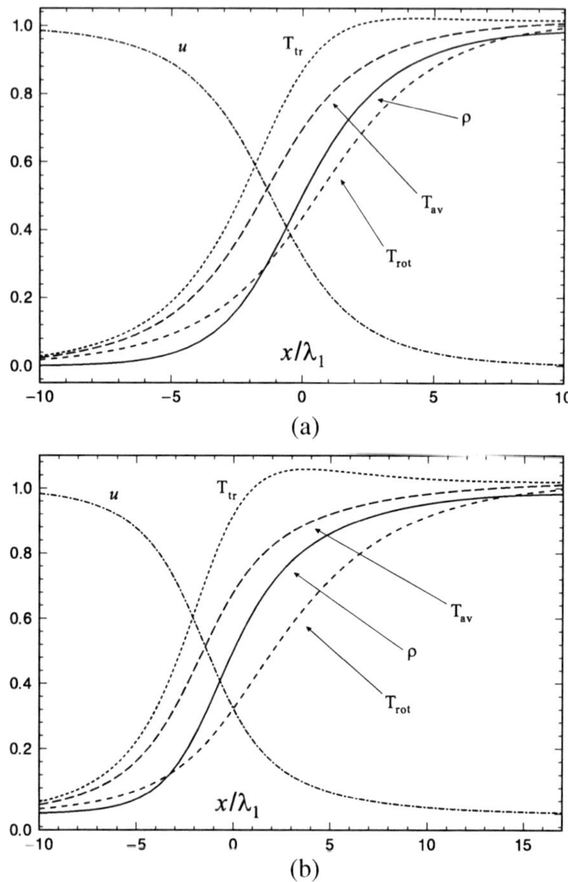

Fig. 8.2 Shock wave problem, Ma $= 1.71$, $Z_{rot} = 5$ (**a**) and $Z_{rot} = 10$ (**b**). The *continuous line* corresponds to ρ, the *dotted line* corresponds to T_{rot}, the temperature T_{tr} is denoted by *dots*, T_{av} is denoted by *long primes*, and u is denoted by *prime-dotted lines*

In Fig. 8.2 ($Z_{rot} = 5$), we see the characteristic profiles in this problem. Namely, the mutual location of curves corresponds to the results of Bird [28], where the profile of the translational temperature T_{tr} has a not large maximum ($T_{tr} = 1.021$) and, moreover, this profile lies to the left from the profile of the rotational temperature T_{rot}, which has no maximum. The inverse width of the profile of the shock wave density is equal to $\lambda_1/\delta = 0.152$. When Z_{rot} increases, the width of the shock wave also increases ($\lambda_1/\delta = 0.136$), and the value of the translational temperature T_{tr} exceeds over its value after the front enlarges approximately by 2 times ($T_{tr} = 1.058$). These results correspond to the data of kinetic computations [28].

The results of computations for Ma $= 7.0$ and Ma $= 12.9$ are presented in Fig. 8.3. They are very close to the results of [129]. In particular, in both cases, the profiles of ρ and T_{rot} lie very close to each other. Moreover, the values of maxima of T_{tr} are also in a good correspondence with the computations of [129], where $T_{tr} = 1.068$ for Ma $= 7.0$ and $T_{tr} = 1.070$ for Ma $= 12.9$. The inverse thickness λ_1/δ of the shock wave for these cases is 0.297 and 0.244, respectively. However,

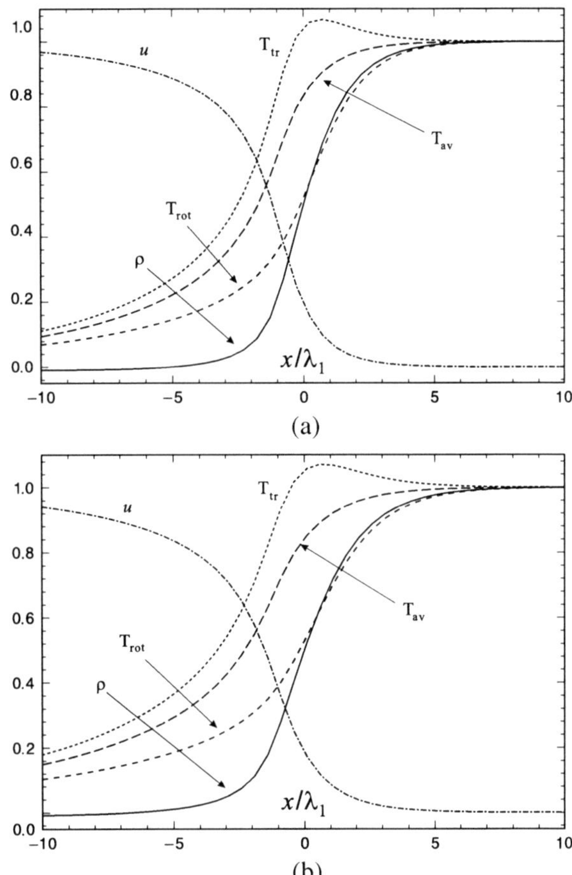

Fig. 8.3 Shock wave problem; Ma $= 7.0$, Z rot $= 5$ (**a**) and Ma $= 12.9$, $Z_{\mathrm{rot}} = 5$ (**b**). The notation is the same as in Fig. 8.2

in a domain before the shock wave, the values of the translational T_{tr} and rotational T_{rot} temperatures computed according to the QGDR model turn out to be excessive as compared with the DSMC data.

The correspondence of the numerical results to the results obtained by the kinetic DSMC algorithm shows a good accuracy of the presented QGDR model for describing the translational–rotational nonequilibrium in the gas.

The use of the QGDR models for computing two-dimensional axially symmetric flows and the comparison of the obtained results with experimental data are presented in [109, 146]. Underexpanded jet flows of CO_2 and N_2 flowing from a sound nozzle of circular sections are considered. In these computations, it is shown that the consistency of computations with experiments is improved when the translational–rotational nonequilibrium is taken into account in the numerical model. The numerical modelling of these problems are also carried out for the QGD and QGDR equations written in the cylindrical coordinate system by using parallel computation systems of cluster type (see [106, 107]).

Chapter 9
Quasi-gas-dynamic Equations for Binary Gas Mixtures

Numerical simulation of gas-mixture flows is of interest for many practical applications. In particular, effective modelling of non-reacting gas flows is a necessary stage preceding the construction of the gas model flows with chemical reactions (see [28, 99]).

To compute gas-mixture flows, two groups of models exist. The first consists of kinetic models, i.e., models based on direct numerical simulation methods DSMC [28] or on the solution of the Boltzmann equation in one or another approximation [128]. The second group consists of systems of moment equations obtained, as a rule, phenomenologically on the basis of the Navier–Stokes equations [99, 135, 198].

Kinetic models ensure a sufficiently exact description of gas flows but lose their efficiency when the Knudsen or Mach number decreases. Moment methods are more economical from the computational viewpoint, but phenomenological methods for their construction lead to a number of problems. In particular, most of the methods are one-fluid approximations, in which one supposes that the velocity and the temperatures of both components of the mixture are equal [99, 135]. Such a simplification does not allow one to analyze the behavior of each of the components and, therefore, it seems to be not sufficiently accurate. There exist few examples of using the two-fluid, two-velocity, or two-temperature approximations, which are mainly constructed for problems of physic of plasma [211]. These methods require the introduction of a number of additional constants whose finding is an independent problem.

In this chapter, we describe a macroscopic model for describing flows of non-reacting binary gas mixture based on the quasi-gas-dynamic equations. This model is a two-fluid approximation, which is a system of equations for the density, the momentum, and the energy of each of the components. Also, we suggest a one-fluid approximation of this model. The constructed system of equations (QGDM) is based on the system of kinetic equations in the relaxation approximation and is a natural generalization of the quasi-gas-dynamic equations to the case of gas mixtures.

The presentation of material is based on [59, 60].

T.G. Elizarova, *Quasi-Gas Dynamic Equations*, Computational Fluid and Solid Mechanics, DOI 10.1007/978-3-642-00292-2_9,
© Springer-Verlag Berlin Heidelberg 2009

9.1 Initial Kinetic Model

In 1954, Bhatnagar, Gross, and Krook [25] published their famous equations, which is the Boltzmann equation with the collision integral in the relaxation form. Despite its simple form, this model preserves the main properties of the initial kinetic equation and, therefore, it finds many applications in analyzing a large number of problems. The generalization of the BGK model to the case of gas mixtures was done by Sirovich in 1962 (see [193]). In 1964, based on the conservation laws, Morse calculated the missing "free parameters" of this model [149]. In what follows, we use this final kinetic model for constructing the moment (QGDM) equations for binary mixture flows. In 1970, Wu and Lee [210] applied the kinetic model mentioned to the computation of a one-dimensional binary mixture flow in a shock tube. Note that these computations correspond to the flow with the Prandtl number $\mathrm{Pr} = 1$, which is determined by the chosen form of the relaxation summand in the initial kinetic model.

Let us present a short description of the mentioned kinetic model in accordance with [210].

Let a mixture consist of two gases a and b with numerical densities n_a and n_b and mass densities $\rho_a = m_a n_a$ and $\rho_b = m_b n_b$, respectively, where m_a and m_b are the masses of molecules of the gases a and b. Let each of the gases be characterized by its temperature T_i and the macroscopic velocity $\boldsymbol{u}_i$, where $i = a, b$. The gas constant is equal to $\mathcal{R}_i = k_B/m_i$, where k_B is the Boltzmann constant.

Then according to [210], the kinetic model for the gas mixture can be written in the form

$$\frac{\partial f_a}{\partial t} + (\boldsymbol{\xi} \cdot \nabla) f_a = \nu_a(F_a - f_a) + \nu_{ab}(\overline{F}_a - f_a), \tag{9.1}$$

$$\frac{\partial f_b}{\partial t} + (\boldsymbol{\xi} \cdot \nabla) f_b = \nu_b(F_b - f_b) + \nu_{ba}(\overline{F}_b - f_b), \tag{9.2}$$

where $f_i(x, \xi, t)$ is the distribution function for the ith specie, $i = a, b$, $\xi = u + c$ is the velocity of the molecule, and c is its heat or thermal velocity. The quantities ν_a and ν_b are the frequencies of mutual collisions between the molecules of the same type, ν_{ab} is the frequency of collisions of molecules of type a with molecules of type b, and ν_{ba} is the frequency of collisions of molecules of type b with molecules of type a. The total number of collisions between the molecules a and b must be balanced, i.e.,

$$n_a \nu_{ab} = n_b \nu_{ba}. \tag{9.3}$$

Let F_a, F_b, $\overline{F}_a$, and $\overline{F}_b$ be the Maxwell distribution functions, which are defined as follows:

$$F_a = \frac{\rho_a}{(2\pi \mathcal{R}_a T_a)^{3/2}} \exp\left(-\frac{(\xi - u_a)^2}{2\mathcal{R}_a T_a}\right), \tag{9.4}$$

$$F_b = \frac{\rho_b}{(2\pi \mathcal{R}_b T_b)^{3/2}} \exp\left(-\frac{(\xi - u_b)^2}{2\mathcal{R}_b T_b}\right), \tag{9.5}$$

$$\overline{F}_a = \frac{\rho_a}{(2\pi \mathcal{R}_a \overline{T}_a)^{3/2}} \exp\left(-\frac{(\xi - \overline{u}_a)^2}{2\mathcal{R}_a \overline{T}_a}\right), \tag{9.6}$$

$$\overline{F}_b = \frac{\rho_b}{(2\pi \mathcal{R}_b \overline{T}_b)^{3/2}} \exp\left(-\frac{(\xi - \overline{u}_b)^2}{2\mathcal{R}_b \overline{T}_b}\right). \tag{9.7}$$

Formulas (9.6) and (9.7) contain "free parameters" or "gas parameters after collisions" denoted by the bars. According to [149], they are calculated through the macroscopic parameters of the gas as follows:

$$\overline{u}_a = \overline{u}_b = \frac{m_a u_a + m_b u_b}{m_a + m_b},$$

$$\overline{T}_a = T_a + \frac{2m_a m_b}{(m_a + m_b)^2}\left(T_b - T_a + \frac{m_b}{6k}(u_b - u_a)^2\right), \tag{9.8}$$

$$\overline{T}_b = T_b + \frac{2m_a m_b}{(m_a + m_b)^2}\left(T_a - T_b + \frac{m_a}{6k}(u_b - u_a)^2\right).$$

These distribution functions are related as follows and define macroscopic parameters of the gases:

$$\int f_i d\xi = \int F_i d\xi = \int \overline{F}_i d\xi = \rho_i, \tag{9.9}$$

$$\int \xi f_i d\xi = \int \xi F_i d\xi = \rho_i u_i, \tag{9.10}$$

$$\int \xi \overline{F}_i d\xi = \rho_i \overline{u}_i, \tag{9.11}$$

$$\int c f_i d\xi = \int c F_i d\xi = \int c \overline{F}_i d\xi = 0, \tag{9.12}$$

$$\int \frac{\xi^2}{2} f_i d\xi = \int \frac{\xi^2}{2} F_i d\xi = \frac{\rho_i u_i^2}{2} + \frac{3 p_i}{2} = E_i, \tag{9.13}$$

$$\int \frac{\xi^2}{2} \overline{F}_i d\xi = \frac{\rho_i \overline{u}_i^2}{2} + \frac{3 \overline{p}_i}{2} = \overline{E}_i. \tag{9.14}$$

In what follows, on the basis of the model presented above, we construct the system of macroscopic equations for describing the flow of binary mixture of gases that do not react with each other.

9.2 Construction of the Moment Equations

To construct the macroscopic or moment equations for a flow of a gas mixture, we use the method that was used for the construction of the QGD equations for a one-component gas.

Assume that the distribution functions of particles a and b are close to the corresponding local Maxwell functions and can be approximately represented in the form of expansions in a small parameter (gradient expansion) in a neighborhood of their equilibrium values as follows:

$$f_a^{QGD} = F_a - \tau(\xi \cdot \nabla)F_a, \tag{9.15}$$

$$f_b^{QGD} = F_b - \tau(\xi \cdot \nabla)F_b, \tag{9.16}$$

where τ is the Maxwell relaxation time for the mixture of gases a and b, which is close in value to the mean collision time for the mixture and is defined as

$$\tau = \frac{\mu}{p}, \tag{9.17}$$

where μ is the viscosity of the mixture and p is the pressure of the mixture, which is the sum of the partial pressures, i.e.,

$$p = p_a + p_b, \quad \text{where} \quad p_a = \rho_a \mathcal{R}_a T_a, \quad p_b = \rho_b \mathcal{R}_b T_b. \tag{9.18}$$

The last two relations are the partial state equations. The method for calculating the mixture viscosity coefficient μ and the relaxation parameter τ is presented below.

The formal replacement of the true values of the distributing functions f_a and f_b by the approximate values (9.15) and (9.16) in the convective terms of Eqs. (9.1) and (9.2) leads to the regularized kinetic equations, which has the following index form:

$$\frac{\partial f_a}{\partial t} + \nabla_i \xi^i F_a - \nabla_i \tau \nabla_j \xi^i \xi^j F_a = v_a(F_a - f_a) + v_{ab}(\overline{F}_a - f_a), \tag{9.19}$$

$$\frac{\partial f_b}{\partial t} + \nabla_i \xi^i F_b - \nabla_i \tau \nabla_j \xi^i \xi^j F_b = v_b(F_b - f_b) + v_{ba}(\overline{F}_b - f_b), \tag{9.20}$$

where the indices i and j correspond to components of the spatial coordinates.

The macroscopic QGDM equations are obtained by the moment averaging of the written equations in the velocity space ξ^i.

The systems of equations for both gases have the same form and, therefore, omitting the index a, we present the procedure for constructing the equations for the gas a.

Let us calculate certain integrals that are used in what follows:

$$\int c^i c^j F \, d\xi = g^{ij} p, \tag{9.21}$$

$$\int c^i c^j c^k F \, d\xi = 0, \tag{9.22}$$

$$\int c_x^4 F \, d\xi = \int c_y^4 F \, d\xi = \int c_z^4 F \, d\xi = 3\frac{p^2}{\rho}, \tag{9.23}$$

$$\int c_x^2 c_y^2 F \, d\xi = \int c_x^2 c_z^2 F \, d\xi = \int c_y^2 c_z^2 F \, d\xi = \frac{p^2}{\rho}, \tag{9.24}$$

$$\int c^i c^j \mathbf{c}^2 F \, d\xi = 5\frac{p^2}{\rho} g^{ij}, \tag{9.25}$$

where g^{ij} is the metric tensor. All integrals are calculated in infinite limits. In their calculation, we have used the relations

$$\int_{-\infty}^{\infty} e^{-y^2} dy = \sqrt{\pi}, \qquad \int_{-\infty}^{\infty} y^2 e^{-y^2} dy = \frac{1}{2}\sqrt{\pi},$$

$$\int_{-\infty}^{\infty} y^4 e^{-y^2} dy = \frac{3}{4}\sqrt{\pi}.$$

Integrating Eq. (9.19) with weight 1 and using relations (9.9), (9.12), and (9.21), we obtain

$$\int \frac{\partial f}{\partial t} \, d\xi = \frac{\partial}{\partial t} \int f \, d\xi = \frac{\partial}{\partial t} \rho,$$

$$\int \nabla_i \xi^i F \, d\xi = \nabla_i \int (u^i + c^i) F \, dc = \nabla_i \rho u^i,$$

$$\int \nabla_i \tau \nabla_j \xi^i \xi^j F \, d\xi = \nabla_i \tau \nabla_j \int (u^i + c^i)(u^j + c^j) F \, dc$$

$$= \nabla_i \tau \nabla_j \left(\rho u^i u^j + \int c^i c^j F \, dc \right) = \nabla_i \tau \nabla_j \left(\rho u^i u^j + g^{ij} p \right).$$

As shown below, the integral of the right-hand side of Eq. (9.19) vanishes.

Therefore, we obtain the following equation for the density:

$$\frac{\partial}{\partial t}\rho + \nabla_i \rho u^i = \nabla_i \tau \nabla_j (\rho u^i u^j + g^{ij} p). \tag{9.26}$$

To obtain the equation for the momentum, we integrate Eq. (9.19) with weight ξ^k using relations (9.10), (9.12), (9.21), and (9.22):

$$\int \frac{\partial f}{\partial t}\xi^k \, d\xi = \frac{\partial}{\partial t}\rho u^k,$$

$$\int \nabla_i \xi^i F \xi^k \, d\xi = \nabla_i (\rho u^i u^k + g^{ik} p),$$

$$\int \nabla_i \tau \nabla_j \xi^i \xi^j F \xi^k \, d\xi = \rho u^i u^j u^k + p(u^k g^{ij} + u^j g^{ik} + u^i g^{jk}).$$

For the collision integral of Eq. (9.19), the quantity ξ is not a summator invariant, since the exchange of momentum between the mixture components is possible, and the integral of the right-hand side is not equal to zero. It is called the exchange term for the energy and is denoted by S^u.

Combining the expressions, we obtain the following equation for ρu^k:

$$\frac{\partial}{\partial t}\rho u^k + \nabla_i (\rho u^i u^k + g^{ik} p)$$
$$= \nabla_i \tau \nabla_j \left[\rho u^i u^j u^k + p(u^k g^{ij} + u^j g^{ik} + u^i g^{jk}) \right] + S^u. \tag{9.27}$$

To obtain the equation for the specific energy E, we average Eq. (9.19) with weight $\xi^2/2$ using relations (9.21), (9.22), (9.23), (9.24), and (9.25):

$$\int \frac{\partial f}{\partial t}\frac{\xi^2}{2}\, d\xi = \frac{\partial}{\partial t}E,$$

$$\int \nabla_i \xi^i F \frac{1}{2}\xi^2 \, d\xi = \nabla_i \frac{1}{2}\int (u^i + c^i) F \xi^2 \, dc = \nabla_i u^i (E + p),$$

$$\int \nabla_i \tau \nabla_j \xi^i \xi^j F \frac{\xi^2}{2}\, d\xi = \nabla_i \tau \nabla_j \left(u^i u^j E + 2 u^i u^j p + \frac{1}{2} u_k u^k g^{ij} p + \frac{5}{2}\frac{p^2}{\rho} g^{ij} \right).$$

For the collision integral of Eq. (9.19), $\xi^2/2$ is not a summator invariant since the exchange of energy between the mixture components is possible, and the integral of the right-hand side of Eq. (9.19) does not vanish. It is called the exchange term and is denoted by S^E.

Combining the expressions obtained and differentiating by parts the last summand, which contains the squared pressure, we obtain

$$\frac{\partial}{\partial t}E + \nabla_i u^i (E + p) = \nabla_i \tau \nabla_j \left(u^i u^j E + 2u^i u^j p + \frac{1}{2} u_k u^k g^{ij} p \right)$$

$$+ \frac{5}{2} \nabla_i \tau \frac{p}{\rho} \nabla_j p g^{ij} + \frac{5}{2} \nabla_i \tau p \nabla_j \frac{p}{\rho} g^{ij} + S^E. \qquad (9.28)$$

The method for obtaining the moment equations presented here leads to expressions for the heat flux with the Prandtl number equal to unity. To generalize the equations to the case of an arbitrary Prandtl number, we multiply the next-to-the-past summand in the energy equation by Pr^{-1}. Here, the energy equation is obtained for a monoatomic gas, which corresponds to $\gamma = 5/3$. A generalization to the case of gases with inner degrees of freedom can be obtained by the formal change of the coefficient $5/3$ by γ in the expression for the total energy (9.13), (9.14) ($3p/2 \rightarrow p/(\gamma - 1)$) and in the last two summands in the energy equation (9.28) ($5/2 \rightarrow \gamma/(\gamma - 1)$).

9.3 Calculation of Exchange Terms

The right-hand sides of the QGDM equations contain the exchange terms that are the moments of the collision integral arising in its averaging in molecule velocities. The use of the relaxation model allows one to calculate these moments and to express them through macroscopic parameters of the gas.

Let us verify that in the density equation (9.26), the exchange terms vanish. Indeed, the direct integration yields

$$\int \nu_a (F_a - f_a)\, d\xi = \nu_a \left(\int F_a\, d\xi - \int f_a\, d\xi \right) = \nu_a (\rho_a - \rho_a) = 0,$$

$$\int \nu_{ab} (\overline{F}_a - f_a)\, d\xi = \nu_{ab} \left(\int \overline{F}_a\, d\xi - \int f_a\, d\xi \right) = \nu_{ab} (\rho_a - \rho_a) = 0.$$

In the relaxation model, it is assumed that the collision frequencies are independent of the molecule velocities.

The integration with weight ξ allows one to calculate the exchange term in Eq. (9.27):

$$\int \nu_a (F_a - f_a)\xi\, d\xi = \nu_a (\rho_a u_a - \rho_a u_a) = 0,$$

$$\int \nu_{ab} (\overline{F}_a - f_a)\xi\, d\xi = \nu_{ab} (\rho_a \overline{u}_a - \rho_a u_a) = S_a^u.$$

Similarly, averaging with weight $\xi^2/2$, we calculate the exchange terms for the energy equation (9.28).

Respectively, for the gases a and b, the exchange terms have the form

$$S_a^u = \nu_{ab}\rho_a(\overline{u}_a - u_a), \qquad S_b^u = \nu_{ba}\rho_b(\overline{u}_b - u_b),$$

$$S_a^E = \nu_{ab}(\overline{E}_a - E_a), \qquad S_b^E = \nu_{ba}(\overline{E}_b - E_b), \tag{9.29}$$

where

$$\overline{E}_a = \frac{\rho_a \overline{u}_a^2}{2} + \frac{\overline{p}_a}{\gamma_a - 1}, \qquad \overline{p}_a = \rho_a \mathcal{R}_a \overline{T}_a,$$

$$\overline{E}_b = \frac{\rho_b \overline{u}_b^2}{2} + \frac{\overline{p}_b}{\gamma_b - 1}, \qquad \overline{p}_b = \rho_b \mathcal{R}_b \overline{T}_b. \tag{9.30}$$

Note that in accordance with the conservation laws of momentum and energy and with account for the balance relation (9.3), we have

$$S_a^u + S_b^u = 0, \qquad S_a^E + S_b^E = 0. \tag{9.31}$$

Therefore, the QGDM model takes into account the momentum and energy exchanges between the mixture components whose intensity is proportional to the collision frequency between different gas particles. The density equations contain no exchange terms which is natural since the mixture components do not react with each other.

9.4 Determination of the Collision Frequencies

To close the system of QGDM equations, we need to specify the collision frequencies ν_{ab} and ν_{ba} and also the relaxation time τ.

According to [28,160,210], the relations for the collision frequencies of molecules of the gas a with each other and those of molecules of the gas a with molecules of gas b can be calculated as follows:

$$\nu_{ab} = \nu_a \left(\frac{d_{ab}}{d_a}\right)^2 \sqrt{\frac{m_a + m_b}{2m_b}} \frac{n_b}{n_a}, \tag{9.32}$$

where d_a is the effective diameter of molecules of the gas a, d_{ab} is the effective diameter of interaction, which can be found (see [28, p. 16]) as $d_{ab} = 0.5(d_a + d_b)$, and n_a and n_b are the numerical densities of the gases a and b, respectively. In turn, the collision frequency ν_a can be related with the gas viscosity. In the approximation of the VHS and VSS particle interaction models (see [28, p. 90]) and using Eq. (2.19), this relation has the following form:

$$v_a = \frac{p_a}{\mu_a} \Omega(\omega_a, \alpha_a),$$

$$\Omega(\omega_a, \alpha_a) = \frac{5(\alpha_a + 1)(\alpha_a + 2)}{\alpha_a(7 - 2\omega_a)(5 - 2\omega_a)}, \tag{9.33}$$

$$\mu_a = \mu_{a\,\mathrm{ref}} \left(\frac{T_a}{T_{a\,\mathrm{ref}}}\right)^{\omega_a}.$$

In what follows, to perform calculations, we set $\alpha_a = 1$, which corresponds to the VHS model (see, e.g., [28, p. 41]). In this case,

$$\Omega(\omega_a, 1) = \Omega(\omega_a) = \frac{30}{(7 - 2\omega_a)(5 - 2\omega_a)}.$$

There also exist other expressions for the collision frequency between the particles of different kinds (see, e.g., [28, p. 96]).

The total number of collisions between the molecules of the gases a and b must be balanced, i.e., relation (9.3) must hold. However, the expressions for collision frequencies in the framework of the VHS model automatically satisfy the above balance relation only in the case of Maxwell molecules ($\omega = 1$) for $T_{a\,\mathrm{ref}} = T_{b\,\mathrm{ref}}$. Therefore, if one of the mutual collision frequencies is found, e.g., according to Eqs. (9.32) and (9.33), then the other must be found from the balance relation (9.3).

Equations (9.26), (9.27), and (9.28) contain the parameter τ, which is defined as the Maxwell relaxation time (9.17) for the mixture. To find the binary mixture viscosity, we have, for example, the Wilke formula (see [99, 207]):

$$\mu = \mu_a \left(1 + G_{ab} \frac{\rho_b}{\rho_a} \frac{\mathcal{M}_a}{\mathcal{M}_b}\right)^{-1} + \mu_b \left(1 + G_{ba} \frac{\rho_a}{\rho_b} \frac{\mathcal{M}_b}{\mathcal{M}_a}\right)^{-1}, \tag{9.34}$$

where

$$G_{ab} = \frac{\left(1 + \sqrt{\mu_a/\mu_b}\sqrt{\mathcal{M}_b/\mathcal{M}_a}\right)^2}{2\sqrt{2}(1 + \mathcal{M}_a/\mathcal{M}_b)},$$

and $\mathcal{M}_a$ and $\mathcal{M}_b$ are the molar masses of the gases a and b, respectively. We calculate G_{ba} similarly to G_{ab} by a cyclic change of indices.

In the literature, there also exist other expressions for finding the viscosity of binary mixtures (see, e.g., [39, p. 275] and [113]). In [162], the viscosity coefficients of separate components are presented.

9.5 Quasi-gas-dynamic Equations for Gas Mixtures

Now we write the final form of the QGDM equations for a binary mixture in the invariant (with respect to coordinates) form. The systems of equations for both gases have the same form and, therefore, we present the system of equations describing the gas a:

$$\frac{\partial}{\partial t}\rho_a + \nabla_i \rho_a u_a^i = \nabla_i \tau (\nabla_j \rho_a u_a^i u_a^j + \nabla^i p_a), \tag{9.35}$$

$$\frac{\partial}{\partial t}\rho_a u_a^k + \nabla_i \rho_a u_a^i u_a^k + \nabla^k p_a$$
$$= \nabla_i \tau \left(\nabla_j \rho_a u_a^i u_a^j u_a^k + \nabla^i p_a u_a^k + \nabla^k p_a u_a^i \right) + \nabla^k \iota \nabla_i p_a u_a^i + S_a'', \tag{9.36}$$

$$\frac{\partial}{\partial t}E_a + \nabla_i u_a^i (E_a + p_a) = \nabla_i \tau \left(\nabla_j (E_a + 2p_a) u_a^i u_a^j + \frac{1}{2}\nabla^i u_{ak} u_a^k p_a \right)$$
$$+ \frac{\gamma_a}{\gamma_a - 1}\nabla_i \tau \frac{p_a}{\rho_a}\nabla^i p_a + \frac{1}{\mathrm{Pr}_a}\frac{\gamma_a}{\gamma_a - 1}\nabla_i \tau p_a \nabla^i \frac{p_a}{\rho_a} + S_a^E, \tag{9.37}$$

where the specific energy of the gas a has the form

$$E_a = \frac{\rho_a u_{ai}^2}{2} + \frac{p_a}{\gamma_a - 1}. \tag{9.38}$$

The exchange terms are calculated according to Eqs. (9.29) and (9.30), the "free parameters" according to Eq. (9.8), and the collision frequencies and the mixture viscosity coefficients can be calculated by using expressions (9.32), (9.33), and (9.34). The relaxation time τ is calculated according to Eq. (9.17). We call attention to the fact that in the right-hand sides of the density equations, there exist terms of diffusion type, as in one-fluid models of the Navier–Stokes type (see, e.g., [135]).

The system written here and complemented by boundary conditions is a closed model for calculation of flows of binary gas mixtures in the two-fluid approximation.

Note that the two-fluid model for gas mixtures using the Chapman–Enskog procedure was written and analyzed in [98]. This model turned out to be very cumbersome and obtained no wide application in the computational practice.

Equations (9.35), (9.36), (9.37), and (9.38) contain no parameters of the gas mixture. However, in what follows, these parameters will be used and, therefore, we present the method for their finding. The mixture parameters (variables without index) are found as follows:

$$n = n_a + n_b, \quad \rho = \rho_a + \rho_b, \quad p = p_a + p_b, \quad u = \frac{\rho_a u_a + \rho_b u_b}{\rho},$$

$$T = \frac{n_a T_a + n_b T_b}{n}, \quad m = \frac{m_a n_a + m_b n_b}{n}, \quad p = \rho \mathcal{R} T, \tag{9.39}$$

$$\mathcal{R} = \frac{\rho_a \mathcal{R}_a + \rho_b \mathcal{R}_b}{\rho} = \frac{k_B}{m}.$$

In the case of one-component gas, system (9.35), (9.36), (9.37), and (9.38) coincides with the QGD system studied earlier.

Similar to the QGD system, Eqs. (9.35), (9.36), and (9.37) can be written as conservation laws. Repeating the reasoning from Sect. 3.7, one can show that the entropy production X for a mixture of gases contains the additional terms proportional to v_{ab} and is nonnegative.

9.6 One-Fluid Approximations

9.6.1 QGDM Model in the One-Fluid Approximation

The two-fluid QGDM system (9.35), (9.36), and (9.37) written here can be simplified and reduced to the one-fluid approximation, which corresponds to the mixture flows in which

$$u_a^i = u_b^i = u^i, \quad T_a = T_b = T. \tag{9.40}$$

In this case, the diffusion velocities of the components w_a^i and w_b^i remain different. The pressures of separate components are calculated as follows:

$$p_a = \rho_a \mathcal{R}_a T, \quad p_b = \rho_b \mathcal{R}_b T. \tag{9.41}$$

Let us construct a one-fluid approximation for the QGDM equations. Assume that $\gamma_a = \gamma_b = \gamma$ and $\mathrm{Pr}_a = \mathrm{Pr}_b = \mathrm{Pr}$ for the components of the mixture. Then the specific energy of the mixture has the form

$$E = E_a + E_b = \frac{(\rho_a + \rho_b)u_i^2}{2} + \frac{p_a + p_b}{\gamma - 1} = \frac{\rho u_i^2}{2} + \frac{p}{\gamma - 1}. \tag{9.42}$$

In passing to the one-fluid approximation, the form of the density equations does not change, except that instead of the velocities u_a^i and u_b^i, the mixture velocity u^i enters them. The equations for the mixture momentum and the specific energy also preserve their general form and are obtained by adding the corresponding two equations for each of the components with account of conditions (9.31) for the exchange terms. The resulting QGDM system takes the form

$$\frac{\partial}{\partial t}\rho_a + \nabla_i \rho_a u^i = \nabla_i \tau (\nabla_j \rho_a u^i u^j + \nabla^i p_a),$$ (9.43)

$$\frac{\partial}{\partial t}\rho_b + \nabla_i \rho_b u^i = \nabla_i \tau (\nabla_j \rho_b u^i u^j + \nabla^i p_b),$$ (9.44)

$$\frac{\partial}{\partial t}\rho u^k + \nabla_i \rho u^i u^k + \nabla^k p$$
$$= \nabla_i \tau (\nabla_j \rho u^i u^j u^k + \nabla^i p u^k + \nabla^k p u^i) + \nabla^k \tau \nabla_i p u^i,$$ (9.45)

$$\frac{\partial}{\partial t}E + \nabla_i u^i (E + p) = \nabla_i \tau \left(\nabla_j (E + 2p) u^i u^j + \frac{1}{2} \nabla^i u_k u^k p \right)$$
$$+ \frac{\gamma}{\gamma - 1} \nabla_i \tau \left(\frac{p_a}{\rho_a} \nabla^i p_a + \frac{p_b}{\rho_b} \nabla^i p_b \right)$$
$$+ \frac{\gamma}{\mathrm{Pr}(\gamma - 1)} \nabla_i \tau \left(p_a \nabla^i \frac{p_a}{\rho_a} + p_b \nabla^i \frac{p_b}{\rho_b} \right).$$ (9.46)

The one-fluid approximation simplifies the QGDM system reducing it to four equations, two equations for densities and the momentum and energy equations, which now contain no exchange terms. Thus, to close the system, it suffices only to find the mixture viscosity coefficient μ, and this model does not require the determination of the mutual collision frequencies ν_{ab} and ν_{ba} and the calculation of the free parameters (9.8). However, in this model, the gas constant $\mathcal{R}$ is no longer a constant and depends on the concentrations of the component. In the case where the values γ and Pr for the mixture components do not coincide, the finding of these quantities for the mixture is also an independent problem.

The finite-difference schemes with artificial viscosity whose differential approximation is similar to the one-fluid approximation of the QGDM equations were written and tested in [42] by examining the reacting gas flow in a neighborhood of a plate.

In the next section, we show the connection of the written one-fluid QGDM model with the traditional approximation of the type of Navier–Stokes equations.

9.6.2 One-Fluid Model for the Navier–Stokes Equations

The one-fluid model for describing flows of gas mixtures without chemical reactions based on the Navier–Stokes equations has the following form (see [99]):

$$\frac{\partial \rho c_a}{\partial t} + \nabla_i \rho c_a u^i = -\nabla_i \mathcal{J}_a^i,$$ (9.47)

$$\frac{\partial \rho c_b}{\partial t} + \nabla_i \rho c_b u^i = -\nabla_i \mathcal{J}_b^i,$$ (9.48)

$$\frac{\partial \rho u^k}{\partial t} + \nabla_i \rho u^i u^k + \nabla^k p = \nabla_i \Pi_{\text{NS}}^{ik}, \tag{9.49}$$

$$\frac{\partial E}{\partial t} + \nabla_i u^i (E + p) = \nabla_i (\Pi_{\text{NS}}^{ik} u^k) - \nabla_i \left(h_a \mathcal{J}_a^i + h_b \mathcal{J}_b^i \right) + \nabla_i \varkappa \nabla^i T, \tag{9.50}$$

where $c_a = \rho_a/\rho$ and $c_b = \rho_b/\rho$ are mass concentrations of the gases a and b, respectively, $\mathcal{J}_a^i$ and $\mathcal{J}_b^i$ are the densities of the diffusion fluxes, and

$$h_a = c_{p_a} T = \frac{\gamma}{\gamma - 1} \frac{p_a}{\rho_a}, \quad h_b = c_{p_b} T = \frac{\gamma}{\gamma - 1} \frac{p_b}{\rho_b} \tag{9.51}$$

are the enthalpies of the gases a and b, respectively. All mixture parameters are found according to Eq. (9.39) and the specific energy is given by

$$E = \frac{\rho u_i^2}{2} + \frac{p}{\gamma - 1},$$

where γ is the specific heat ratio for the mixture.

According to [99], the expressions for the diffusion fluxes in the binary mixture are written as follows:

$$\mathcal{J}_a^i = -\rho c_a D \left(\nabla^i \ln c_a + \frac{m_b - m_a}{m} c_b \nabla^i \ln p \right) - \rho c_a D_a^T \nabla^i \ln T,$$
$$\mathcal{J}_b^i = -\rho c_b D \left(\nabla^i \ln c_b + \frac{m_a - m_b}{m} c_a \nabla^i \ln p \right) - \rho c_b D_b^T \nabla^i \ln T, \tag{9.52}$$

where D is the diffusion coefficient of the binary mixture and D_a^T and D_b^T are the thermodiffusion coefficients of the gases a and b, respectively. Under the absence of mass forces, in the model presented, diffusion can arise for the following three reasons: under the action of the gradient of concentrations (mass diffusion, the first term in Eq. (9.52)), under the action of the gradient of pressures (barodiffusion, the second term), and by the action of the gradient of temperatures (thermodiffusion, the third term in Eq. (9.52)). For mixtures of light and heavy components, the contribution of the thermodiffusion to the mass transportation can become sufficiently appreciable. The expression for the thermodiffusion coefficient is presented, for example, in [39]. However, one considers [135] that for many flows, the thermodiffusion is a second-order effect as compared with the mass diffusion, and its influence is neglected.

The diffusion coefficient is connected with the mixture viscosity coefficient by the following relation [99]:

$$D = \frac{\mu}{\rho \, \text{Sc}}, \tag{9.53}$$

where Sc is the Schmidt number, which is close to 1 for gases.

The value of the diffusion coefficient for the binary mixture can be also found by the following expression (see [162]):

$$D = 1.8826 \times 10^{-22} \sqrt{\frac{T^3(\mathcal{M}_a + \mathcal{M}_b)}{\mathcal{M}_a \mathcal{M}_b}} \frac{1}{p\sigma^2 \Omega^{(1.1)*}(T^*)}, \qquad (9.54)$$

where $\mathcal{M}_a$ and $\mathcal{M}_b$ are molar masses of the gases, p is the pressure, $\sigma = 0.5(\sigma_a + \sigma_b)$, where σ_a and σ_b (m) are effective collision diameters, $T^* = k_B T/\varepsilon$ is the characteristic temperature, ε/k_B is the molecule potential energy parameter characterizing the interaction of molecules of types a and b, $\varepsilon = \sqrt{\varepsilon_a \varepsilon_b}$, and ε_a and ε_b are the parameters of the molecular interaction potential function. $\Omega^{(1.1)*}(T^*)$ is the dimensionless collision integral for the mass transport, which expresses the measure of deviation from the model considering the gas molecules as hard spheres for which $\Omega^{(1.1)*} = 1$.

As in the QGDM model, the mixture viscosity coefficient of the gas can be found by different methods, for example, from relation (9.34). In this case, the viscosity coefficient for each of the components can be found according to Eq. (9.33) or from other relations, for example, according to [162]:

$$\mu_a = 26.69 \times 10^{-27} \frac{\sqrt{\mathcal{M}_a T_a}}{\sigma_a^2 \Omega^{(2.2)*}(T_a^*)}, \qquad (9.55)$$

where $\Omega^{(2.2)*}(T_a^*)$ is the dimensionless collision integral for the momentum transport expressing the measure of deviation from the model considering the gas molecules as hard spheres for which $\Omega^{(2.2)*} = 1$ and $T_a^* = k_B T/\varepsilon_a$ is the characteristic temperature. The collision integrals $\Omega^{(1.1)*}(T^*)$ and $\Omega^{(2.2)*}(T^*)$ are calculated in [113] by using the Lennard–Jones potential and can be found by the following approximate formulas with accuracy sufficient for many applications:

$$\begin{aligned}
\Omega^{(1.1)*}(T^*) &= 1.074(T^*)^{-0.1604}, \\
\Omega^{(2.2)*}(T^*) &= 1.157(T^*)^{-0.1472}.
\end{aligned} \qquad (9.56)$$

9.6.3 QGDM and Navier–Stokes One-Fluid Approximations

The QGDM system in the one-fluid approximation (9.43), (9.44), (9.45), and (9.46) can be reduced to the form (9.47), (9.48), (9.49), and (9.50) by using some simplifications.

Equation (9.43) is represented in the form (9.47), where the diffusion flux is found by the expression

$$\mathcal{J}_a^i = -\rho_a w_a^i = -\tau(\nabla_j \rho c_a u^i u^j + \nabla^i p_a). \qquad (9.57)$$

Neglecting the term with squared velocity in the right-hand side, writing $p_a = \rho c_a \mathcal{R}_a T$, and then differentiating this expressions by parts, we obtain the following expression with the thermodiffusion coefficient $D_a^T = 0$, which is similar to Eq. (9.52):

$$\mathcal{J}_a^i = -\rho c_a D \frac{m}{m_a} (\nabla^i \ln c_a + \nabla^i \ln p), \tag{9.58}$$

where the coefficient D is defined in Eq. (9.53) for Sc $= 1$.

Similarly, Eq. (9.44) reduces to the form (9.48), where

$$\mathcal{J}_b^i = -\rho c_b D \frac{m}{m_b} (\nabla^i \ln c_b + \nabla^i \ln p). \tag{9.59}$$

The dissipative summand of the QGD equations can be represented as the sum of the dissipative terms of the Navier–Stokes equations and an addition of the first order with respect to the Knudsen number. Neglecting these additions, we immediately reduce the momentum equation (9.45) to the form (9.49).

The energy equation (9.46) can be reduced to the form (9.50) by extracting the dissipative terms and the heat flux of the form $\nabla_i (\Pi_{\text{NS}}^{ik} u^k + \varkappa \nabla^i T)$ from the right-hand side of Eq. (9.46) and neglecting the remaining terms summands with powered velocity. In this case, the terms with gradients of pressure preserve and are rewritten in the form

$$\frac{\gamma}{\gamma - 1} \nabla_i \tau \left(\frac{p_a}{\rho_a} \nabla^i p_a + \frac{p_b}{\rho_b} \nabla^i p_b \right) = -\nabla_i (h_a \mathcal{J}_a^i + h_b \mathcal{J}_b^i),$$

where h_a and h_b are determined by relations (9.51) and the diffusion fluxes have the form (9.58) and (9.59).

The one-fluid QGDM system is thus reduced and leads to the known one-fluid approximation for describing binary mixtures based on the Navier–Stokes equations.

9.7 QGDM System for One-Dimensional Flows

Let us write the system of quasi-gas-dynamic equations generalized to the case of binary gas mixture (9.35), (9.36), and (9.37) in the one-dimensional case for a plane-parallel flow (the index a denotes the parameters of one gas and the index b the parameters of the other):

$$\frac{\partial \rho_a}{\partial t} + \frac{\partial}{\partial x} \rho_a u_a = \frac{\partial}{\partial x} \tau \frac{\partial}{\partial x} (\rho_a u_a^2 + p_a), \tag{9.60}$$

$$\frac{\partial \rho_b}{\partial t} + \frac{\partial}{\partial x}\rho_b u_b = \frac{\partial}{\partial x}\tau \frac{\partial}{\partial x}(\rho_b u_b^2 + p_b), \tag{9.61}$$

$$\frac{\partial \rho_a u_a}{\partial t} + \frac{\partial}{\partial x}(\rho_a u_a^2 + p_a) = \frac{\partial}{\partial x}\tau \frac{\partial}{\partial x}(\rho_a u_a^3 + 3p_a u_a) + S_a^u, \tag{9.62}$$

$$\frac{\partial \rho_b u_b}{\partial t} + \frac{\partial}{\partial x}(\rho_b u_b^2 + p_b) = \frac{\partial}{\partial x}\tau \frac{\partial}{\partial x}(\rho_b u_b^3 + 3p_b u_b) + S_b^u, \tag{9.63}$$

$$\frac{\partial E_a}{\partial t} + \frac{\partial}{\partial x}u_a(E_a + p_a) = \frac{\partial}{\partial x}\tau \frac{\partial}{\partial x}u_a^2(E_a + 2.5 p_a) + \frac{\gamma_a}{\gamma_a - 1}\frac{\partial}{\partial x}\tau \frac{p_a}{\rho_a}\frac{\partial p_a}{\partial x}$$
$$+ \frac{\gamma_a}{\gamma_a - 1}\frac{1}{\mathrm{Pr}_a}\frac{\partial}{\partial x}\tau p_a \frac{\partial}{\partial x}\frac{p_a}{\rho_a} + S_a^E, \tag{9.64}$$

$$\frac{\partial E_b}{\partial t} + \frac{\partial}{\partial x}u_b(E_b + p_b) = \frac{\partial}{\partial x}\tau \frac{\partial}{\partial x}u_b^2(E_b + 2.5 p_b) + \frac{\gamma_b}{\gamma_b - 1}\frac{\partial}{\partial x}\tau \frac{p_b}{\rho_b}\frac{\partial p_b}{\partial x}$$
$$+ \frac{\gamma_b}{\gamma_b - 1}\frac{1}{\mathrm{Pr}_b}\frac{\partial}{\partial x}\tau p_b \frac{\partial}{\partial x}\frac{p_b}{\rho_b} + S_b^E. \tag{9.65}$$

The specific energy for the gases a and b is written in the form

$$E_a = \frac{\rho_a u_a^2}{2} + \frac{p_a}{\gamma_a - 1}, \quad E_b = \frac{\rho_b u_b^2}{2} + \frac{p_b}{\gamma_b - 1}. \tag{9.66}$$

According to Eq. (9.29), the exchange terms in the momentum equations for the gases a and b have the form

$$S_a^u = (\rho_a \bar{u}_a - \rho_a u_a)v_{ab}, \quad S_b^u = (\rho_b \bar{u}_b - \rho_b u_b)v_{ba}, \tag{9.67}$$

and in the energy equations, they are

$$S_a^E = (\bar{E}_a - E_a)v_{ab}, \quad S_b^E = (\bar{E}_b - E_b)v_{ba}. \tag{9.68}$$

In turn, v_{ab} is calculated according to Eqs. (9.32) and (9.33) for $\alpha_a = 1$, and v_{ba} is found from the balance relation (9.3).

Formulas (9.67) and (9.68) contain the "free parameters" denoted by the bars. According to Eq. (9.8), these parameters are calculated as follows:

$$\bar{u}_a = \bar{u}_b = \frac{m_a u_a + m_b u_b}{m_a + m_b},$$
$$\bar{T}_a = T_a + \frac{2m_a m_b}{(m_a + m_b)^2}\left(T_b - T_a + \frac{m_b}{6k}(u_b - u_a)^2\right), \tag{9.69}$$
$$\bar{T}_b = T_b + \frac{2m_a m_b}{(m_a + m_b)^2}\left(T_a - T_b + \frac{m_a}{6k}(u_b - u_a)^2\right).$$

The specific energy of the gases in relations (9.68) is given by

$$\overline{E}_a = \frac{\rho_a \overline{u_a^2}}{2} + \rho_a \frac{R_a}{\gamma_a - 1}\overline{T}_a, \quad \overline{E}_b = \frac{\rho_b \overline{u_b^2}}{2} + \rho_b \frac{R_b}{\gamma_b - 1}\overline{T}_b.$$

The parameter τ is found according to Eqs. (9.17) and (9.34).

9.7.1 Dimensionless Form

We solve system (9.60), (9.61), (9.62), (9.63), (9.64), and (9.65) in the dimensionless variables. As dimensional scales, we take the following characteristics of the gas a: the density $\rho_{a\,\mathrm{ref}}$, the sound speed $a_{a\,\mathrm{ref}} = \sqrt{\gamma_a R_a T_{a\,\mathrm{ref}}}$ under the temperature $T_{a\,\mathrm{ref}}$, and the free path $\lambda_{a\,\mathrm{ref}}$. The mean free path of the molecule can be calculated as follows (see [28]):

$$\lambda = \frac{4\mu}{\rho\sqrt{RT}}\frac{1}{\sqrt{2\pi}\,\Omega(\omega, 1)} = \frac{\mu}{\rho\sqrt{RT}}\frac{2(7 - 2\omega)(5 - 2\omega)}{15\sqrt{2\pi}}. \tag{9.70}$$

Then the relations between the dimensional and dimensionless characteristics have the following form (all parameters of the gas b are scaled with respect to the parameters of the gas a):

$$\rho = \tilde{\rho}\rho_{a\,\mathrm{ref}}, \quad a = \tilde{a}a_{a\,\mathrm{ref}}, \quad u = \tilde{u}a_{a\,\mathrm{ref}}, \quad p = \tilde{p}\rho_{a\,\mathrm{ref}}a_{a\,\mathrm{ref}}^2,$$

$$m = \tilde{m}\rho_{a\,\mathrm{ref}}\lambda_{a\,\mathrm{ref}}^3, \quad T = \tilde{T}\frac{a_{a\,\mathrm{ref}}^2}{\gamma_a R_a} = \tilde{T}T_{a\,\mathrm{ref}}, \quad x = \tilde{x}\lambda_{a\,\mathrm{ref}},$$

$$t = \tilde{t}\frac{\lambda_{a\,\mathrm{ref}}}{a_{a\,\mathrm{ref}}}, \quad \tau = \tilde{\tau}\frac{\lambda_{a\,\mathrm{ref}}}{a_{a\,\mathrm{ref}}}, \quad n = \tilde{n}\frac{1}{\lambda_{a\,\mathrm{ref}}^3},$$

$$\mu = \tilde{\mu}\lambda_{a\,\mathrm{ref}}\rho_{a\,\mathrm{ref}}a_{a\,\mathrm{ref}}\frac{15\sqrt{2\pi}}{2\sqrt{\gamma_a}(7 - 2\omega_a)(5 - 2\omega_a)} = \tilde{\mu}\mu_{a\,\mathrm{ref}}.$$

After reducing to the dimensionless form, Eqs. (9.60), (9.61), (9.62), (9.63), (9.64), and (9.65) do not change their form. Let us write the relations between the gas parameters (connection equation) used in calculations:

$$\tilde{a}_a = \sqrt{\tilde{T}_a}, \quad \tilde{a}_b = \sqrt{\frac{\gamma_b}{\gamma_a}\frac{R_b}{R_a}\tilde{T}_b}, \quad \tilde{T}_a = \frac{\gamma_a \tilde{p}_a}{\tilde{\rho}_a}, \quad \tilde{T}_b = \frac{\gamma_a \tilde{p}_b}{\tilde{\rho}_b}\frac{R_a}{R_b},$$

$$\tilde{\mu}_a = \tilde{T}_a^{\omega_a}, \quad \tilde{\mu}_b = \frac{\mu_{b\,\mathrm{ref}}}{\mu_{a\,\mathrm{ref}}}\left(\frac{T_{a\,\mathrm{ref}}}{T_{b\,\mathrm{ref}}}\right)^{\omega_b}\tilde{T}_b^{\omega_b},$$

where $\mu_{a\,\mathrm{ref}}$, $T_{a\,\mathrm{ref}}$, $\mu_{b\,\mathrm{ref}}$, and $T_{b\,\mathrm{ref}}$ are the values of the viscosity coefficients and the corresponding temperatures of the gases a and b used in the viscosity variation law (9.33),

$$\tilde{\lambda}_a = \frac{(\gamma_a \tilde{p}_a)^{\omega_a - 0.5}}{\tilde{\rho}_a^{\omega_a + 0.5}},$$

$$\tilde{\lambda}_b = \frac{(\gamma_a \tilde{p}_b)^{\omega_b - 0.5}}{\tilde{\rho}_b^{\omega_b + 0.5}} \frac{\mu_{b\,\text{ref}}}{\mu_{a\,\text{ref}}} \left(\frac{T_{a\,\text{ref}}}{T_{b\,\text{ref}}}\right)^{\omega_b} \frac{(7 - 2\omega_b)(5 - 2\omega_b)}{(7 - 2\omega_a)(5 - 2\omega_a)} \left(\frac{\mathcal{R}_a}{\mathcal{R}_b}\right)^{\omega_b}.$$

The values of mean free paths are not directly used in calculations, but they are necessary for the choice of the spatial step and the graphical representations of the results obtained.

The mean free path time between collisions is calculated for each of the components as follows:

$$\tilde{\tau}_a = \frac{(\gamma_a \tilde{p}_a)^{\omega_a - 1}}{\tilde{\rho}_a^{\omega_a}} \frac{15\sqrt{2\pi\gamma_a}}{2(7 - 2\omega_a)(5 - 2\omega_a)},$$

$$\tilde{\tau}_b = \frac{\mu_{b\,\text{ref}}}{\mu_{a\,\text{ref}}} \left(\frac{T_{a\,\text{ref}}}{T_{b\,\text{ref}}}\right)^{\omega_b} \frac{(\gamma_a \tilde{p}_b)^{\omega_b - 1}}{\tilde{\rho}_b^{\omega_b}} \frac{15\sqrt{2\pi\gamma_a}}{2(7 - 2\omega_a)(5 - 2\omega_a)} \left(\frac{\mathcal{R}_a}{\mathcal{R}_b}\right)^{\omega_b}.$$

The dimensionless form of the mean free path time $\tilde{\tau}_{ab}^c$ is written as follows:

$$\tilde{\tau}_{ab}^c = \frac{\tilde{\tau}_a}{\Omega(\omega_a)} \left(\frac{2d_a}{d_a + d_b}\right)^2 \sqrt{\frac{2m_b}{m_a + m_b} \frac{\rho_a m_b}{\rho_b m_a}}.$$

The value of $\tilde{\tau}_{ba}^c$ is found by using the balance relation (9.3). Moreover, $\tau_{ab}^c = 1/\nu_{ab}$ and $\tau_{ba}^c = 1/\nu_{ba}$ and τ_{ba}^c is found as follows:

$$\tau_{ba}^c = \tau_{ab}^c \frac{\rho_b}{\rho_a} \frac{m_a}{m_b}. \tag{9.71}$$

9.8 Structure of Shock Waves in the Mixture of Helium and Xenon

9.8.1 Problem Formulation

As the first example of using the QGDM equations, we consider the problem on the structure of stationary shock waves in the mixture of helium and xenon (He is the gas a and Xe is the gas b). The density profiles of these gases measured by using an electric gun and a laser interferometer can be found in [97]. The measurements are carried out for the following variants of gas mixtures:

V1 98.5% He and 1.5% Xe, i.e., $n_a/n = 0.985$ and $n_b/n = 0.015$;
V2 97% He and 3% Xe, i.e., $n_a/n = 0.97$ and $n_b/n = 0.03$;
V3 94% He and 6% Xe, i.e., $n_a/n = 0.94$ and $n_b/n = 0.06$;
V4 91% He and 9% Xe, i.e., $n_a/n = 0.91$ and $n_b/n = 0.09$.

Table 9.1 Dimensional parameters for the mixture components

	V1		V2		V3		V4	
	He	Xe	He	Xe	He	Xe	He	Xe
ρ (kg/m^3) $\times 10^5$	5.15	2.57	5.16	2.22	4.91	10.3	4.57	14.8
p (Pa)	33.14	0.51	33.21	1.02	31.62	2.02	29.42	2.91
T (K)	310							
u (m/s)	3076.76		2882.6		2672.8		2530.3	
Ma	2.97	17.01	2.78	15.93	2.58	14.78	2.44	13.99

For the second variant of percent concentration (V2), there exists the computation of this problem by using the direct Monte Carlo simulation (DSMC, see [28]), which can be considered as standard and practically coinciding with experimental data.

The mixture parameters before the shock wave chosen in accordance with the experiment data of [97] and the computation in [28] are presented in Table 9.1.

In Table 9.2, we present the physical parameters of helium and xenon according to [28] needed for performing the computation by using the QGDM model. The Prandtl number Pr for the gases mentioned above is constant and equal to 2/3.

Table 9.2 Tabulated values for the mixture components

	He	Xe
m (kg)	6.65×10^{-27}	218.0×10^{-27}
$\mathcal{R}$ (J/(kg·K))	2076.2	63.33
$\mathcal{M}$ (kg/mol)	4.0	131.4
d (m)	2.30×10^{-10}	5.65×10^{-10}
γ	1.66	1.66
ω	0.66	0.85
μ_{ref} (N/(m·s)) for $T = 273$K	2.03×10^{-5}	2.34×10^{-5}

Note that the molecular masses of the gases considered differ from one another by more than 30 times.

9.8.2 Computation on the Two-Fluid QGDM Model

The computation was carried out in dimensionless variables, all quantities were reduced to the dimensionless form with respect to the parameters of helium He (gas A) in the incident flow.

To construct the boundary conditions on the right and left boundaries, we use the Rankin–Hugoniot conditions for a stationary shock wave in the gas mixture. In this case, the quantities to the right from the discontinuity are calculated as follows:

Table 9.3 Dimensionless parameters for variant V1

	Gas A (He)	Gas B (Xe)	Mixture
ρ	1.	0.499	1.499
T	1.	1.	1.
a	1.	0.175	0.823
λ	1.	11.09	1.151
p	0.6	0.0091	0.609
Ma	2.97	17.01	3.61

Table 9.4 Dimensionless parameters for variant V2

	Gas A (He)	Gas B (Xe)	Mixture
ρ	1.	1.011	2.011
T	1.	1.	1.
a	1.	0.175	0.715
λ	1.	5.485	1.134
p	0.6	0.0185	0.618
Ma	2.78	15.93	3.89

Table 9.5 Dimensionless parameters for variant V3

	Gas A (He)	Gas B (Xe)	Mixture
ρ	1.	2.095	3.095
T	1.	1.	1.
a	1.	0.175	0.587
λ	1.	2.646	1.167
p	0.6	0.0383	0.638
Ma	2.58	14.77	4.4

Table 9.6 Dimension-free parameters for variant V4

	Gas A(He)	Gas B (Xe)	Mixture
ρ	1.	3.245	4.245
T	1.	1.	1.
a	1.	0.175	0.509
λ	1.	1.708	1.064
p	0.6	0.0594	0.659
Ma	2.44	13.99	4.8

$$\rho_2 = \rho_1 \frac{(\gamma + 1)\,\mathrm{Ma}^2}{2 + (\gamma - 1)\,\mathrm{Ma}^2}, \qquad p_2 = p_1 \frac{2\gamma\,\mathrm{Ma}^2 - \gamma + 1}{\gamma + 1},$$

$$u_2 = u_1 \frac{2 + (\gamma - 1)\,\mathrm{Ma}^2}{(\gamma + 1)\,\mathrm{Ma}^2}, \qquad T_2 = \frac{\gamma p_2}{\rho_2}, \tag{9.72}$$

where the subscripts 1 and 2 correspond to the Rankin–Hugoniot conditions for the mixture before and after the shock wave, respectively, and Ma is the Mach number for the mixture.

To calculate the parameters of separate components, we assume that their temperatures and velocities before and after the shock wave become equal, whereas the percent concentrations remain fixed in passing through the shock front. Then on the basis of conditions (9.72), the parameters of each of the mixture components before and after the shock wave are found from the relations

$$\rho_{a1} = m_a n_a, \qquad \rho_{b1} = m_b n_b, \qquad T_{a_1} = T_{b_1} = T_1, \qquad u_{a_1} = u_{b_1} = u_1,$$
$$\rho_{a2} = \rho_{a1} \frac{\rho_2}{\rho_1}, \qquad \rho_{b2} = \rho_{b1} \frac{\rho_2}{\rho_1}, \qquad T_{a_2} = T_{b_2} = T_2, \qquad u_{a_2} = u_{b_2} = u_2. \tag{9.73}$$

The initial conditions are the following discontinuity at the point $x = 0$:

$$\text{for } x \le 0: \quad \rho_a = \rho_{a1}, \quad \rho_b = \rho_{b1}, \quad T_a = T_b = T_1, \quad u_a = u_b = u_1,$$
$$\text{for } x \ge 0: \quad \rho_a = \rho_{a2}, \quad \rho_b = \rho_{b2}, \quad T_a = T_b = T_2, \quad u_a = u_b = u_2. \tag{9.74}$$

The same values are used as the boundary conditions.

To solve system (9.60), (9.61), (9.62), (9.63), (9.64), and (9.65), we apply the explicit finite-difference scheme. The steady-state solution was obtained as the limit of a time-evolving process. All spatial derivatives, including the convective terms, were approximated by central finite differences.

The problem is solved on the uniform spatial grid for the chosen accuracy $\varepsilon_{\rho a} = 10^{-5}$. In condensing the grid by 2 times, the difference in the results of computations are very small, which allows us to make a conclusion on the attained convergence with respect to the grid. As an example, in Table 9.7, we present the parameters of numerical computation for variant V2.

The profiles of gas-dynamic parameters (velocity, density, and temperature) in the shock wave are presented in the normalized form on the basis of the Rankin–Hugoniot conditions up and down with respect to the flow. In this case, $\rho \to (\rho - \rho_1)/(\rho_2 - \rho_1)$ and similarly for temperature. For the velocity, we have $u \to (u - u_2)/(u_1 - u_2)$.

Let us dwell on the results of computing variant V2 in more detail. In Figs. 9.1, 9.2, and 9.3, we present the profiles of gas-dynamic parameters on the shock wave front in comparison with the corresponding results obtained in [28] on the basis of the DSMC method. The curves corresponding to the results of computations with respect to the DSMC method are imposed on the data of the QGDM model in such a way that the values of the mixture mean density coincide for $x = 0$.

In Fig. 9.1, we present the profiles of densities and temperature of helium and xenon. In Fig. 9.2, we present the distributions of the mean temperature and density of the mixture. As in the DSMC model, the mean temperature of the mixture is very

Table 9.7 Parameters of computations for variant V2

	Grid 601	Grid 1201
Grid step h	0.5	0.25
Step in time Δt	4.8×10^{-3}	1.2×10^{-3}
Number of iterations N_{iter}	90251	360450

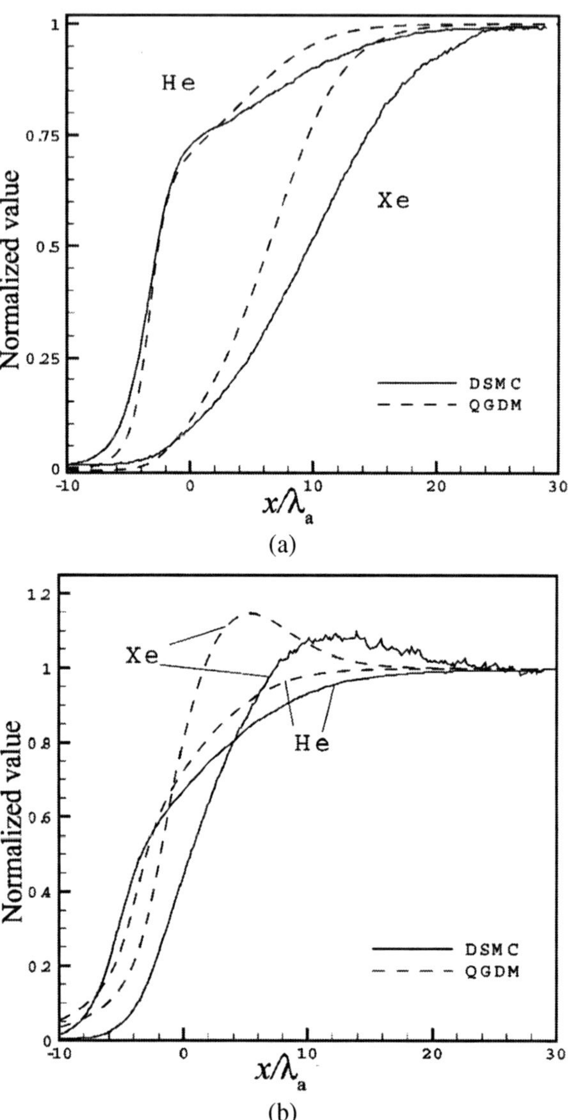

Fig. 9.1 Profiles of densities (**a**) and temperatures (**b**) in the mixture He–Xe

close to the temperature of helium, and the temperature of xenon exceeds its value after the shock wave by $\sim 10\%$.

In Fig. 9.2(b), we show the diffusion velocities for the mixture components relative to the velocity of the unperturbed flow. The diffusion velocities of the components u_{da} and u_{db} are found according to [28] as follows:

$$u_{da} = u_a - u, \qquad u_{db} = u_b - u, \tag{9.75}$$

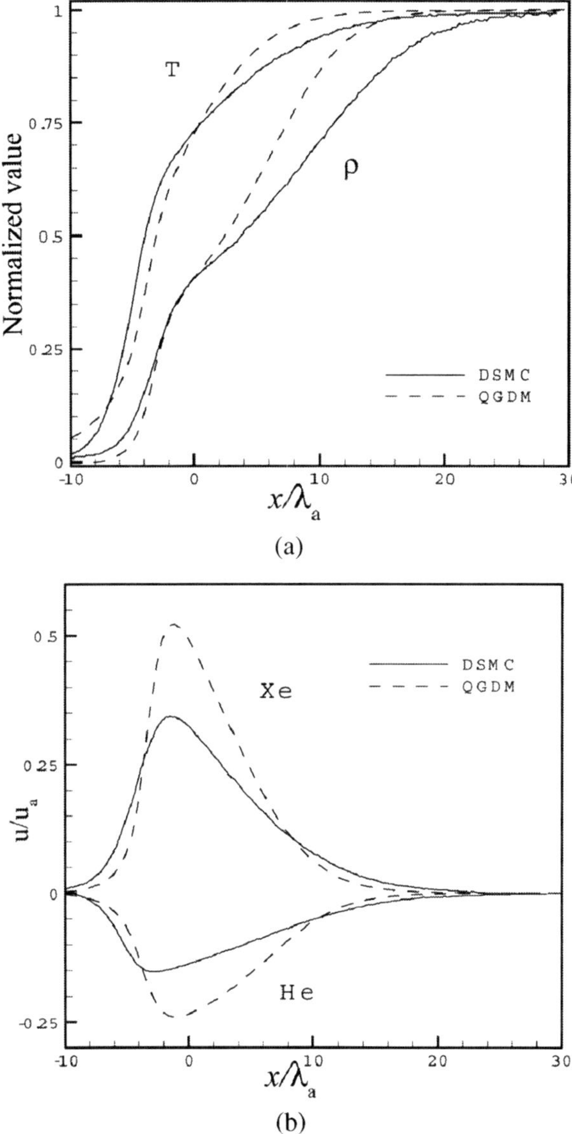

Fig. 9.2 Profiles of density and mean temperature (**a**) and diffusion velocities (**b**) in the mixture He–Xe

where u is the velocity of the mixture motion found in accordance with Eq. (9.39). For such definition of diffusion velocity, we have $u_{da}\rho_a + u_{db}\rho_b = 0$.

In Fig. 9.3(a), we present the change of the concentration of xenon. It is seen that the values of the concentration of xenon on the shock wave front is less by 2 times than those in the unperturbed flow domain.

Fig. 9.3 Concentration of Xe
(**a**) and relative thickness of
shock waves (**b**) in the
mixture He–Xe

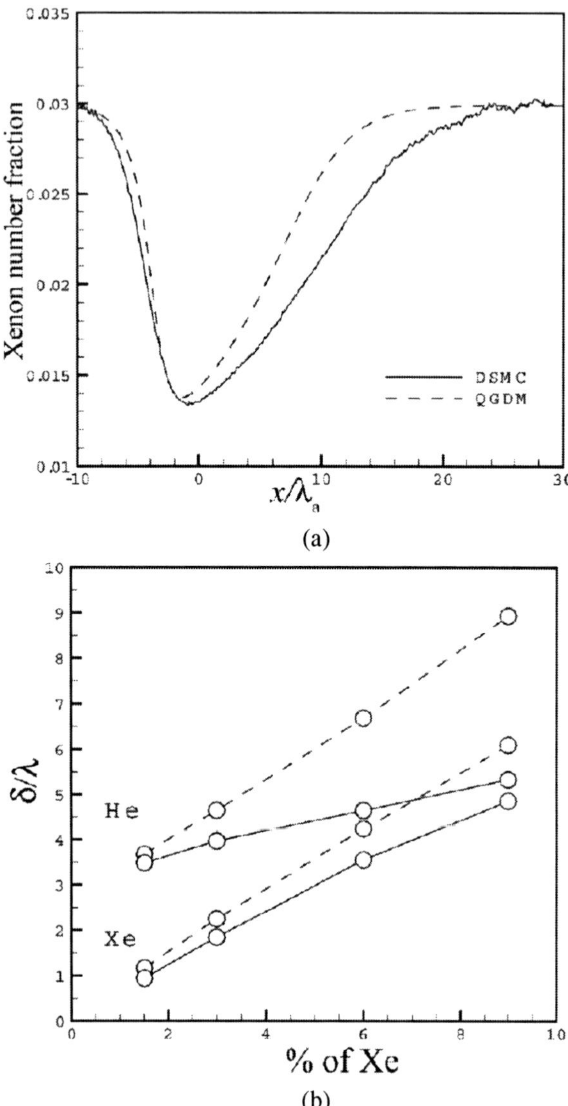

(a)

(b)

The curves presented testify that the QGDM model qualitatively, as well as quantitatively, reflects the main peculiarities of the flow considered: the mutual location
and their form correspond to the standard results.

In Fig. 9.4, we present the density profiles of He and Xe in the shock wave
for variants V1–V4 corresponding to the experiments (see [97]). The results show
that varying the composition of the mixture, the results of numerical computation
qualitatively correspond to the data of the natural experiments in the limits of its
accuracy.

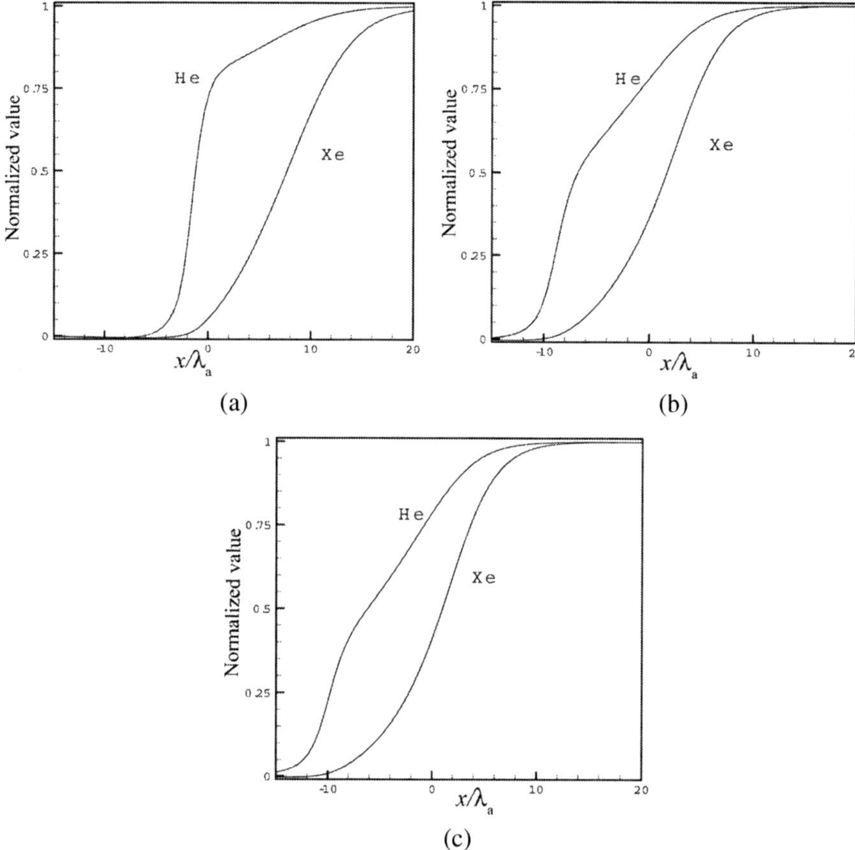

Fig. 9.4 Profiles of density in the mixture He–Xe: (a) 1.5% Xe, (b) 6% Xe, (c) 9% Xe

On the basis of computing variants V1–V4, in Fig. 9.3(b), we present the dependencies of the relative shock wave thickness of both components δ_{He}/λ_{He} and δ_{Xe}/λ_{Xe} depending on the concentration of Xe in the mixture before the shock wave in comparison with the results of [97]. In this case, the shock wave thickness is calculated as

$$\delta = \frac{\rho_2 - \rho_1}{\max\limits_{x} \dfrac{\partial \rho}{\partial x}}. \tag{9.76}$$

The mean free path of molecules for each of the components is calculated in accordance with Eq. (9.70) for parameters of each of the gas components before the shock wave. The continuous line denotes the results of experiment and the dotted line the results of computations with respect to the QGDM model. All curves are presented in the form similar to [97].

It is seen that for variant V1 corresponding to the minimal concentration of Xe the data of computation and experiment practically coincide. When the concentration of Xe grows, the thickness of the shock wave obtained in the computation exceeds the experimental values. In this case, the qualitative character of dependencies corresponds to the data of experiment and the known theoretical results in accordance to which, the relative thickness of the shock wave for the chosen value of ω increases when the Mach number decreases. Note that the thickness of the shock wave is a very sensitive characteristic, and its calculation on the basis of the moment equations for a one-component gas is a sufficiently complicated problem.

9.8.3 Computation in the One-Fluid Approximation

In the framework of the Navier–Stokes equations, there are no developed models for computing the gas-mixture flows in the two-fluid approximation. Therefore, to compare the QGDM equations with the known macroscopic approaches, we choose the comparison of the one-fluid approximation of the QGDM equations (9.43), (9.44), (9.45), and (9.46) with the Navier–Stokes-type model (9.47), (9.48), (9.49), and (9.50) by examining the numerical modelling of the problem on the shock wave structure (variant V2). In this case, the QGDM equations (9.43), (9.44), (9.45), and (9.46) are written for a planar one-dimensional flow. The method for solving these equations and the finding of all necessary constants completely coincide with those described in Sect. 9.8.2.

When using the Navier–Stokes-type model (9.47), (9.48), (9.49), and (9.50), the mixture viscosity coefficient is found by relation (9.34). Moreover, the viscosity coefficients of separate components are found by the following two methods: relations (9.33) and (9.55), which practically has no influence on the results of computations. To find the viscosity coefficient (9.55) and the diffusion coefficient in the expressions for diffusion flows (9.52), it turns out that the values presented in Table 9.2 are insufficient, and it is necessary, in addition, to find a number of constants. The parameters of the molecular interaction potential are found according to [162]: $\varepsilon_a/k_B = 10.22\,\mathrm{K}$ and $\varepsilon_b/k_B = 231.0\,\mathrm{K}$. The effective collision diameters $\sigma_a = 2.551 \times 10^{-10}\,\mathrm{m}$ and $\sigma_b = 4.047 \times 10^{-10}\,\mathrm{m}$ are presented therein. The coefficients D_a^T and D_b^T determining the thermodiffusion are assumed to be equal to zero.

As in the previous case, to solve the Navier–Stokes-type system (9.47), (9.48), (9.49), and (9.50) numerically, we use the explicit in time finite-difference scheme with central differences. However, to ensure the stability of the finite-difference algorithm, it is necessary to approximate the terms with $\ln p$ entering expressions (9.52) for diffusing flows by using up-wind finite-difference derivatives of the form

$$\frac{\partial \ln p}{\partial x} \sim \frac{\ln p_{i+1} - \ln p_i}{\Delta x}, \tag{9.77}$$

which introduces an additional scheme dissipation of order $O(h)$ (here i is the coordinate of a node of the computational grid). For system (9.47), (9.48), (9.49), (9.50), (9.51), and (9.52), the following balance relations hold:

$$\rho c_a + \rho c_b = \rho, \quad \mathcal{J}_a + \mathcal{J}_b = 0.$$

In numerically solving the above system, the quantities ρc_b and $\mathcal{J}_b$ are not directly computed, but they are found from the above balance relations by using the computed values of ρc_a and $\mathcal{J}_a$.

As the initial conditions, we use conditions (9.74). The same conditions are also posed on the left boundary. In contrast to the QGDM equations, for Navier–Stokes-type system on the right boundary, we pose the soft boundary conditions $\partial\psi/\partial x = 0$, where $\psi = (\rho_a, \rho_b, u, E)$. This gives a possibility to oscillations arising in the process of numerically solving the problem and spreading along the flow to leave the computational domain through its right boundary without obstructions. Note that in the use of the QGDM equations, such oscillations do not arise.

The parameters of computation for this variant are close to those presented in Table 9.7. The number of steps in time up to the convergence is slightly greater than that in Table 9.7. For example, for the grid with the number of steps equal to 601, we have $N_{\text{iter}} = 140000$ for the same accuracy of $\epsilon_{\rho a} = 10^{-5}$.

Therefore, the stability of the numerical algorithm for Navier–Stokes-type model is substantially lesser than that for the QGDM model, which is expressed by the appearance of oscillations; for damping these oscillations, it is necessary to use the approximation of the first-order accuracy of the form (9.77) and the "soft" boundary conditions on the right boundary.

In Figs. 9.5 and 9.6, we present the gas-dynamic parameters in the shock wave computed according to two one-fluid models. The notation and the normalization in the figures are the same as in the previous section.

In Fig. 9.5(a), similarly to Fig. 9.2(a), we present the mixture densities, and in Fig. 9.5(b), similarly to Fig. 9.3(a), we present the mixture mean temperature and density. Note that this approach does not allow us to find the temperatures of separate mixture components. It is seen that the results obtained according to both models are sufficiently close to each other and, at the same time, they notably differ from the standard results of the DSMC model in the form of the curves. In Fig. 9.6(a), analogous to Fig. 9.2(b), and in Fig. 9.6(b), analogous to Fig. 9.3(a), we present the dimensionless profiles of diffusion velocities for xenon and helium and also the concentration of xenon on the shock wave, respectively. It is impossible to compute the diffusion velocity for the one-fluid QGDM model on the basis of relation (9.75). For the QGDM equations, the diffusion component velocities are calculated analogously to the model of the Navier–Stokes type through the diffusing fluxes (9.57) in the form

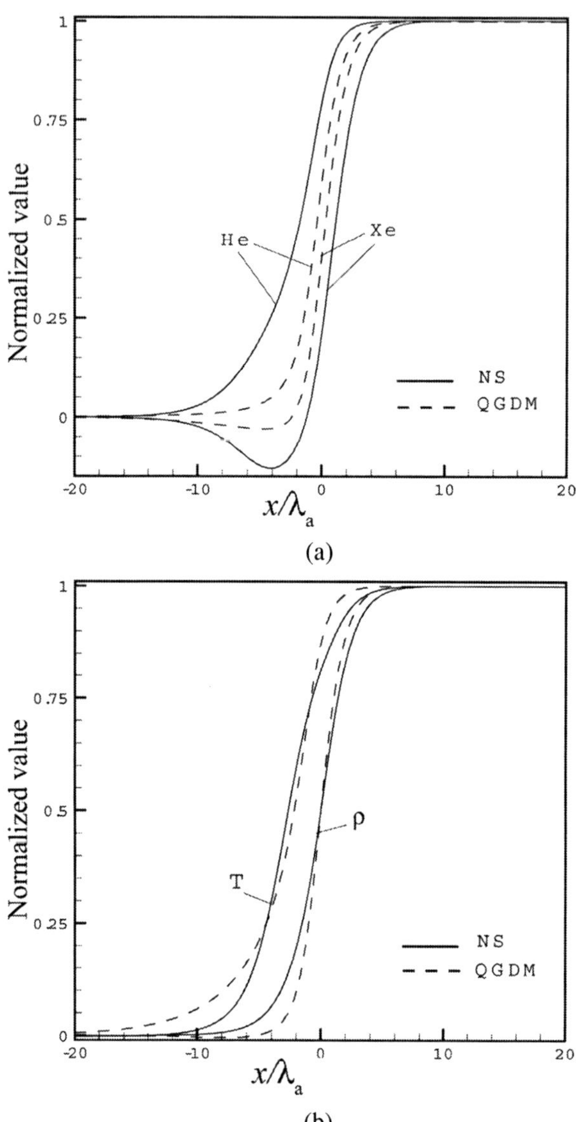

Fig. 9.5 Profiles of density components (**a**) and the mean temperature (**b**) in the mixture He–Xe. One-fluid models

$$u_{da} = -w_a = -\frac{\tau}{\rho_a}\frac{\partial}{\partial x}(\rho_a u^2 + p_a),$$

$$u_{db} = -w_b = -\frac{\tau}{\rho_b}\frac{\partial}{\partial x}(\rho_b u^2 + p_b). \tag{9.78}$$

For the Navier–Stokes model, the diffusion velocities of the components are found as follows:

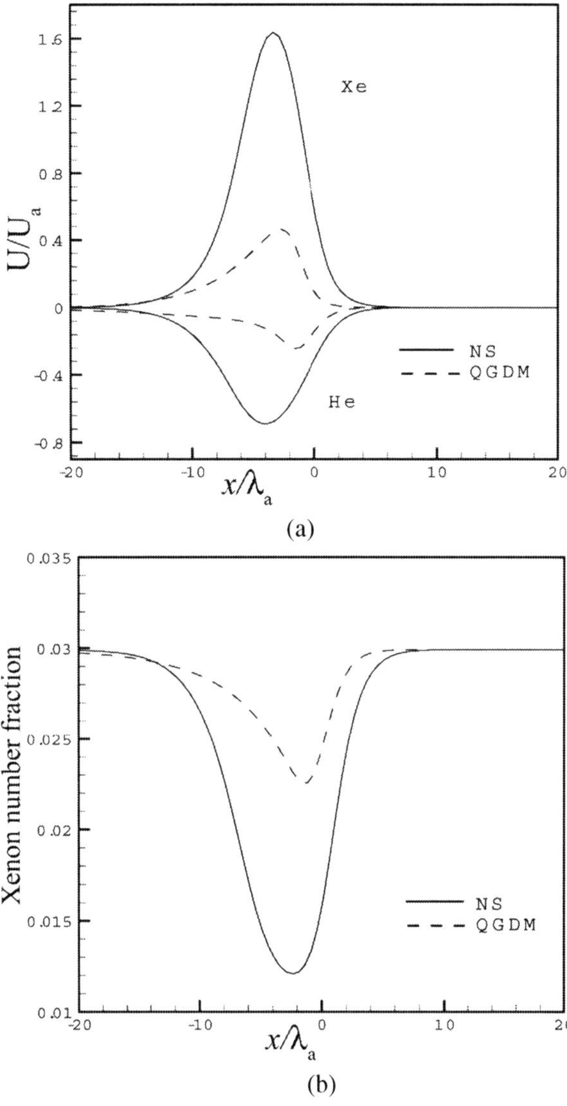

Fig. 9.6 Diffusion velocities (**a**) and the concentration of Xe (**b**) in the mixture He–Xe. One-fluid models

$$u_{da} = \frac{\mathcal{J}_a}{\rho c_a}, \quad u_{db} = \frac{\mathcal{J}_b}{\rho c_b}. \tag{9.79}$$

The model of the Navier–Stokes equations type substantially overstates the values of both diffusion velocities given that the concentration of Xe in the shock wave is close to the standard values, and, on the contrary, the QGDM model yields more precise values of diffusion velocities overstating the concentration of Xe.

Therefore, it follows from the presented computation that both one-fluid models are close to each other, although they turn out to be notably lesser exact than the two-fluid QGDM approximation and, the more so, than the results of the DSMC method.

9.9 Diffusion Problem of Argon and Helium

As the second example of approbation of the two-fluid QGDM equations, we consider the mass diffusion problem of helium and argon in the statement corresponding to the computation according to the DSMC on [28].

In the distance $L = 1$ m from each other, let two reservoirs be placed, and let they be filled in by gases He (gas a, to the right) and Ar (gas b, to the left). The numerical molecule density in the reservoirs is constant and equal to $n = 2.8 \times 10^{20}$ m^{-3}. It is assumed that the gases in the reservoirs have the same temperature $T = 273$ K and the same velocity equal to zero.

The constants for helium and argon necessary in the computation are presented in Table 9.8 in accordance with [28]. Note that, in contrast to the previous variant, the molecular masses of gases differ by 10 times.

Using these constants, we obtain the missing initial conditions: the helium density $\rho_a = nm_a = 1.862 \times 10^{-6}$ kg/m^3, the sound speed $a_a = \sqrt{\gamma_a \mathcal{R}_a T_a} = 971.9$ m/s, and the mean free path of the molecules calculated by formula (9.70), $\lambda_a = 1.479 \times 10^{-2}$ m. The density of argon is $\rho_b = nm_b = 1.856 \times 10^{-5}$ kg/m^3, the sound speed is $a_b = \sqrt{\gamma_b \mathcal{R}_b T_b} = 307.81$ m/s, and the free path of molecules defined by formula (9.70) is $\lambda_b = 4.63 \times 10^{-3}$ m.

As in the previous section, the calculation is carried out in dimensionless variables and, moreover, all quantities are normalized relative to the parameters of the gas a (helium) in the reservoir. The corresponding dimensionless parameters are presented in Table 9.9.

We consider the planar one-dimensional flow described by Eqs. (9.60), (9.61), (9.62), (9.63), (9.64), and (9.65). As the boundary conditions, we use the following dimensionless relations:

Table 9.8 Table values for the mixture components

	He	Ar
m (kg)	6.65×10^{-27}	66.3×10^{-27}
$\mathcal{R}$ (J/(kg·K))	2076.2	208.24
$\mathcal{M}$ (kg/mol)	4.0	39.926
d (m)	2.30×10^{-10}	4.17×10^{-10}
γ	1.66	1.66
ω	0.66	0.81
Pr	0.666	0.666
μ_{ref} (N/(m·s)) for $T = 273$ K	1.865×10^{-5}	2.117×10^{-5}

Table 9.9 Dimensionless parameters

	Gas a (He)	Gas b (Ar)
ρ	1.	9.969
T	1.	1.
a	1.	0.316
λ	1.	0.313
p	0.6	0.60

$$\rho_a = 1 - 10^{-10}, \quad \rho_b = 10^{-10}, \quad T_a = T_b = 1, \quad \frac{\partial u_a}{\partial x} = \frac{\partial u_b}{\partial x} = 0$$

on the left boundary and

$$\rho_b = 1 - 10^{-10}, \quad \rho_a = 10^{-10}, \quad T_a = T_b = 1, \quad \frac{\partial u_a}{\partial x} = \frac{\partial u_b}{\partial x} = 0$$

on the right boundary, i.e., we assume that in each of the gases, $\sim 10^{-10}\%$ molecules of the other gas are present. At the initial instant of time, we assume that the density of components between the reservoirs linearly varies:

$$\rho_a(x) = \frac{\rho_{a_{x=L}} - \rho_{a_{x=0}}}{L} x + \rho_{a_{x=0}}, \quad \rho_b(x) = \frac{\rho_{b_{x=L}} - \rho_{b_{x=0}}}{L} x + \rho_{b_{x=0}}.$$

We use the same numerical algorithm as in the previous section for solving the QGDM equations. The problem is solved on the uniform spatial grid consisting of 339 points with the spatial step 0.2, which corresponds to $0.2\lambda_a$ and $0.64\lambda_b$.

In Fig. 9.7(a), we present the variation of numerical densities of Ar and He between the reservoirs relative to the numerical density in the reservoirs. The diffusion velocities of both gases are shown in Fig. 9.7(a) in the dimensional form. In both figures, we present the comparison with the corresponding computational results of [28]. The diffusion velocities are computed in accordance with Eq. (9.75). We see the qualitative and quantitative similarity of the results obtained by using both methods. Precisely, a point of equal concentration is displaced from the middle of the domain to the right close to the reservoir with the heavier gas. The velocity of helium diffusion is greater than that of argon, and the qualitative behavior of diffusion velocity of components is different: the helium diffusion velocity has a weakly expressed minimum at the middle of the computation domain.

In Fig. 9.8, we present the distribution of pressures and velocities of components between the reservoirs. The pressure distribution repeats the distribution of numerical densities (Fig. 9.8(a)). The velocity of each of the components is very small near the reservoir of this component and considerably increases near the opposite reservoir (Fig. 9.8(b)).

The calculation presented demonstrates that, despite the use of relatively simple expressions for the mixture viscosity coefficient and the particle collision frequency, the QGDM model yields a good description of mixture flows even if molecules in them are strongly different in mass. From the computational viewpoint, the QGDM

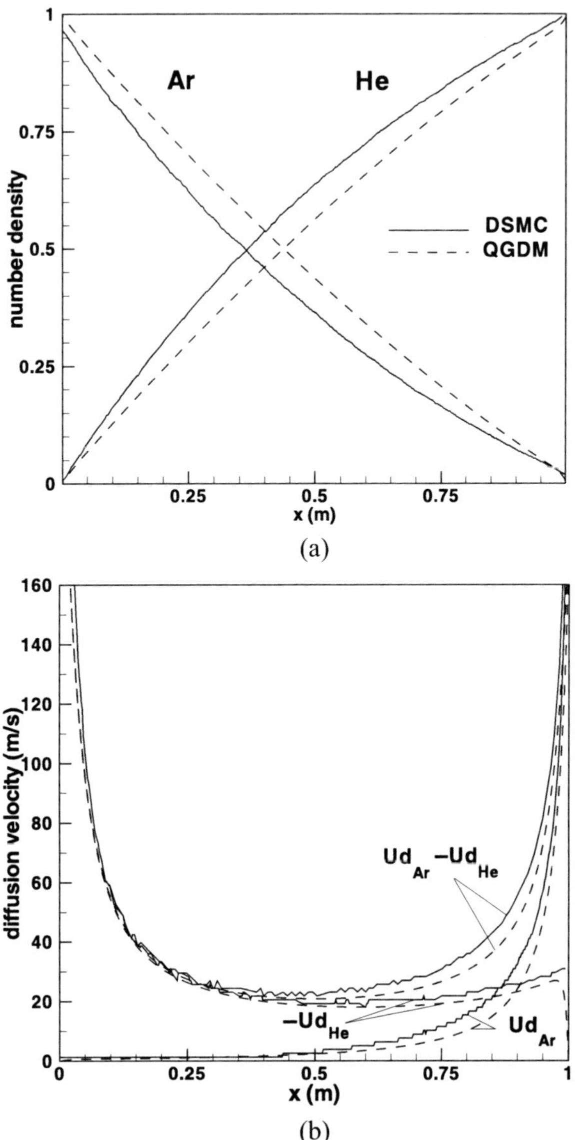

Fig. 9.7 Numerical density (**a**) and diffusion velocities (**b**) in the mixture Ar–He

algorithms are more stable than the corresponding algorithms constructed on the basis of the traditional conservation equations of the Navier–Stokes type. If necessary, in the QGDM model, we can include an artificial dissipation in the form of an addition to the coefficient τ of the form $\alpha h/c$.

The QGDM system is written in the invariant form, which allows us to use it for various spatial problems. To enlarge the exactness of the model, it is appropriate to use more precise expressions for the collision frequencies. The approach used in

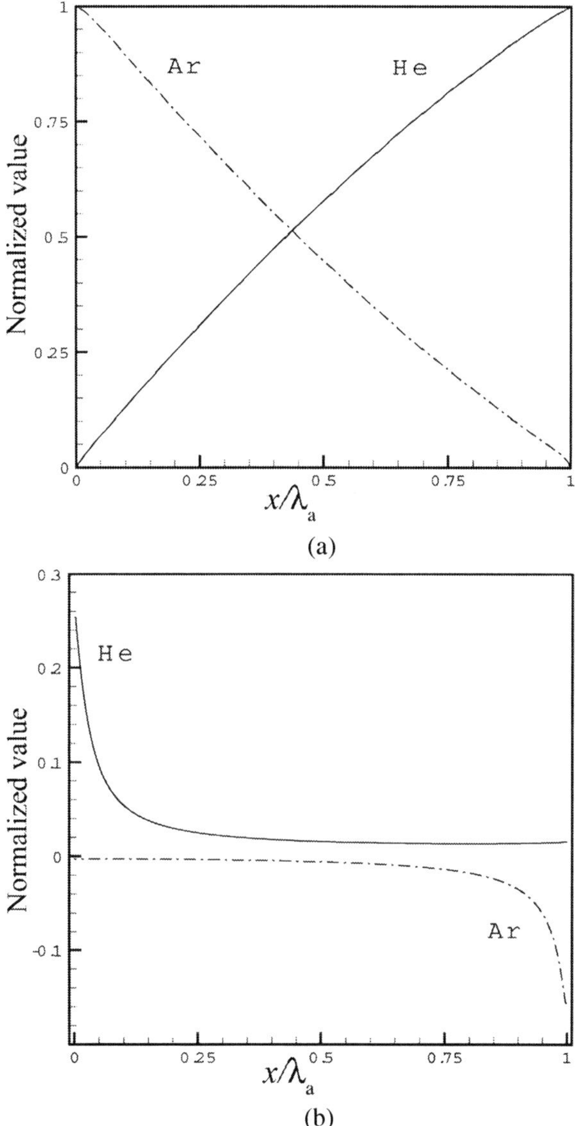

Fig. 9.8 Pressures (**a**) and velocities (**b**) of components in the mixture Ar–He

deducing the QGDM equations can be useful in constructing the models for describing gas flows with weighted particles.

In concluding this chapter, we note that the QGDM system can be represented in the form of conservation laws (3.13), (3.14), (3.15), and (3.16). For this system, one succeeds in deriving the equation for the entropy (of the form Eq. (1.57)) with nonnegative entropy production X.

Appendix A
Example of Constructing Quasi-gas-dynamic Equations

As an example, we present a detailed deduction of the quasi-gas-dynamic equations by using the kinetic model.

Consider a planar one-dimensional flow along the axis x. In this case, the velocities of molecules $\xi = (\xi_x, \xi_y, \xi_z)$ are connected with the heat velocities $c = (c_x, c_y, c_z)$ by the relations

$$\xi_x = u + c_x, \quad \xi_y = c_y, \quad \xi_z = c_z,$$

where u is the macroscopic gas velocity along the axis x. In this case, the regularized kinetic equation (3.4) takes the form

$$\frac{\partial f}{\partial t} + \xi_x \frac{\partial f^{(0)}}{\partial x} - \xi_x \frac{\partial}{\partial x}\left(\tau \xi_x \frac{\partial f^{(0)}}{\partial x}\right) = \mathcal{I}(f, f). \tag{A.1}$$

The macroscopic equations are constructed by sequentially multiplying Eq. (A.1) by the summator invariants

$$h(\xi) = 1, \ \xi_x, \ \frac{\xi^2}{2}$$

and by averaging in all particle velocities ξ. The conservation laws of mass, momentum, and energy are expressed by the following expression for the collision integral:

$$\int \mathcal{I}(f, f) h(\xi) d\xi = 0. \tag{A.2}$$

Therefore, the terms with the collision integral disappear from the resulting equations.

A.1 Equation of Continuity

Let us integrate Eq. (A.1) in all particle velocities. In integrating the first term, we obtain

$$\int \frac{\partial f}{\partial t} d\xi = \frac{\partial}{\partial t} \int f d\xi = \frac{\partial \rho}{\partial t}.$$

The second term transforms as follows:

$$\int \frac{\partial f^{(0)}}{\partial x} \xi_x \, d\xi = \frac{\partial}{\partial x} \int (u + c_x) \, f^{(0)} \, d\xi = \frac{\partial}{\partial x} \int u \, f^{(0)} \, d\xi = \frac{\partial}{\partial x} \rho u. \qquad (A.3)$$

Here, we use relations (3.7) in the form

$$\int c_x f^{(0)} d\xi = 0, \qquad (A.4)$$

together with the definition of ρ,

$$\rho = \int f d\xi = \int f^{(0)} \, d\xi.$$

The second term in the left-hand side of Eq. (A.1) transforms as follows:

$$\begin{aligned}
\int \xi_x \frac{\partial}{\partial x} \left(\tau \xi_x \frac{\partial f^{(0)}}{\partial x} \right) d\xi &= \frac{\partial}{\partial x} \left(\tau \frac{\partial}{\partial x} \int \xi_x^2 \, f^{(0)} \, d\xi \right) \\
&= \frac{\partial}{\partial x} \left(\tau \frac{\partial}{\partial x} \int (u + c_x)^2 \, f^{(0)} \, d\xi \right) \\
&= \frac{\partial}{\partial x} \tau \frac{\partial}{\partial x} (\rho u^2 + p). \qquad (A.5)
\end{aligned}$$

Here, we use relations (3.7) and (A.4), together with the definition of the pressure p in the form

$$p = \frac{1}{3} \int c^2 \, f^{(0)} \, d\xi = \frac{1}{3} \int (c_x^2 + c_y^2 + c_z^2) f^{(0)} \, d\xi = \int c_x^2 f^{(0)} \, d\xi. \qquad (A.6)$$

Uniting expressions (A.2), (A.3), and (A.5), we obtain the first equation of the QGD system in the form

$$\frac{\partial \rho}{\partial t} + \frac{\partial \rho u}{\partial x} = \frac{\partial}{\partial x} \tau \frac{\partial}{\partial x} (\rho u^2 + p). \qquad (A.7)$$

A.2 Equation for the Momentum

Let us multiply Eq. (A.1) by ξ_x and integrate over all velocities ξ.

Taking into account the definition of the density and Eqs. (3.7), (A.4), and (A.6), we write the first and second terms of the kinetic equation as follows:

$$\int \xi_x \frac{\partial f}{\partial t} d\xi = \frac{\partial}{\partial t} \int \xi_x f d\xi = \frac{\partial}{\partial t} \rho u,$$

$$\int \xi_x{}^2 \frac{\partial}{\partial x} f^{(0)} d\xi = \frac{\partial}{\partial x} \int (u^2 + 2 c_x u + c_x^2) f^{(0)} d\xi = \frac{\partial}{\partial x} (\rho u^2 + p). \qquad \text{(A.8)}$$

The latter term transforms as follows:

$$\int \xi_x{}^2 \frac{\partial}{\partial x} \tau \xi_x \frac{\partial f^{(0)}}{\partial x} d\xi = \frac{\partial}{\partial x} \tau \frac{\partial}{\partial x} \int \xi_x{}^3 f^{(0)} d\xi = \frac{\partial}{\partial x} \tau \frac{\partial}{\partial x} \int (u + c_x)^3 f^{(0)} d\xi$$

$$= \frac{\partial}{\partial x} \tau \frac{\partial}{\partial x} \int (u^3 + 3u^2 c_x + 3u c_x{}^2 + c_x{}^3) f^{(0)} d\xi$$

$$= \frac{\partial}{\partial x} \tau \frac{\partial}{\partial x} (\rho u^3 + 3 p u). \qquad \text{(A.9)}$$

Here, we use expressions (3.7), (A.4), and (A.6), together with the formula

$$\int c_x^3 f^{(0)} d\xi = 0. \qquad \text{(A.10)}$$

In a more general case, because of symmetry, the following relation holds:

$$\int c_i c_j^2 f^{(0)} d\xi = 0. \qquad \text{(A.11)}$$

Uniting Eqs. (A.2), (A.8), and (A.9), we obtain the second equation of the QGD system in the form

$$\frac{\partial}{\partial t} \rho u + \frac{\partial}{\partial x} (\rho u^2 + p) = \frac{\partial}{\partial x} \tau \frac{\partial}{\partial x} (\rho u^3 + 3 p u). \qquad \text{(A.12)}$$

A.3 Equation for the Energy

Averaging Eq. (A.1) with the weight $\xi^2/2$, for the first two terms of the model kinetic equation, we obtain

$$\int \frac{\xi_x{}^2 + \xi_y{}^2 + \xi_z{}^2}{2} f d\xi = \frac{\partial E}{\partial t},$$

$$\int \frac{\partial}{\partial x} f^{(0)} \xi_x \frac{\xi_x^{2} + \xi_y^{2} + \xi_z^{2}}{2} \, d\xi = \frac{\partial}{\partial x} \int u \frac{\xi_x^{2} + \xi_y^{2} + \xi_z^{2}}{2} \, f^{(0)} \, d\xi$$

$$+ \frac{\partial}{\partial x} \int c_x \frac{(c_x + u)^2 + c_y^{2} + c_z^{2}}{2} \, f^{(0)} \, d\xi$$

$$= \frac{\partial}{\partial x} u E + \frac{\partial}{\partial x} u p. \tag{A.13}$$

Here, we use the definitions of ρ and p, together with Eq. (A.11), and the definition of the total energy in the form

$$E = \frac{1}{2} \int \xi^2 f d\xi. \tag{A.14}$$

The last term of the kinetic equation transforms into the terms with the second spatial derivatives in the equation for the total energy:

$$\int \frac{\xi^2}{2} \xi_x \frac{\partial}{\partial x} \tau \xi_x \frac{\partial f^{(0)}}{\partial x} \, d\xi = \frac{\partial}{\partial x} \int \frac{\xi^2}{2} \xi_x^{2} f^{(0)} \, d\xi$$

$$= \frac{\partial}{\partial x} \tau \frac{\partial}{\partial x} \int (u + c_x)^2 \frac{(c_x + u)^2 + c_y^{2} + c_z^{2}}{2} \, f^{(0)} \, d\xi$$

$$= \frac{\partial}{\partial x} \tau \frac{\partial}{\partial x} \int u^2 \frac{(c_x + u)^2 + c_y^{2} + c_z^{2}}{2} \, f^{(0)} \, d\xi$$

$$+ \frac{\partial}{\partial x} \tau \frac{\partial}{\partial x} \int 2u c_x \frac{(c_x + u)^2 + c_y^{2} + c_z^{2}}{2} \, f^{(0)} \, d\xi$$

$$+ \frac{\partial}{\partial x} \tau \frac{\partial}{\partial x} \int c_x^{2} \frac{(c_x + u)^2 + c_y^{2} + c_z^{2}}{2} \, f^{(0)} \, d\xi$$

$$= \frac{\partial}{\partial x} \tau \frac{\partial}{\partial x} \left(u^2 E + 2u^2 p + \frac{u^2 p}{2} + \int c_x^{2} \frac{c_x^{2} + c_y^{2} + c_z^{2}}{2} f^{(0)} \, d\xi \right).$$

To calculate the latter integral, we perform the change of variables

$$x = \frac{c_x}{\sqrt{2RT}}, \quad y = \frac{c_y}{\sqrt{2RT}}, \quad z = \frac{c_z}{\sqrt{2RT}}.$$

Taking into account the formulas

$$\int e^{-x^2} dx = \sqrt{\pi}, \quad \int x^2 e^{-x^2} dx = \frac{\sqrt{\pi}}{2},$$

$$\int^4 e^{-x^2} dx = \frac{3}{4}\sqrt{\pi},$$

we obtain

$$\int c_x{}^2 \frac{c_x{}^2 + c_y{}^2 + c_z{}^2}{2}\, f^{(0)}\, dc_x\, dc_y\, dc_z = \frac{5p^2}{2\rho}. \tag{A.15}$$

In accordance with the relations obtained above, the desired terms in the energy equation have the form

$$\frac{\partial}{\partial x} \tau \frac{\partial}{\partial x} \left(u^2 \left(E + \frac{5}{2} p \right) + \frac{5p^2}{2\rho} \right). \tag{A.16}$$

Uniting formulas (A.2), (A.13), and (A.16), we obtain the last equation of the QGD system in the form

$$\frac{\partial}{\partial t} E + \frac{\partial}{\partial x} u(E + p)$$
$$= \frac{\partial}{\partial x} \tau \frac{\partial}{\partial x} u^2 \left(E + \frac{5}{2} p \right) + \frac{5}{2} \left(\frac{\partial}{\partial x} \tau \frac{p}{\rho} \frac{\partial p}{\partial x} + \frac{\partial}{\partial x} \tau p \frac{\partial}{\partial x} \frac{p}{\rho} \right). \tag{A.17}$$

A.4 Resulting System of Equations

Equations (A.7), (A.12), and (A.17) compose the resulting system of equations. In the final form of this system, we transform the right-hand side of the energy equation (A.17) as follows. In it, we distinguish the term describing the heat flux into which we introduce the Prandtl number Pr and identify the coefficient 5/3 with the parameter γ as $5/2 = \gamma/(\gamma - 1)$. The obtained system of the QGD equations takes the form

$$\frac{\partial \rho}{\partial t} + \frac{\partial \rho u}{\partial x} = \frac{\partial}{\partial x} \tau \frac{\partial}{\partial x} (\rho u^2 + p), \tag{A.18}$$

$$\frac{\partial}{\partial t} \rho u + \frac{\partial}{\partial x} (\rho u^2 + p) = \frac{\partial}{\partial x} \tau \frac{\partial}{\partial x} (\rho u^3 + 3pu), \tag{A.19}$$

$$\frac{\partial}{\partial t} E + \frac{\partial}{\partial x} u(E + p) = \frac{\partial}{\partial x} \tau \frac{\partial}{\partial x} \left(u^2(E + 2p) + \frac{u^2}{2} p \right)$$
$$+ \frac{\gamma}{\gamma - 1} \frac{\partial}{\partial x} \tau \frac{p}{\rho} \frac{\partial p}{\partial x} + \frac{1}{\mathrm{Pr}} \frac{\gamma}{\gamma - 1} \frac{\partial}{\partial x} \tau p \frac{\partial}{\partial x} \frac{p}{\rho}. \tag{A.20}$$

Expression (A.16) can be equivalently rewritten in the form

$$\frac{\partial}{\partial x} \tau \frac{\partial}{\partial x} u^2(E + 2p) + \frac{\partial}{\partial x} \tau \frac{\partial}{\partial x} \frac{p}{\rho}(E + p).$$

This form was used in the first versions of kinetic-consistent finite-difference schemes of the form (2.39).

Appendix B
Flows of Viscous Compressible Gas in Microchannels

In this appendix, we consider the problem on flows of viscous compressible gas in microchannels with isothermal walls. For such flows, we construct approximate formulas for calculating the mass-flow rate. We show that the QGD models complemented by the Maxwell-type slip boundary conditions for the velocity predict the existence of the minimum mass-flow rate in the channel, the so-called Knudsen minimum. We suggest corrections that allow us to write the approximate formulas for the mass-flow rate, which hold for any Knudsen numbers. The results presented are based on [53, 58, 85].

B.1 Introduction

In flows of moderately rarefied gases, effects arise that are difficult to be explained in the framework of the classical Navier–Stokes equations, even if one uses the velocity-slip and the temperature-jump boundary conditions (see [28, 34, 127, 160, 195], etc). In particular, such effects are observed in gas flows in long isothermal microchannels. In these problems, one observes the increase of the mass-flow rate as compared to the mass-flow rate calculated by the Navier–Stokes equations and also the so-called Knudsen effect closely related to this phenomenon.

The experiments of Knudsen carried out in the 1900s[1] show that in gas flows in a long cylindrical channel, the normalized mass-flow rate through the section does not depend monotonically on the Knudsen number, and for the Knudsen numbers $\mathrm{Kn} \sim 1$, one observes the minimum of the specific mass-flow rate. These classical experiments are mentioned, e.g., in [34, Chap. 7] and [160, Sect. 4.3]. Also, it is stressed there that the existing macroscopic models do not allow one to obtain the Knudsen minimum. However, as was shown in [33, 34], if the self-diffusion processes are taken into account in the macroscopic equations, then this result can be obtained.

When using kinetic approaches, the Knudsen effect was obtained in a number of works by using variational approaches to solving the Boltzmann equation in the

---

[1] M. Knudsen, *Ann. Phys.*, **28**, 75 (1909).

BGK approximation [16, 33–37, 127, 132, 178]. The application of Monte Carlo methods to such problems is problematic, which is related to the necessity of damping statistical fluctuations in computing slow-velocity flows, and leads to very large computational rates (see [132]).

On the basis of macroscopic equations, one succeeds in describing the Knudsen effect by using some variants of the Barnett equations [132] or the Navier–Stokes system complemented by velocity-slip boundary conditions of an artificial form [44, 46, 116].

In this appendix, we consider flows in microchannels on the basis of the QGD/QHD models with the classical slip boundary conditions. We have shown above that for flows with moderate Knudsen numbers, which are the flows in thin cracks, holes, capillaries, and microchannels, where the coefficient τ is not very small, additional terms in the QGD/QHD equations can influence the solution of the problem considered. Furthermore, on the basis of QGD/QHD equations with the classical slip boundary conditions, the Maxwell-type conditions, we construct approximate analytical formulas describing the increase of the mass-flow rate and the Knudsen effect in microchannels. We suggest corrections that allow us to obtain the unified formulas for the mass-flow rates for planar and cylindrical channels that hold in the whole range of Knudsen numbers.

B.2 Poiseuille Flows in Planar Channels

B.2.1 Formulas for the Mass-Flow Rate

Consider a gas flow in a planar channel of length L and width H. In the input and output of the channel, let the pressures be equal to p_1 and p_2, where $p_1 > p_2$. Following [134, Chap. 2, Sect. 18, Problem 6], we assume that the gradient of the pressure along the channel is not large, and for a small channel length dx, we can assume that the gas density ρ is constant. We seek the solution of the problem in the form

$$u_x = u(y), \quad u_y = 0, \quad p = p(x), \quad T = T_0. \tag{B.1}$$

In this case, both QGD and QHD systems are reduced to a system of two equations. Namely, it follows from the continuity equation that

$$\frac{\partial}{\partial x} \tau \frac{\partial p(x)}{\partial x} = 0, \tag{B.2}$$

and from the equations of motion for both systems, we obtain

$$\frac{\partial p(x)}{\partial x} = \frac{\partial}{\partial y} \mu \frac{\partial u(y)}{\partial y} + 2u(y) \frac{\partial}{\partial x} \tau \frac{\partial p(x)}{\partial x}. \tag{B.3}$$

These two equations are equivalent to one equation

$$\frac{dp(x)}{dx} = \mu_0 \frac{d^2 u(y)}{dy^2}. \tag{B.4}$$

The same equation is obtained under the substitution of relations (B.1) in the Navier–Stokes system. Here $\mu(T) = \mu(T_0) = \mu_0$.

Using the Maxwell-type slip conditions

$$\left(u - \frac{2-\sigma}{\sigma} \lambda \frac{du}{dy} \right) \Big|_{y=0} = 0, \quad \left(u + \frac{2-\sigma}{\sigma} \lambda \frac{du}{dy} \right) \Big|_{y=H} = 0$$

for the velocity as the boundary conditions (see [3, 144] and Sect. 2.7.1), we find the velocity profile, which has the form of the modified Poiseuille parabola

$$u_x = -\frac{1}{2\mu_0} \frac{dp(x)}{dx} \left[y(H-y) + \frac{2-\sigma}{\sigma} \lambda H \right],$$

where σ is the accommodation coefficient for the velocity, or a part of diffusionally reflected molecules, which is close to unity, and λ is the mean free path of particles, which is related to the viscosity coefficient as follows:

$$\lambda = A \frac{\mu}{p} \sqrt{\mathcal{R} T}. \tag{B.5}$$

The coefficient A is equal to $A = \sqrt{\pi/2}$ for the Chapman formula (see [3]) and $A = 2(7 - 2\omega)(5 - 2\omega)/15\sqrt{2\pi}$ for the Bird formula (see [28]).

We also succeed in obtaining the velocity distribution in the channel for the case where the accommodation coefficients on the upper and lower walls of the channel are different. On the lower wall, let the accommodation coefficient be equal to σ_1, and on the upper wall, let it be equal to σ_2. Then the boundary conditions for the velocity take the form

$$\left(u - \frac{2-\sigma_1}{\sigma_1} \lambda \frac{du}{dy} \right) \Big|_{y=0} = 0, \quad \left(u + \frac{2-\sigma_2}{\sigma_2} \lambda \frac{du}{dy} \right) \Big|_{y=H} = 0,$$

and the velocity distribution can be written as follows:

$$u_x = -\frac{1}{2\mu_0} \frac{dp(x)}{dx} \left[y(H\beta - y) + \frac{2-\sigma_2}{\sigma_2} \lambda H\beta \right], \tag{B.6}$$

where the coefficient β has the form

$$\beta = \frac{H + 2\lambda(2-\sigma_2)/\sigma_2}{H + \lambda(2-\sigma_1)/\sigma_1 + \lambda(2-\sigma_2)/\sigma_2}.$$

For $\sigma_1 = \sigma_2 = \sigma$, the latter formula passes to the relation obtained above.

For the Navier–Stokes equations, the mass-flow density is $j_{mx} = \rho u_x$, and the gas mass-flow rate with respect to the channel width unity in an arbitrary section is calculated as follows:

$$J_{\mathrm{NS}} = \int_0^H j_{mx}\,dy = \int_0^H \rho u_x\,dy = -\frac{H^3}{8\mu_0 \mathcal{R} T_0}\left[\frac{2}{3}p\frac{dp}{dx} + 4\frac{2-\sigma}{\sigma}p\frac{dp}{dx}\frac{\lambda}{H}\right].$$
(B.7)

In calculating the integral, following [134], we perform the change $\rho = p/\mathcal{R} T_0$. A similar formula for the mass-flow rate in a long channel with account for the velocity-slip conditions is presented in [160] for a cylindrical channel and is called the modified Poiseuille formula for the mass-flow rate.

For the QGD and QHD equations, we have $j_{mx} = \rho(u_x - w_x)$ and, moreover, for the problem considered, the quantity w_x is the same for both models and is given by

$$w_x = -\frac{\tau}{\rho}\frac{dp}{dx} = -\frac{\mu}{p\,\mathrm{Sc}}\frac{1}{\rho}\frac{dp}{dx}.$$

Within the framework of the QGD/QHD approach, the mass-flow rate through the section of the channel is given by

$$J^{xy} = \int_0^H \rho(u_x - w_x)\,dy = \int_0^H \rho u_x\,dy - \frac{\mu}{\mathrm{Sc}}\int_0^H \frac{1}{p}\frac{dp(x)}{dx}\,dy$$

$$= -\frac{H^3}{8\mu_0\mathcal{R} T_0}\left[\frac{2}{3}p\frac{dp}{dx} + 4\frac{2-\sigma}{\sigma}p\frac{dp}{dx}\frac{\lambda}{H} + \frac{8}{A^2\,\mathrm{Sc}}p\frac{dp}{dx}\left(\frac{\lambda}{H}\right)^2\right].\quad \text{(B.8)}$$

The last term in this formula is obtained as follows:

$$\frac{\mu}{\mathrm{Sc}}\int_0^H \frac{1}{p}\frac{dp}{dx}\,dy = \frac{\mu}{\mathrm{Sc}}\frac{1}{p}\frac{dp}{dx}H = \frac{\mu^2}{\mathrm{Sc}}\frac{1}{\mu_0}\frac{1}{p}\frac{dp}{dx}H,$$

where the viscosity coefficient μ is expressed through the mean free path (B.5).

The first term in Eq. (B.8) corresponds to the mass-flow rate determined by the Poiseuille parabola with adhesion conditions, the second term describes the increase of the mass-flow rate that accounts for the slip boundary conditions for the velocity, and the third term describes the mass-flow rate that accounts for self-diffusion processes. The last term has order $O(\tau^2)$ or $O(\mathrm{Kn}^2)$, where the Knudsen number is $\mathrm{Kn} = \lambda/H$. For stationary flows, precisely this distinction between the Navier–Stokes equations and the QGD/QHD models exists.

B.2.2 Minimum of the Mass-Flow Rate, or the Knudsen Effect

In Knudsen's experiments, the mass-flow rate was calculated in the normalized form, where the normalizing factor was determined by the mass-flow rate in the channel for a free molecular flow. The mass-flow rate in a section of a long channel for a free molecular flow is determined by diffusion processes. For the gas consisting of hard-sphere molecules, for $\sigma = 1$, this mass-flow rate is written as, e.g., in [160]. For a planar channel, it is given by

$$J_0^{xy} = \frac{4H^2\sqrt{2}}{3\sqrt{\pi \mathcal{R}T_0}} \frac{dp}{dx}, \tag{B.9}$$

where H is the width of the channel.

We call attention to the distinction in the formulas for the mass-flow rates written within the framework of the dense gas model (the first term in Eqs. (B.7) and (B.8)) and those in the free molecular flow approximation (B.9). In the first case, the mass-flow rate is proportional to H^3, the gradient of the pressure, and the mean pressure. In the second case, the mass-flow rate is proportional to H^2 and the gradient of the pressure and is independent of the mean pressure in the channel.

Expressing the viscosity coefficient through the mean free path (B.5), we calculate the normalized value of the mass-flow rate (B.8):

$$Q_{xy} = \frac{J^{xy}}{J_0^{xy}} = \frac{3\sqrt{\pi} A}{8\sqrt{2}} \left[\frac{\mathrm{Kn}^{-1}}{6} + \frac{2-\sigma}{\sigma} + \frac{2}{A^2 \mathrm{Sc}} \mathrm{Kn} \right]. \tag{B.10}$$

This implies that Q has a minimum for the Knudsen number given by

$$\mathrm{Kn}_m = \frac{A}{2} \sqrt{\frac{\mathrm{Sc}}{3}}.$$

The location of the minimum is independent of the accommodation coefficient σ. For $\mathrm{Sc} = 1$ and $A = \sqrt{\pi/2}$, we have $Kn_m = 0.36$.

In [34, 36, 127], for the hard-sphere molecules ($\omega = 0.5$), the mass-flow rate in the planar channel is calculated in the BGK approximation. The obtained results are not expressed analytically and are presented in the form of tables and graphs. For small numbers Kn (Kn $\to$ 0), these results are presented in the form of the approximate formula

$$Q_{\mathrm{cer}} = \frac{\mathrm{Kn}^{-1}}{6} + \sigma + (2\sigma^2 - 1)\,\mathrm{Kn}. \tag{B.11}$$

For $\sigma = 1$, expressions (B.10) and (B.11) differ from one another only by the numerical coefficient in the term $\sim \mathrm{Kn}$ and the factor of order 1 of the bracket. This factor is related to the normalization chosen in [34, 36] for which $A = \sqrt{2}$ and

$$J_0^{xy} = \frac{H^2}{\sqrt{2\mathcal{R}T_0}} \frac{dp}{dx}. \tag{B.12}$$

Numerous data and asymptotic formulas for the mass-flow rate obtained by using various forms of the BGK equation are collected in [178]. These formulas differ from Eq. (B.11) by coefficients close to 1 in the second and third terms and contain an additional fourth term $\sim -\mathrm{Kn}^2$.

B.2.3 Dependence of the Mass-Flow Rate on the Pressure Overfall

As a rule, in experiments, one measures the mass-flow rate in a microchannel depending on the pressure overfall at its ends. Let us obtain such a formula. For this purpose, we return to expression (B.8) and calculate the integral mass-flow rate with respect to the channel length L.

In Eq. (B.8), we set $\lambda = \lambda_2 p_2/p$, where λ_2 is the mean free path on the channel output. Integrate the obtained expression over the closed interval $[0, L]$. Assume that the channel width D is much greater than its height H. Then neglecting the influence of the lateral walls, we calculate the mass-flow rate:

$$Q_{xy}^L = \frac{D}{L} \int_0^L J^{xy} dx = D \frac{H^3 p_2^2}{8\mu_0 \mathcal{R} T_0 L}$$
$$\times \left[\frac{(p_1/p_2)^2 - 1}{3} + 4\frac{2-\sigma}{\sigma} \mathrm{Kn}_2 \left(\frac{p_1}{p_2} - 1 \right) + \frac{8}{A^2 \mathrm{Sc}} \mathrm{Kn}_2^2 \ln \left(\frac{p_1}{p_2} \right) \right],$$
$$\tag{B.13}$$

where $\mathrm{Kn}_2 = \lambda_2/H$. A similar formula for the Navier–Stokes system with velocity slip conditions has the form

$$Q_{NS}^L = D \frac{H^3 p_2^2}{8\mu_0 \mathcal{R} T_0 L} \left[\frac{(p_1/p_2)^2 - 1}{3} + 4\frac{2-\sigma}{\sigma} \mathrm{Kn}_2 \left(\frac{p_1}{p_2} - 1 \right) \right]. \tag{B.14}$$

The expression

$$Q_{NS}^{L2} = D \frac{H^3 p_2^2}{8\mu_0 \mathcal{R} T_0 L}$$
$$\times \left[\frac{(p_1/p_2)^2 - 1}{3} + 4\frac{2-\sigma}{\sigma} \mathrm{Kn}_2 \left(\frac{p_1}{p_2} - 1 \right) + 9 \mathrm{Kn}_2^2 \ln \left(\frac{p_1}{p_2} \right) \right], \tag{B.15}$$

which differs from Eq. (B.13) only by the coefficient of the last term, was obtained from the Navier–Stokes equations by using the so-called slip conditions of the second order (see [15, 46]):

$$u_s = \frac{2-\sigma}{\sigma} \lambda \left(\frac{\partial u}{\partial y} \right) + \frac{9\lambda^2}{8} \left(\frac{\partial^2 u}{\partial y^2} \right),$$

where the mean free path λ is calculated by the Chapman formula. Formulas (B.13), (B.14), and (B.15) for the mass-flow rate were compared with experimental data for nitrogen and helium.

In [44], the results of experimental measurements of mass-flow rate in isothermal helium and nitrogen flows in microchannels of length L, height H, and width D are presented. The most appropriate for comparison is the microchannel pattern of sizes $L = 5 \times 10^{-3}$ m, $H = 0.545 \times 10^{-6}$ m, and $D = 5 \times 10^{-5}$ m. Here, the condition $H \ll D$ holds, and the channel can be assumed to be planar.

In helium flows, the effect of the increase of the mass-flow rate is sharply expressed if $T_0 = 294.2$ K and $p_2 = 0.75 \times 10^5$ Pa. The value of helium dynamic viscosity for the above temperature taken from reference books is $\mu_0 = 19.6 \times 10^{-6}$ Pa $\cdot$ s, the gas constant is $\mathcal{R} = 2.0785 \times 10^3$ J/(kg $\cdot$ K), and the Schmidt number is Sc $= 0.77$. The corresponding graphs in comparison with the experimental points are shown in Fig. B.1. In the first case, to calculate the mean free path, one used the Chapman formula, and in the second case, the Bird formula with the exponent $\omega = 0.66$ (see [28]). The part σ of diffusionally reflected molecules is assumed to be equal to unity. It is seen that the calculations using the QGD/QHD equation describe the experiment most precisely. The Navier–Stokes equations with the traditional Maxwell slip boundary conditions yield a not completely satisfactory result.

Similar data for nitrogen are presented in Fig. B.2. The distinctions in the problem parameters as compared with the previous variant are as follows: $\mu_0 = 17.8 \times 10^{-6}$ Pa $\cdot$ s, $\mathcal{R} = 2.962 \times 10^2$ J/(kg$\cdot$K), Sc $= 0.74$, $\omega = 0.74$, and $\sigma = 0.93$. The last quantity corresponds to the data of [44]. The curve corresponding to the QGD/QHD models is in an ideal correspondence with the experiment, although the effect of the increase of the mass-flow rate is not sharply expressed in this case.

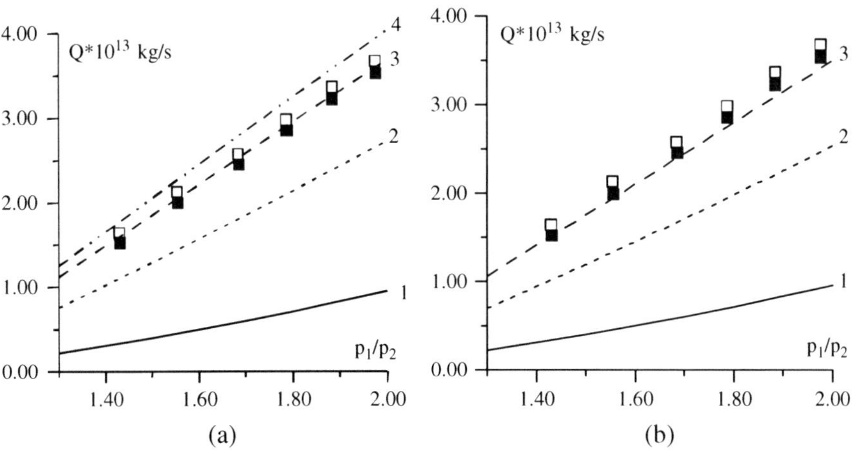

Fig. B.1 Results for helium. (a) The Chapman formula; (b) the Bird formula. *Line 1*: the Navier–Stokes system with adhesion conditions; *line 2*: the Navier–Stokes system with the Maxwell slip conditions; *line 3*: the QGD/QHD equations with the Maxwell slip conditions; *line 4*: the Navier–Stokes system with the slip condition of the second order. The white and black squares correspond to the spread of the experimental data (see [44])

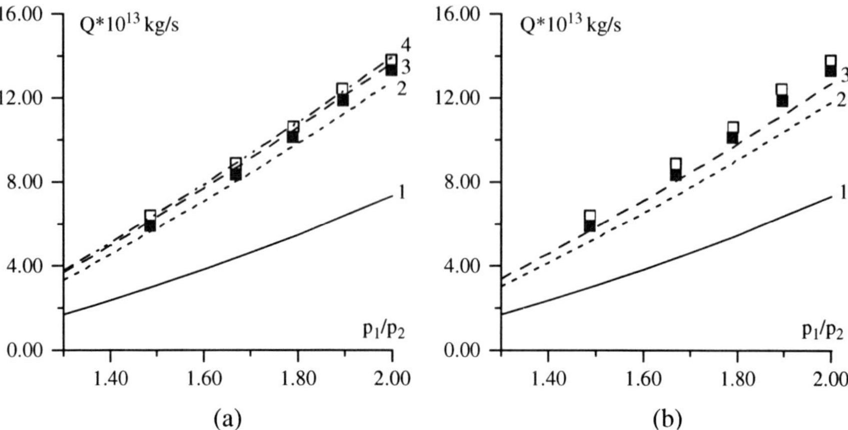

Fig. B.2 Results for nitrogen. (**a**) The Chapman formula; (**b**) the Bird formula. *Line 1*: the Navier–Stokes system with adhesion conditions; *line 2*: the Navier–Stokes system with the Maxwell slip conditions; *line 3*: the QGD/QHD equations with the Maxwell slip conditions; *line 4*: the Navier–Stokes system with the slip conditions of the second order. The white and black squares correspond to the spread of the experimental data (see [44])

The mass-flow rate for argon can be additionally compared with the data of [132], and for helium and nitrogen, with the data of [147].

In the variants considered, the Knudsen number in the channel output does not exceed 0.5. For $Kn_2 \leq 0.1$, the mass-flow rate is adequately described by the Navier–Stokes system with the Maxwell-type slip conditions, since the term of the second order of smallness in Kn_2 can be omitted in formula (B.13).

B.3 Poiseuille Flows in Circular Tubes

Repeating the above reasoning, let us present the formulas for the mass-flow rate in the case of a gas flow in a circular tube. For this purpose, we consider a gas flow in a long isothermal channel of length L and radius H. We seek the solution of the problem in the form

$$u_z = u(r), \quad u_r = 0, \quad p = p(z), \quad T = T_0. \tag{B.16}$$

Substituting Eq. (B.16) in the Navier–Stokes system and the QGD/QHD systems, we obtain the following equation connecting the gradient of the pressure and the longitudinal velocity:

$$\frac{dp(z)}{dz} = \mu_0 \frac{1}{r}\frac{d}{dr}\left(r\frac{du(r)}{dr}\right). \tag{B.17}$$

Using the slip boundary conditions for the velocity,

$$\left(u + \frac{2-\sigma}{\sigma}\lambda\frac{du}{dr}\right)\bigg|_{r=H} = 0,$$

we obtain the velocity profile of the form

$$u(r) = -\frac{1}{4\mu_0}\frac{dp(z)}{dz}\left[(H^2 - r^2) + 2\frac{2-\sigma}{\sigma}\lambda H\right].$$

For the Navier–Stokes equations, we have $j_{mr} = \rho u_z$. Replacing $\rho = p/\mathcal{R}T_0$, we calculate

$$J_{\text{NS}} = 2\pi \int_0^H r\rho u_z\, dr = -\frac{\pi H^4}{8\mu_0\mathcal{R}T_0}\left[p\frac{dp}{dz} + 4\frac{2-\sigma}{\sigma}p\frac{dp}{dz}\frac{\lambda}{H}\right]. \qquad (B.18)$$

With accuracy up to the notation, such a formula for the hard-sphere gas in a long cylindrical channel with account for the velocity-slip conditions was presented in [160] and was called the modified Poiseuille formula.

For the QGD/QHD systems, we have $j_{mz} = \rho(u_z - w_z)$. For the problem considered, the additions for both models are the same and are equal to

$$w_z = -\frac{\tau}{\rho}\frac{dp}{dz} = -\frac{\mu}{p\,\text{Sc}}\frac{1}{\rho}\frac{dp}{dz}.$$

The mass-flow rate in the section of the channel is given by

$$\begin{aligned}
J^{rz} &= 2\pi \int_0^H r\rho(u_z - w_z)dr \\
&= 2\pi \int_0^H r\rho u_z dr - 2\pi\frac{\mu}{\text{Sc}}\int_0^H r\frac{1}{p}\frac{dp(z)}{dz}dr \\
&= -\frac{\pi H^4}{8\mu_0\mathcal{R}T_0}\left[p\frac{dp}{dz} + 4\frac{2-\sigma}{\sigma}p\frac{dp}{dz}\frac{\lambda}{H} + \frac{8}{A^2\,\text{Sc}}p\frac{dp}{dz}\left(\frac{\lambda}{H}\right)^2\right]. \quad (B.19)
\end{aligned}$$

According to [160, Chap. 4-3, formula (4-43)], the mass-flow rate in the section of the cylindrical channel calculated under the assumption that the flow is free molecular and the condition of the complete velocity accommodation on the wall $\sigma = 1$ holds is as follows for the hard-sphere gas:

$$J_0^{rz} = \frac{4H^3}{3}\sqrt{\frac{2\pi}{\mathcal{R}T_0}}\frac{dp}{dz}.$$

In [160], it was noted that in the experiments for sufficiently rarefied gases, the presented formula describes the mass-flow rate in the channel with accuracy up to 1%.

The normalized value of the mass-flow rate is

$$Q_{rz} = \frac{J^{rz}}{J_0^{rz}} = \frac{3\sqrt{\pi}}{8} \frac{A}{\sqrt{2}} \left[\frac{\mathrm{Kn}^{-1}}{4} + \frac{2-\sigma}{\sigma} + \frac{2}{A^2 \mathrm{Sc}} \mathrm{Kn} \right]. \qquad (B.20)$$

It follows from Eq. (B.20) that Q_{rz} has the minimum for

$$\mathrm{Kn}_m = \frac{A}{2} \sqrt{\frac{\mathrm{Sc}}{2}}.$$

For $\mathrm{Sc} = 1$ and $A = \sqrt{\pi/2}$, we have $\mathrm{Kn}_m = 0.42$.

Expression (B.20) was compared with the computation of [34, 37, 127] whose results are presented in the table form. Here, for small numbers Kn ($\mathrm{Kn} \to 0$), a simplified formula for the mass-flow rate is presented:

$$Q_{cer} = \frac{\mathrm{Kn}^{-1}}{4} + \sigma + (2\sigma^2 - 1)\,\mathrm{Kn}. \qquad (B.21)$$

For $\sigma = 1$, formulas (B.20) and (B.21) differ by the factor $A/\sqrt{2} = 0.89$ and the Schmidt number Sc. According to the computations in [34, 35], the minimum value of the mass-flow rate is observed for $\mathrm{Kn} \sim 3$. Note that in [34], $A = \sqrt{2}$ and the mass-flow rate is normalized to the quantity

$$J_0^{rz} = \frac{\pi H^3}{\sqrt{2\mathcal{R}T_0}} \frac{dp}{dz}. \qquad (B.22)$$

Using Eq. (B.19), the mass-flow rate is calculated as follows depending on the pressure overfall at the channel ends:

$$Q_{rz}^L = \frac{1}{L} \int_0^L J(z)dz = \frac{\pi H^4 p_2^2}{8\mu_0 \mathcal{R} T_0 L}$$

$$\times \left[\frac{(p_1/p_2)^2 - 1}{2} + 4\frac{2-\sigma}{\sigma} \mathrm{Kn}_2 \left(\frac{p_1}{p_2} - 1 \right) + \frac{8}{A^2 \mathrm{Sc}} \mathrm{Kn}_2^2 \ln\left(\frac{p_1}{p_2} \right) \right], \qquad (B.23)$$

where $\mathrm{Kn}_2 = \lambda_2/H$.

The comparison of the mass-flow rate in the section of planar and cylindrical channels with experimental data in the BGK approximation is presented below. Here, we only note that the analysis of the results shows that for the Knudsen numbers $0.1 < \mathrm{Kn} < 0.5$, the mass-flow rate formulas obtained on the basis of the QGD/QHD models in long isothermal channels correspond better to the existing data than the formulas based on the Navier–Stokes equations. In both problems, the classical Maxwell velocity-slip conditions are posed. However, for Knudsen numbers $Kn > 0.5$, the difference between the QGD/QHD results and the kinetic-model data sharply increases. Below, we suggest a method for eliminating this effect by introducing a correction into the coefficient τ.

B.4 Computation of the Mass-Flow Rate for Rarefied Flows

B.4.1 Correction of the Parameter τ for Rarefied Flows

For locally steady-state flows of a sufficiently dense gas, the smoothing parameter τ is

$$\tau = \frac{\mu}{p\,\mathrm{Sc}} = \frac{\gamma}{\mathrm{Sc}}\frac{\mu}{\rho c_s^2} = \frac{\lambda}{\mathrm{Sc}\,A\sqrt{\mathcal{R}T}}, \tag{B.24}$$

where γ is the specific heat ratio, Sc is the Schmidt number, which is close to unity for gases, and c_s is the sound speed. For the ideal gas, the quantity c_s is of order of mean thermal velocity $\langle c \rangle \sim \sqrt{2\mathcal{R}T}$ of the heat motion of molecules and the coefficient μ is equal to $\lambda\langle c \rangle$ in order, where λ is the mean free path of particles. Therefore, it follows from Eq. (B.24) that τ is equal to the mean free time $\lambda/\sqrt{2\mathcal{R}T}$ of particles in the gas with accuracy up to a coefficient of unit order. The computations of dense and moderately rarefied gas flows justify the true choice of the smoothing parameter in the form (B.24).

We have $\tau \sim 1/\rho$, and under the increase of the rarefaction of the gas (Knudsen number Kn), this quantity unboundedly grows. For flows of sufficiently rarefied gases, when $\lambda \geq H$, i.e., $Kn = \lambda/H \geq 1$, it is natural to limit the smoothing parameter and, in addition, to relate it with the characteristic size of the problem. For this purpose, we relate the smoothing parameter with Kn replacing it by the value

$$\tau \to \tau(Kn) = \frac{\tau}{1 + Kn} = \frac{\lambda}{\mathrm{Sc}(1 + Kn)A\sqrt{\mathcal{R}T}}. \tag{B.25}$$

As $Kn \to 0$, expression (B.25) degenerates into formula (B.24).

Let us estimate the influence of the correction in the limit case of strongly rarefied gas flows. Substituting the expression for λ of the form (B.5), for large Knudsen numbers $Kn \gg 1$, we obtain

$$\tau = \frac{\lambda}{\mathrm{Sc}(1 + Kn)A\sqrt{\mathcal{R}T}} \sim \frac{\lambda}{\mathrm{Sc}\,Kn\,A\sqrt{\mathcal{R}T}} = \frac{H}{\mathrm{Sc}\,A\sqrt{\mathcal{R}T}}. \tag{B.26}$$

Therefore, for rarefied flows, $\tau \sim H/\sqrt{\mathcal{R}T}$ is of order of the characteristic time of the free path of molecules between collisions with the boundaries of the domain considered.

Using the relaxation time τ calculated in such a way, we can find the modified coefficients of diffusion, viscosity, and heat conductivity in a rarefied gas. Then the viscosity coefficient takes the form

$$\mu(\mathrm{Kn}) = \tau(\mathrm{Kn})p\,\mathrm{Sc} \sim \frac{\rho H \sqrt{\mathcal{R}T}}{A}. \tag{B.27}$$

The diffusion coefficient becomes

$$D(\mathrm{Kn}) = \frac{\tau(\mathrm{Kn})p\,\mathrm{Sc}}{\rho\,\mathrm{Sc}} \sim \frac{H\sqrt{\mathcal{R}T}}{\mathrm{Sc}\,A}. \tag{B.28}$$

The heat conduction coefficient for an ideal polytropic gas ($\varepsilon = c_v T,\ c_v - \mathcal{R}/(\gamma - 1)$) is

$$\varkappa(\mathrm{Kn}) = \frac{\gamma \mathcal{R}\mu(\mathrm{Kn})}{(\gamma - 1)\,\mathrm{Pr}} \sim \frac{\gamma \mathcal{R}}{(\gamma - 1)\,\mathrm{Pr}} \frac{\rho H \sqrt{\mathcal{R}T}}{A}. \tag{B.29}$$

The transition coefficients of a close structure for flows in long channels are suggested earlier in [156].

As an example of using the coefficient thus obtained, we consider free molecular gas flows in a planar layer. The expressions for the mass flux, the friction force, and the heat flux for these problems obtained by methods of kinematic theory are known (see, e.g., [28, 45, 127] for $\sigma = 1$).

B.4.1.1 Diffusion Problem

Consider two infinite reservoirs filled in by a gas with densities ρ_1 and ρ_2. Let the reservoirs be separated by a thin partition with a hole of unit area. Let the mean free path of particles be sufficiently large ($\lambda \gg 1$). For definiteness, let $\rho_2 > \rho_1$. The temperature of the gas is constant and equal to T. According to [28, 127], the diffusion flux between the reservoirs is calculated as follows:

$$j_D{}^{\mathrm{kin}} = \rho_2 \frac{\sqrt{8\mathcal{R}T}}{\sqrt{\pi}} - \rho_1 \frac{\sqrt{8\mathcal{R}T}}{\sqrt{\pi}} = \frac{\sqrt{8\mathcal{R}T}}{\sqrt{\pi}}(\rho_2 - \rho_1), \tag{B.30}$$

where $\sqrt{8\mathcal{R}T/\pi}$ is the mean velocity of particles calculated on the basis of Maxwell distribution function (2.18) (see [160]).

Using the diffusion coefficient in the form (B.28), we obtain

$$j_D = D\frac{d\rho}{dy} = D\frac{\rho_2 - \rho_1}{H} = \frac{\sqrt{\mathcal{R}T}}{A\,\mathrm{Sc}}(\rho_2 - \rho_1). \tag{B.31}$$

Relations (B.30) and (B.31) coincide with accuracy up to the numerical coefficient ~ 1. In both cases, the mass flux through the hole is proportional to the difference of the densities and the heat velocity of particles.

B.4.1.2 Couette Problem

Let us calculate the friction force arising between two infinite parallel laminas separated from one another by distance H. Let one of the laminas be fixed and the other move with speed U. A gas in the layer between the laminas has temperature T and density ρ. According to [28], for a free molecular flow, the friction force between the laminas is calculated as follows:

$$F_t{}^{\mathrm{kin}} = \frac{\rho U \sqrt{\mathcal{R}T}}{\sqrt{2\pi}}.$$

(B.32)

The expression presented in [122] differs from Eq. (B.32) by the coefficient 2 in the numerator. In constant to the case of a dense gas, where the friction force is determined by the Newton law $F_t = \mu\, du/dy \sim \mu U/H$, for free molecular flows, the friction force is proportional to the density of particles and is independent of the distance between the laminas.

Using the viscosity coefficient (B.27), on the basis of the Newton formula, we obtain

$$F_t = \mu \frac{du}{dy} = \frac{\rho U \sqrt{\mathcal{R}T}}{A}.$$

(B.33)

Formulas (B.32) and (B.33) coincide with accuracy up to a numerical coefficient of order 1.

B.4.1.3 Heat Transfer Problem

Kinetic theory yields a formula for the heat flow between two infinite laminas of temperatures T_1 and T_2 ($T_2 > T_1$) for a gas with Kn $\gg$ 1 in the following form (see [28, 45]):

$$q^{\mathrm{kin}} = \frac{\gamma + 1}{\sqrt{2\pi}(\gamma - 1)} \frac{\rho \sqrt{\mathcal{R}T_1 T_2}}{(\sqrt{T_1} + \sqrt{T_2})} \mathcal{R}(T_2 - T_1).$$

(B.34)

Note that, according to [127], the first numerical factor is equal to $\sqrt{8/\pi}$ for this problem.

Substituting the heat conduction coefficient (B.29) in the Fourier law

$$q = \varkappa \frac{dT}{dy} = \varkappa \frac{T_2 - T_1}{H},$$

we immediately obtain

$$q = \frac{\gamma}{\Pr A(\gamma - 1)} \rho \sqrt{\mathcal{R}T} \mathcal{R}(T_2 - T_1).$$ (B.35)

The last two relations differ from each other by the method of finding the mean temperature T between the laminas as a function of T_1 and T_2. If one defines the mean temperature between the laminas as $T = T_1 T_2/(\sqrt{T_1} + \sqrt{T_2})^2$, then the formulas differ by a numerical coefficients of order 1.

Therefore, the suggested form of modification of the smoothing parameter τ allows us to obtain the values of diffusion mass flux, the friction force, and the heat flux for mentioned problems that correspond to the results of kinetic approach to the free molecular flows.

B.4.2 Unified Formulas for Calculating the Mass-Flow Rate

Here we use the modified formula for τ for the calculation of the values of the mass-flow rate in the microchannels presented in the previous sections of the present work. In this case, we take into account the fact that the coefficient τ enters only the diffusion component of the mass flux, which is described by the last terms in the corresponding formulas.

In finding the number Kn, there exists a definite arbitrariness related to the choice of the characteristic scale H and the finding of the characteristic length. Therefore, it is convenient to introduce a gauge coefficient α into Eq. (B.25); as a result, we obtain

$$\tau = \frac{\mu}{p \, \mathrm{Sc}(1 + \alpha \, \mathrm{Kn})}.$$ (B.36)

Then the formula for the local mass-flow rate in the section of the planar channel (B.8) takes the form

$$J^{xy} = -\frac{H^3}{8\mu_0 \mathcal{R}T_0}$$
$$\times \left[\frac{2}{3} p \frac{dp}{dx} + 4\frac{2-\sigma}{\sigma} p \frac{dp}{dx} \frac{\lambda}{H} + \frac{8}{A^2 \, \mathrm{Sc}} p \frac{dp}{dx} \left(\frac{\lambda}{H}\right)^2 \frac{1}{1 + \alpha\lambda/H} \right].$$ (B.37)

The normalized value of the mass-flow rate (an analog of formula (B.10)) is

$$Q_{xy} = \frac{J^{xy}}{J_0^{xy}} = \frac{3\sqrt{\pi}}{8} \frac{A}{\sqrt{2}} \left[\frac{\mathrm{Kn}^{-1}}{6} + \frac{2-\sigma}{\sigma} + \frac{2}{A^2 \, \mathrm{Sc}} \frac{\mathrm{Kn}}{1 + \alpha \, \mathrm{Kn}} \right].$$ (B.38)

The normalized mass-flow rate has the minimum for the Knudsen number

$$\mathrm{Kn}_m = \frac{A}{2}\sqrt{\frac{\mathrm{Sc}}{3}}\left(1 - \alpha\frac{A}{2}\sqrt{\frac{\mathrm{Sc}}{3}}\right)^{-1},$$

where Kn_m is the positive root of the corresponding quadratic equation. For $\mathrm{Sc} = 1$, $A = \sqrt{\pi/2}$, and $\alpha = 1$, we have $\mathrm{Kn}_m = 0.56$.

The existence condition for the minimum of the Knudsen number $\mathrm{Kn}_{xy} > 0$ imposes the following restriction on α:

$$\alpha < \frac{A\sqrt{\mathrm{Sc}}}{2\sqrt{3}} \sim 3.$$

For $\mathrm{Kn} \gg 1$, the mass-flow rate in the section of the channel is equal to that for the free molecular flow, and Eq. (B.38) takes the form

$$Q_{xy} = \frac{3\sqrt{\pi}}{8}\frac{A}{\sqrt{2}}\left[\frac{2-\sigma}{\sigma} + \frac{2}{\alpha A^2\,\mathrm{Sc}}\right] = 1. \tag{B.39}$$

From this, we can uniquely find the coefficient α. For $\sigma = 1$, we have

$$\alpha = \frac{6\sqrt{\pi}}{A\,\mathrm{Sc}(8\sqrt{2} - 3A\sqrt{\pi})}.$$

If $A = \sqrt{\pi/2}$ and $\mathrm{Sc} = 1$, then $\alpha = 1.82$.

The formula (B.19) for the local mass-flow rate in the section of the cylindrical channel takes the form

$$J^{rz} = -\frac{\pi H^4}{8\eta_0\mathcal{R}T_0}$$
$$\times\left[p\frac{dp}{dx} + 4\frac{2-\sigma}{\sigma}p\frac{dp}{dx}\frac{\lambda}{H} + \frac{8}{A^2\,\mathrm{Sc}}p\frac{dp}{dx}\left(\frac{\lambda}{H}\right)^2\frac{1}{1+\alpha\lambda/H}\right]. \tag{B.40}$$

The normalized value of the mass-flow rate takes the form (an analog of formula (B.20))

$$Q_{rz} = \frac{J^{rz}}{J_0^{rz}} = \frac{3\sqrt{\pi}}{8}\frac{A}{\sqrt{2}}\left[\frac{\mathrm{Kn}^{-1}}{4} + \frac{2-\sigma}{\sigma} + \frac{2}{A^2\,\mathrm{Sc}}\frac{\mathrm{Kn}}{1+\alpha\,\mathrm{Kn}}\right]. \tag{B.41}$$

Q_{rz} has the minimum for the Knudsen number

$$\mathrm{Kn}_{rz} = \frac{A}{2}\sqrt{\frac{\mathrm{Sc}}{2}}\left(1 - \alpha\frac{A}{2}\sqrt{\frac{\mathrm{Sc}}{2}}\right)^{-1} \sim 0.72$$

for $\mathrm{Sc} = 1$, $A = \sqrt{\pi/2}$, and $\alpha = 1$.

For Kn $\gg$ 1, the mass-flow rate in the channel is equal to that in the free-molecular flow. Moreover, Eq. (B.41) coincides with Eq. (B.39) and allows us to find the value of the coefficient α.

The comparison of Eqs. (B.38) and (B.41) with the results of [34] is presented in Fig. B.3 for $\sigma = 1$, $A = \sqrt{\pi/2}$, and Sc $= 1$. The lower curve corresponds to the Navier–Stokes equations with the adhesion condition for the velocity, line 1 corresponds to the Navier–Stokes equation with the slip conditions for the velocity, line 2 corresponds to the QGD/QHD models for $\alpha = 0$, line 3 corresponds to $\alpha = 1$, and line "Cer" is the data of [34]. In Fig. B.3, line 4 corresponds to $\alpha = 2$.

Note that as Kn $\to \infty$, the BGK model for a planar channel has the asymptotics $Q_{xy} \sim \log$ Kn, and for a cylindrical channel, it is $Q_{rz} \sim$ const (see [34]). The QGD/QHD estimates obtained by a simple method have the asymptotics $Q \sim$ const (Fig. B.3) in both cases.

The graphs presented demonstrate that for the Knudsen numbers not exceeding Kn $\sim$ 0.1, the Navier–Stokes equations and QGD/QHD systems with the slip conditions for the velocity are in a good agreement with the results of computations according to the BGK model. For Kn $<$ 0.5, the QGD/QHD models are in a good agreement with the standard results, but further, when the number Kn grows, the QGD/QHD data rapidly lose their accuracy. When using the correction of QGD/QHD, the results are in a very good agreement with the data of the BGK approximation in the whole range of the Knudsen numbers. Moreover, $\alpha = 1$ for planar channels and $\alpha = 2$ for cylindrical channels.

We present the integral formulas for the mass-flow rate depending on the pressure overfall for the plane and cylindrical channels.

An analog of formula (B.13) takes the form

$$
Q_{xy}^L = D \frac{H^3 p_2^2}{8\mu_0 \mathcal{R} T_0 L} \left[\frac{(p_1/p_2)^2 - 1}{3} + 4 \frac{2-\sigma}{\sigma} \mathrm{Kn}_2 \left(\frac{p_1}{p_2} - 1 \right) \right.
$$
$$
\left. + \frac{8}{A^2 \, \mathrm{Sc}} \, \mathrm{Kn}_2^2 \ln \left(\frac{p_1/p_2 + \alpha \, \mathrm{Kn}_2}{1 + \alpha \, \mathrm{Kn}_2} \right) \right], \tag{B.42}
$$

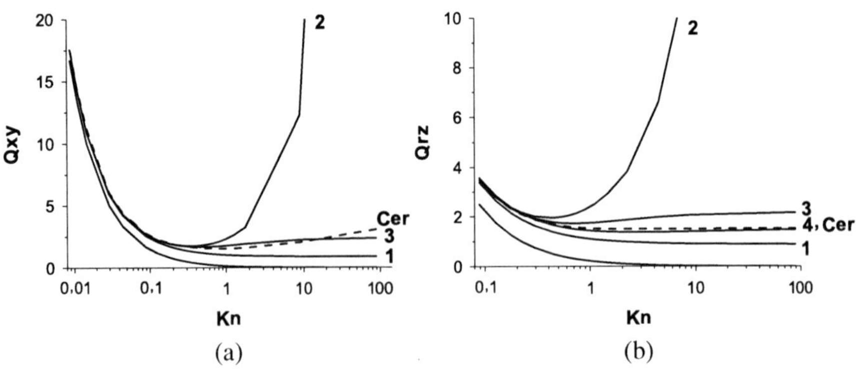

Fig. B.3 Dependence of the specific mass-flow rate Q on the Knudsen number Kn in the planar channel (**a**) and the cylindrical channel (**b**)

where $\mathrm{Kn}_2 = \lambda_2/H$. In the limit of large Knudsen numbers, the last term becomes linear in the number Kn and can be rewritten in the form

$$\frac{8\,\mathrm{Kn}_2}{A^2\,\mathrm{Sc}\,\alpha}\left(\frac{p_1}{p_2}-1\right).$$

An analog of formula (B.23) with correction has the form

$$Q_{rz}^L = \frac{\pi H^4 p_2^2}{8\eta_0 \mathcal{R}\, T_0 L}\left[\frac{(p_1/p_2)^2 - 1}{2} + 4\frac{2-\sigma}{\sigma}\,\mathrm{Kn}_2\left(\frac{p_1}{p_2}-1\right)\right.$$
$$\left. + \frac{8}{A^2\,\mathrm{Sc}}\,\mathrm{Kn}_2^2 \ln\left(\frac{p_1/p_2 + \alpha\,\mathrm{Kn}_2}{1 + \alpha\,\mathrm{Kn}_2}\right)\right]. \tag{B.43}$$

B.5 Comparison with Experimental Data

Let us compare the obtained formulas for the mass-flow rate with the experimental data presented in [147]. Here, the results obtained in [13,44] and in the author work [147] for helium and nitrogen flows in planar silicon microchannels are collected. The measured values of the mass-flow rate are represented in the dimensionless form by normalizing the value for the Poiseuille flow to the mass-flow rate (B.7):

$$J_0 = \frac{H^3 p}{12\mu_0 \mathcal{R} T_0}\frac{dp}{dx} \tag{B.44}$$

and are represented in the form of dependence on the Knudsen number in the range $0 < \mathrm{Kn} < 1$. The Knudsen number is found in [147] as $\mathrm{Kn} = \lambda/H$, where the mean free path is calculated as follows:

$$\lambda = \frac{k_B T}{\sqrt{2}\pi p a^2};$$

here, a is the diameter of the molecule and k_B is the Boltzmann constant. This definition of the mean free path corresponds to the Bird formula for the hard-sphere gas (B.5). This can be seen by using the Bird formula (B.5) and the viscosity coefficient in the form (see [28])

$$\mu = \frac{15}{8}\sqrt{\frac{\pi}{2}}\frac{m\sqrt{2k_B T/m}}{(2-\alpha)^\alpha\,\Gamma(4-\alpha)\sigma_{\mathrm{ref}}}\left(\frac{T}{T_{\mathrm{ref}}}\right)^\alpha,$$

where $\sigma_{\mathrm{ref}} = \pi a^2$ is the interaction cross-section for a molecule and $\alpha = \omega - 1/2$. For the hard-sphere gas, $\omega = 1/2$, $\alpha = 0$, $p = \rho\mathcal{R}T$, and $\mathcal{R} = k_B/m$ (recall that $\Gamma(4) = 3! = 6$).

To compare with the experimental data, we normalize the mass-flow rate (B.8) to the mass-flow rate value in the Poiseuille flow (B.44):

$$S = \frac{J_{xy}}{J_0} = 1 + 6\frac{2-\sigma}{\sigma}\,\mathrm{Kn} + 12\frac{1}{A^2\,\mathrm{Sc}}\,\mathrm{Kn}^2. \tag{B.45}$$

In [147], the analytical formula for the dependence of the mass-flow rate value on Kn is represented in the form

$$S = 1 + 6A_1\,\mathrm{Kn} + 12A_2\,\mathrm{Kn}^2,$$

where $A_1 = (2-\sigma)/\sigma$ and the coefficient A_2 is found by matching to the experiment and is assumed to depend on σ and Kn. The accommodation coefficient is also assumed to depend on the Knudsen number, and for the correspondence with the experiment, in a number of cases, it is chosen to be greater than one, $\sigma > 1$. The same assumptions for calculating the mass-flow rate are made in [145].

The mass-flow rate formula (B.45) obtained for the QGD/QHD models has the same structure as that predicted in [145, 147]. Moreover, the coefficient A_2 is determined by the coefficient A in Eq. (B.5) and by the value of the Schmidt number $A_2 = 1/(A^2\,\mathrm{Sc})$, i.e., by the gas properties.

According to Eq. (B.5), the value of the coefficient A in the Bird formula (B.5) varies in the following limits: for Maxwell molecules, $\omega = 1$ and $A = 0.798$; for nitrogen, $\omega = 0.74$ and $A = 1.034$; for helium, $\omega = 0.66$ and $A = 1.112$; and for the hard-sphere gas, $\omega = 0.5$ and $A = 16/(5\sqrt{2\pi}) = 1.28$. For the Chapman formula, $A = \sqrt{\pi/2} = 1.252$ is a constant value.

For the Schmidt number Sc, there exist approximate formulas; for example, according to [28], the Schmidt number can be calculated as $\mathrm{Sc} = 5/(7 - 2\omega)$. In accordance with this, $\mathrm{Sc} = 0.883$ for helium and $\mathrm{Sc} = 0.905$ for nitrogen. At the same time, for many gases, the Schmidt number is measured in experiments with isotopes and is known sufficiently exactly. In [28], the experimental values are presented: $\mathrm{Sc} = 0.757$ for helium and $\mathrm{Sc} = 0.746$ for nitrogen.

In Fig. B.4, we present the comparison of expression (B.45) with experimental data of [147] for nitrogen and helium. Here, $A = 16/(5\sqrt{2\pi}) = 1.28$ is calculated by the Bird formula (B.5) for the hard-sphere gas, $\sigma = 1$, $\mathrm{Sc} = 0.88$ and 0.75.

It is seen that the QGD/QHD data are in a sufficiently good correspondence with the experiment up to $\mathrm{Kn} \sim 0.4\text{--}0.5$. Line 1 represents the dependence $1 + 6\,\mathrm{Kn}$, the mass-flow rate calculated within the framework of the Navier–Stokes model. Here, the coincidence with the experiment is attained for $\mathrm{Kn} < 0.1$.

For $\mathrm{Kn} \sim 0.5$, the influence of the dependence of solution on the number Sc appears. When Sc grows, the value of the mass-flow rate diminishes and becomes close to the experiment. We succeed in obtaining a concordance with the experiment for $\mathrm{Sc} > 1$, which does not correspond to real values of this quantity. For $\mathrm{Kn} > 0.5$, the correspondence with the experiment is attained by introducing a correction into the smoothing parameter (B.25) and selecting the coefficient α. In this case, an analog of formula (B.45) has the form

$$S = 1 + 6\frac{2-\sigma}{\sigma}\,\mathrm{Kn} + 12\frac{1}{A^2\,\mathrm{Sc}}\frac{\mathrm{Kn}^2}{1 + \alpha\,\mathrm{Kn}}. \tag{B.46}$$

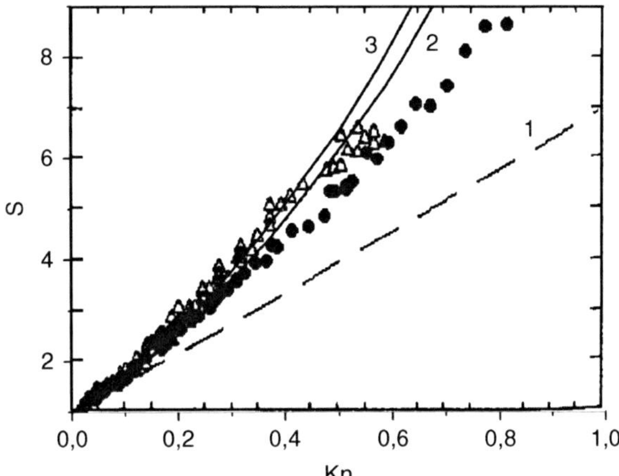

Fig. B.4 Dependence of S on Kn. *Line 1*: the Navier–Stokes model with the slip condition; *line 2*: the QGD/QHD model for Sc = 0.88; *line 3*: the QGD/QHD model for Sc = 0.75. Experimental data [147] are represented by triangles for nitrogen and by points for helium

The comparison of the mass-flow rate calculated according to Eq. (B.46) for $\sigma = 1$, $A = 16/(5\sqrt{2\pi}) = 1.28$, and Sc = 0.75 with the experimental data is shown in Fig. B.5. We see a good correspondence for nitrogen when $\alpha = 1$ and for helium when $\alpha = 2$ for QGD/QHD models up to Kn = 1.

Note that the insufficient accuracy in finding the coefficients makes it difficult to perform an accurate comparison of computational and experimental data. The value of the accommodation coefficient σ whose finding is an independent problem influences the value of the mass-flow rate. In this consideration, we set $\sigma = 1$.

In [116], to calculate the mass-flow rate in long microchannels, the artificial boundary condition for the velocity slip was introduced as

$$u_s = \frac{2 - \sigma}{\sigma} \frac{\text{Kn}}{(1 - b\,\text{Kn})} \lambda \frac{\partial u}{\partial y},$$

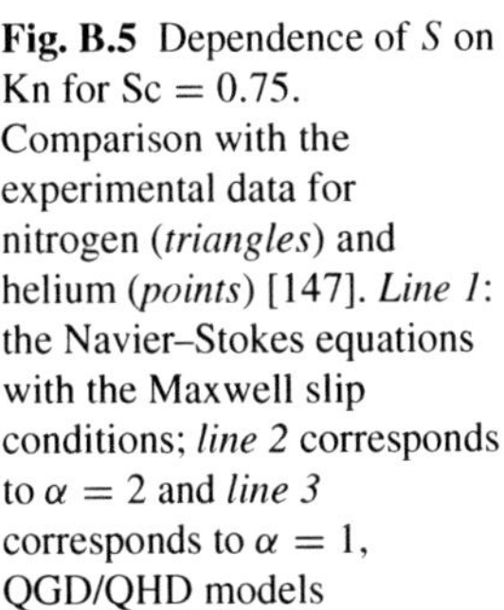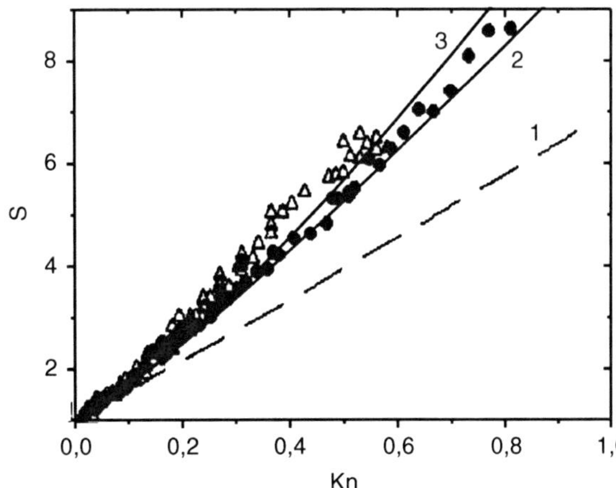

Fig. B.5 Dependence of S on Kn for Sc = 0.75. Comparison with the experimental data for nitrogen (*triangles*) and helium (*points*) [147]. *Line 1*: the Navier–Stokes equations with the Maxwell slip conditions; *line 2* corresponds to $\alpha = 2$ and *line 3* corresponds to $\alpha = 1$, QGD/QHD models where the coefficient of the derivative of velocity depends on the Knudsen number and a certain variable coefficient b. The resulting formulas for the mass-flow rate practically coincide with the QGD/QHD formula (B.46), except that of the quantities A and Sc, which depend on the type of gas in our consideration, while the expression of [116] involves numerical coefficients found from additional assumptions. The comparison of formulas from [116] with kinetic computation data and those of experiment demonstrates their good accuracy, which implicitly justifies the accuracy of the obtained universal formulas (B.42), (B.43), and (B.46) for large Knudsen numbers.

B.5.1 Concluding Remarks

The results presented demonstrate that, based on the Navier–Stokes equations with the classical Maxwell conditions, one succeeds in constructing the formulas for the mass-flow rates in long microchannels, which hold for the Knudsen number satisfying $\mathrm{Kn} < 0.1$.

The approximate formulas for the mass-flow rate obtained within the framework of the QGD/QHD equations predict the Knudsen effect and allow one to compute the value of the mass-flow rate with a sufficient accuracy up to the Knudsen numbers of order $\mathrm{Kn} \sim 0.5$.

The introduction of a correction into the relaxation parameter of the QGD/QHD model allows one to obtain the value of the mass-flow rate which coincides well with the results of kinetic theory in the whole range of Knudsen numbers.

Using the above approach, it is not possible to obtain analytical expressions for calculating gas flows in nonisothermal channels. These problems can be solved by using simplified gas-dynamic equations. Such equations are the Prandtl equations and the family of the so-called parabolized equations. The Prandtl equations for the two-dimensional planar flows have the form

$$\frac{\partial \rho}{\partial t} + \frac{\partial (\rho u_x)}{\partial x} + \frac{\partial (\rho u_y)}{\partial y} = 0, \tag{B.47}$$

$$\frac{\partial (\rho u_x)}{\partial t} + \frac{\partial (\rho u_x^2)}{\partial x} + \frac{\partial (\rho u_x u_y)}{\partial y} + \frac{\partial p}{\partial x} = \frac{\partial}{\partial y}\left(\mu \frac{\partial u_x}{\partial y}\right), \tag{B.48}$$

$$\frac{\partial p}{\partial y} = 0, \tag{B.49}$$

$$\frac{\partial}{\partial t}\left[\rho\left(\frac{u_x^2}{2} + \varepsilon\right)\right] + \frac{\partial}{\partial x}\left[\rho u_x\left(\frac{u_x^2}{2} + \varepsilon + \frac{p}{\rho}\right)\right]$$
$$+ \frac{\partial}{\partial y}\left[\rho u_y\left(\frac{u_x^2}{2} + \varepsilon + \frac{p}{\rho}\right)\right] = \frac{\partial}{\partial y}\left(\mu u_x \frac{\partial u_x}{\partial y}\right) + \frac{\partial}{\partial y}\left(\varkappa \frac{\partial T}{\partial y}\right). \tag{B.50}$$

The variants of parabolic approximations for the Navier–Stokes equations and the QGD/QHD equations for stationary flows can be obtained by neglecting second-order terms containing gradients in the longitudinal direction [45, 99, 187]. These simplified equations for stationary flows can be effectively solved by using marching algorithms, as the marching coordinate, the coordinate along the main direction of the flow, is chosen. The variants of such algorithms for Navier–Stokes models are presented, e.g., in [7, 99, 164].

Appendix C
Numerical Modelling of the Stationary Shock Wave Structure

We present the results of numerical simulation of the problem on the stationary shock wave structure in argon and nitrogen obtained on the basis of the QGD equations and the Navier–Stokes equations. The obtained data improve the existing ideas of applicability limits of macroscopic equations for monoatom and two-atom rarefied gas flows. We obtain that the density profile in the shock wave in argon and nitrogen in the range of Mach numbers from 1.5 ip to 10 is essentially closer to the experimental data as considered early. The distinction of experiment in the profile width is of order 30%, and, moreover, the density profiles themselves coincide with experiment sufficiently well. In computations for a two-atom gas, the agreement with the experiment is attained under the account for the second viscosity coefficient. It is shown that the Navier–Stokes equations and the QGD equations in this problem yield very close results.

The material for this appendix is based on [57, 62, 68, 70, 89, 90].

C.1 Introduction

The problem on the stationary shock wave structure is a classical problem by which examining the limited character of the Navier–Stokes model for describing rarefied gas flows is demonstrated. In particular, in numerous books devoted to kinetic models for describing the rarefied gas flows (see [28, 35, 122], etc.), the comparison of the shock wave width computed according to the Navier–Stokes model with the experimental data for argon and nitrogen is presented; among them, the most complete are data of [6]. Among the first numerical computations of the shock wave structure, which are cited in comparison with the experiment, we mention [172] and [143]. The obtained data show that the results based on the Navier–Stokes system adequately describe the shock wave width up to Mach numbers less than 2. For large Mach numbers, the density profile width in computation exceeds the values of experimental data by 1.5–2 times. This fact serves as a permanent stimulus for improving the Navier–Stokes model for describing moderately rarefied gas flows, which is reflected in a large number of scientific publications. As an example, we mention [202] and the bibliography therein.

The computations of the shock wave structure on the basis of the Navier–Stokes system described above were carried out sufficiently long ago by using a number of approximation and lower-power computation tools. In what follows, we present the results of computing this problem on the modern level in the framework of the Navier–Stokes and QGD equations by applying various numerical algorithms with improved constants depending on the viscosity and the temperature. The results of computations are compared with each other and with the experimental data for two gases: argon, which is a monoatom gas having no inner degrees of freedom, and nitrogen, which is a two-atom gas having additional degrees of freedom related to the rotational energy.

The molecular gas properties are chosen in accordance with [28]. As a standard, we take the experimental data of [6]. These data are obtained in a shock tube, where the shock wave form can be considered to be plane with a high accuracy, which allows us to solve the problem in the one-dimensional approximation. In these experiments, the temperature before the shock wave is of order $300°$K, which yields a possibility to use a power dependence of the viscosity coefficient on the temperature.

In [6], by using an electric probe, the density in the shock wave is measured. In particular, the density profiles and the dependencies of the inverse width of the shock wave on the Mach number are presented. The temperature and velocity profiles are not measured in these experiments. The comparison of numerical models with the experiment are carried out only for the density distributions. In comparison of the numerical and experimental data, the reduction to the dimension-free form is used [6].

C.2 Statement of the Problem

The mathematical model for solving the problem is the gas-dynamic system of equations for a plane one-dimensional flow. This system is presented in Sect. 5.7 and has the form

$$\frac{\partial \rho}{\partial t} + \frac{\partial j_m}{\partial x} = 0, \tag{C.1}$$

$$\frac{\partial (\rho u)}{\partial t} + \frac{\partial (j_m u)}{\partial x} + \frac{\partial p}{\partial x} = \frac{\partial \Pi}{\partial x}, \tag{C.2}$$

$$\frac{\partial E}{\partial t} + \frac{\partial (j_m H)}{\partial x} + \frac{\partial q}{\partial x} = \frac{\partial (\Pi u)}{\partial x}. \tag{C.3}$$

Here, ρ is the gas density, u is the velocity, $p = \rho \mathcal{R} T$ is the pressure, T is the temperature, γ is the specific heat ratio, $\mathcal{R}$ is the gas constant, and E and H are the total energy of the volume unit and the total specific enthalpy, respectively, which are calculated by the formulas $E = \rho u^2/2 + p/(\gamma - 1)$ and $H = (E + p)/\rho$. The method for calculating the quantities j_m, Π, and q entering the system is presented in Sect. 5.7. The dynamic viscosity coefficient μ, the heat conduction coefficient $\varkappa$, and the relaxation parameter τ are connected by the relations

$$\mu = \mu_\infty \left(\frac{T}{T_\infty}\right)^\omega, \quad \varkappa = \frac{\gamma \, \mathcal{R}}{(\gamma - 1)\,\mathrm{Pr}}\mu, \quad \tau = \frac{\mu}{p\,\mathrm{Sc}},$$

where Pr and Sc are the Prandtl and Schmidt numbers, respectively.

As measurement units x, t, ρ, u, p, T, E, and H we use the quantities l, l/c_∞, ρ_∞, c_∞, $\rho_\infty c_\infty^2$, T_∞, $\rho_\infty c_\infty^2$, and c_∞^2. Here, l is the linear size, c_∞ is the sound speed, ρ_∞ is the density, and T_∞ is the temperature (the last three quantities are calculated in the unperturbed flow before the shock wave front). The characteristic size l in this problem is chosen to be equal to the mean free path λ_∞ in the unperturbed flow, which, according to [28], is calculated as follows:

$$\lambda = \mu/(\rho\sqrt{2\pi \mathcal{R}T} \cdot \Omega/4), \tag{C.4}$$

where $\Omega(\omega) = 30/((7 - 2\omega)(5 - 2\omega))$. In this problem, the mean free path chosen for the deduction to the dimensionless form is calculated according to Eq. (C.4) for $\Omega(0.5) = 5/4$ (hard-sphere gas) and is given by

$$\lambda_\infty = \frac{16}{5} \frac{\mu_\infty}{\rho_\infty \sqrt{2\pi \mathcal{R}T_\infty}}. \tag{C.5}$$

Precisely this definition of the mean free path is applied for representation of results in [6] and in other works devoted to the study of the shock wave structure.

The reduction to the dimensionless form does not change the structure of equations. The relaxation parameter and the coefficients of viscosity and heat conduction are calculated as follows in the dimensionless form:

$$\mu = \gamma^{\omega-0.5}\frac{5\sqrt{2\pi}}{16}\left(\frac{p}{\rho}\right)^\omega, \quad \varkappa = \frac{\mu}{\mathrm{Pr}(\gamma - 1)}, \quad \tau = \frac{\mu}{p\,\mathrm{Sc}}$$

and, moreover, $p = \rho T/\gamma$. In the dimensionless form, the Mach number satisfies $\mathrm{Ma} = u_\infty$ and the sound speed is equal to $c = \sqrt{T}$.

The system of Eqs. (C.1), (C.2), and (C.3) is complemented by the initial and boundary conditions. The initial conditions are a jump at the point $x = 0$. In this case, to the left from the discontinuity, we have $\rho = \rho^{(1)} = 1$, $u = u^{(1)} = \mathrm{Ma}$, and $p = p^{(1)} = 1/\gamma$, whereas to the right from the discontinuity, the values are found from the Rankin–Hugoniot conditions:

$$\rho^{(2)} = \rho^{(1)}\frac{(\gamma + 1)\,\mathrm{Ma}^2}{2 + (\gamma - 1)\,\mathrm{Ma}^2}, \quad u^{(2)} = u^{(1)}\frac{2 + (\gamma - 1)\,\mathrm{Ma}^2}{(\gamma + 1)\,\mathrm{Ma}^2},$$

$$p^{(2)} = p^{(1)}\frac{2\gamma\,\mathrm{Ma}^2 - \gamma + 1}{\gamma + 1}.$$

The values on the boundaries of the computational domain are fixed and are determined by the relations written above.

For each value of the Mach number, we find the inverse thickness λ_∞/δ of the front, where δ is the width of the density front. The dimensionless value of δ is calculated according to the maximum value of the derivative $\partial\rho/\partial x$:

$$\frac{1}{\delta} = \max\left(\frac{\partial\rho}{\partial x}\right)\frac{1}{\rho^{(2)} - \rho^{(1)}}.$$

To solve the initial-boundary-value problem (C.1), (C.2), and (C.3) numerically, we define the computational domain as $-L \leq x \leq L$. Introduce the uniform grid x_i, $i = 0, \ldots, N - 1$. Let us use the explicit finite-difference scheme with approximation of all spatial derivatives by central differences and that of the derivatives in time by forward differences of the first order. Such a scheme is written in Sect. 5.7. The obtained algorithm allows us to find a solution for the Navier–Stokes equations (for $\tau = 0$), as well as for the QGD system. We choose the step h of the spatial mesh to be sufficiently small as compared with the shock wave thickness, which ensures the stability of the numerical algorithm without introduction of an artificial viscosity.

The stationary solution is found by using the ascertainment, or time-evolution method. The criterion of stopping the computations has the form

$$\max(\widehat{\rho} - \rho)/\Delta t < \varepsilon = 10^{-3}.$$

The dimensionless value of the reciprocal shock thickness is calculated as follows:

$$\frac{1}{\delta} = \max_i\left(\frac{\rho_{i+1} - \rho_{i-1}}{2h}\right)\frac{1}{\rho^{(2)} - \rho^{(1)}}.$$

In what follows, we present the results of computations for argon, which are obtained by the ascertainment method.

C.3 Results of Computations: Ascertainment Method

The parameter values for argon (one-atom gas) are as follows: $\gamma = 5/3$, $\omega = 0.81$, $Sc = 0.752$, and $Pr = 2/3$. The degree exponent in the viscosity law and the Schmidt number is taken from [28].

For computations, we use the QGD system (C.1), (C.2), and (C.3) and also the Navier–Stokes system, which is obtained from Eqs. (C.1), (C.2), and (C.3) for $\tau = 0$. The number of grid points is $N = 1200$, and the grid step is $h = 0.25$. The step in time is $\Delta t = \beta h/\max(\sqrt{T} + |u|)$, $\beta = 0.001$.

The values of the front inverse width obtained in the computations are presented in Fig. C.1. In the same figure, we depict the experimental data from [6]. The results obtained by the author [6] are labelled by the sign $\triangledown$.

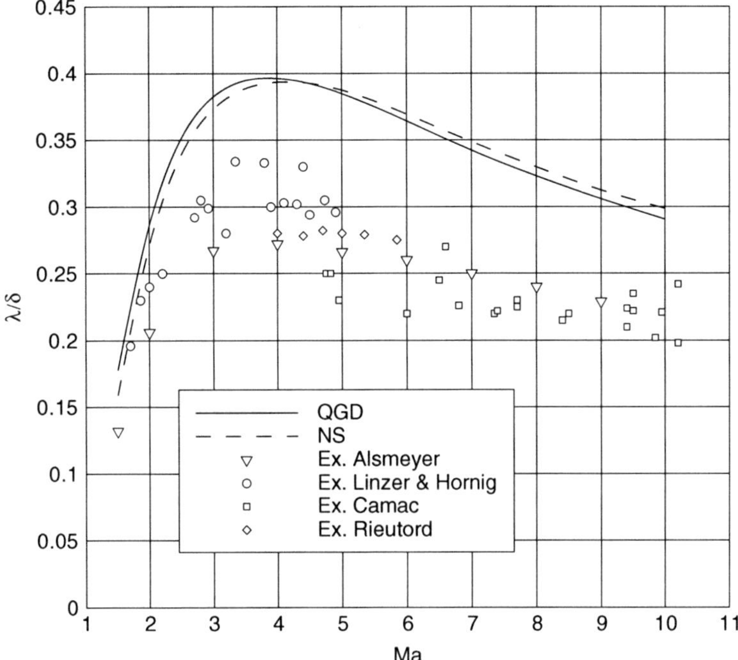

Fig. C.1 Reciprocal shock wave thickness vs the Mach number for argon

In Fig. C.1, we see that in the whole range of Mach numbers, the QGD and Navier–Stokes models yield very close results. Thus, the data obtained according to both models turn out to be essentially closer to the experimental data than was considered early.

A close accordance of the results is obtained in [84] in computing this problem for idealized monoatomic gases, the rigid sphere gases ($\omega = 0.5$), Maxwell molecules ($\omega = 1$) in comparison with the DSMC data, and also in [57] in comparing the QGD calculations with those according to kinetic theory for the monoatom gas with the degree exponent in the viscosity law being equal to $\omega = 0.72$.

These results change the ideas presented in a number of publications (see, e.g., [28, 34, 35, 122]) according to which for Mach numbers greater than 2, the computation results by the Navier–Stokes equations differ from the experimental data and the kinetic computations by 1.5–2.

In Fig. C.2, we depict the profiles of density, temperature, and velocity in the shock wave in argon for Ma $= 9$. Here and in what follows, the profiles are presented in the normalized form, i.e., $f_\rho = (\rho - \rho^{(1)})/(\rho^{(2)} - \rho^{(1)})$, where f_ρ is the density value in the figure, $\rho^{(1)}$, $\rho^{(2)}$ are the values on the boundary; the things are analogous for the temperature. For the velocity, we have $f_u = (u - u^{(2)})/(u^{(1)} - u^{(2)})$. The experimental points for the velocity profile are taken from [6]. The experimental data for the density profile are close to the results of computations by the QGD model and are satisfactory for the computations by the Navier–Stokes model.

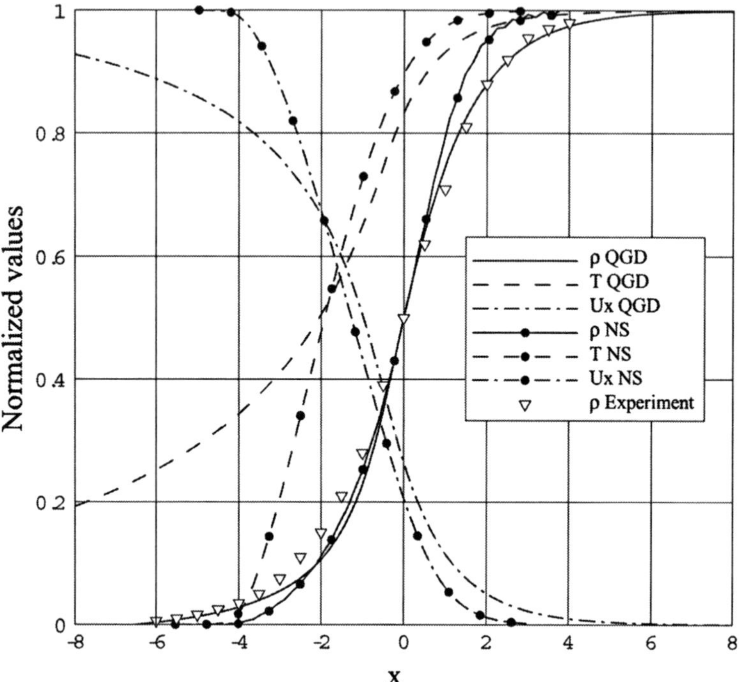

Fig. C.2 Profiles of density, temperature, and velocity of the shock wave in argon for Ma $= 9$

The temperature and velocity profiles for the QGD model and the Navier–Stokes equations are considerably different in the domain before the shock wave ($x < -2$): in the QGD model, these profiles are more smoothed. Precisely this distinction is observed in the comparison of computations in the BGK approximations with those by the Navier–Stokes model [122]. The comparisons with the experiments for these profiles are of great difficulty and are not presented here.

The number of steps in time to attain the steady-state solution for the QGD model is 3×10^5–3×10^6 for Mach numbers 1.5–10, respectively. For the Navier–Stokes system, the number of iteration steps vary in the limits 1×10^6–2×10^7. That is, the computations for the Navier–Stokes equations require the number of steps in time be 3–10 times more for the ascertainment. In Fig. C.3, we depict the density profiles in the shock wave in argon in the enlarged scale. It is seen that the algorithm based on the Navier–Stokes equations detects a computational instability in the zone after the shock wave, oscillations of the numerical solutions with the period equal to the spatial mesh step. At the same time, the QGD algorithm yields a smooth curve. This effect explains a slow convergence of the numerical solution for the Navier–Stokes equations as compared with the QGD algorithm.

The ascertainment method is studied with respect to the convergence in the grid. For this purpose, on the basis of the QGD system (C.1), (C.2), and (C.3), we carry out the computation for argon with Ma $= 10$, $\beta = 0.0001$, $N = 2400$, and the

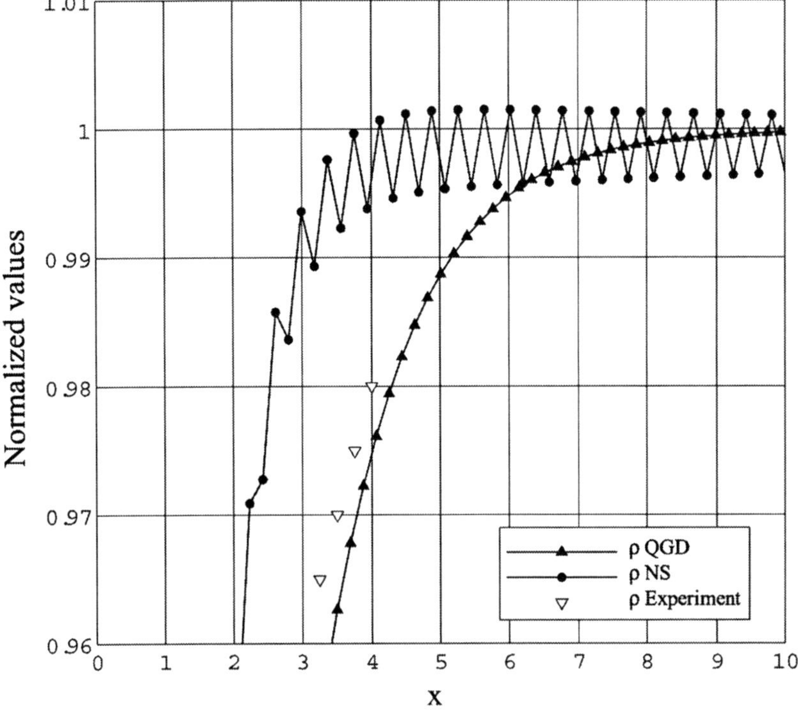

Fig. C.3 Density profiles in the shock wave in argon for Ma $= 9$

grid step is 2 times diminished, $h = 0.125$. In this case, the quantities λ/δ differ in the third decimal place and the number of steps in time is $N_t = 6.6415 \times 10^7$. Therefore, we can assert that in the presented computations, the convergence in the grid is attained.

The solution of the problem on the stationary shock wave structure in the framework of the Navier–Stokes equations can be obtained by a more effective method reducing the system of partial differential equations to a system of two ordinary differential equations. The corresponding algorithm is presented in the next section.

C.4 Solution of the Steady-State Navier–Stokes Equations

In the case where $\tau = 0$, which corresponds to the Navier–Stokes system, for the stationary flow, we succeed in the simplification of the system (C.1), (C.2), and (C.3) reducing it to the Cauchy problem for a system of two ordinary differential equations. The Navier–Stokes system for the problem considered has the form

$$\frac{d(\rho u)}{dx} = 0, \tag{C.6}$$

$$\frac{d(\rho u^2)}{dx} + \frac{dp}{dx} = \frac{d\Pi}{dx}, \tag{C.7}$$

$$\frac{d(\rho u H)}{dx} + \frac{dq}{dx} = \frac{d(\Pi u)}{dx}. \tag{C.8}$$

Here, $\Pi = \Pi_{NS} = 4\mu/3(du/dx)$. Integrating the system (C.6), (C.7), and (C.8) one time, we arrive at the system

$$\rho u = c_0,$$
$$\rho u^2 + p = \Pi + c_1,$$
$$\rho u H + q = \Pi u + c_2,$$

where the constants c_0, c_1, and c_2 are found from the boundary conditions. Substituting the expressions for the quantities entering here, we can reduce the obtained system to the following system of two equations for p and ρ:

$$\frac{d\rho}{dx} = \frac{3\rho^2}{4\mu c_0} \left(c_1 - p - \frac{c_0^2}{\rho} \right), \tag{C.9}$$

$$\frac{dp}{dx} = \frac{d\rho}{dx} \cdot \frac{p}{\rho} - \frac{\mathcal{R}\rho}{\varkappa} \left(c_2 - \frac{4\mu c_0^2}{3\rho^3}\frac{d\rho}{dx} - \frac{c_0^3}{2\rho^2} - \frac{c_0 p}{\rho} \cdot \frac{\gamma}{\gamma - 1} \right). \tag{C.10}$$

Let us reduce the equations to the dimension-free form, as is done in the previous section. In this case, Eqs. (C.9) and (C.10) take the form

$$\frac{d\rho}{dx} = \frac{3\rho^2}{4\mu c_0} \left(c_1 - p - \frac{c_0^2}{\rho} \right), \tag{C.11}$$

$$\frac{dp}{dx} = \frac{d\rho}{dx} \cdot \frac{p}{\rho} - \frac{\rho}{\varkappa\gamma} \left(c_2 - \frac{4\mu c_0^2}{3\rho^3}\frac{d\rho}{dx} - \frac{c_0^3}{2\rho^2} - \frac{c_0 p}{\rho} \cdot \frac{\gamma}{\gamma - 1} \right). \tag{C.12}$$

The dimensionless coefficients μ and $\varkappa$ are calculated as follows:

$$\mu = \gamma^{\omega - 0.5} \frac{5\sqrt{2\pi}}{16} \left(\frac{p}{\rho} \right)^{\omega}, \qquad \varkappa = \frac{\mu}{\mathrm{Pr}(\gamma - 1)}.$$

The problem consists in searching for a solution of Eqs. (C.11) and (C.12) satisfying the asymptotic conditions (Rankin–Hugoniot conditions) at infinity of the form

$$\rho(-\infty) = \rho_L = 1, \quad p(-\infty) = p_L = \frac{1}{\gamma}, \tag{C.13}$$

$$\rho(+\infty) = \rho_R = \frac{(\gamma + 1)\,\mathrm{Ma}^2}{2 + (\gamma - 1)Ma^2}, \quad p(+\infty) = p_R = \frac{2\gamma\,\mathrm{Ma}^2 - (\gamma - 1)}{\gamma(\gamma + 1)}. \tag{C.14}$$

The integration constants entering the system of equations are expressed from the boundary conditions (C.13) and (C.14) and from the relations $d\rho/dx = dp/dx = 0$

$$c_0 = \mathrm{Ma}, \quad c_1 = \mathrm{Ma}^2 + \frac{1}{\gamma}, \quad c_2 = \frac{\mathrm{Ma}^3}{2} + \frac{\mathrm{Ma}}{\gamma - 1}.$$

The numerical solution of system (C.11) and (C.12) with the boundary conditions (C.13) and (C.14) is complicated by the following two circumstances:

1. the equations imply that if functions $\rho(x)$ and $p(x)$ are solutions of the problem, then the functions $\tilde{\rho}(x) = \rho(x + c)$ and $\tilde{p}(x) = p(x + c)$, where c is an arbitrary constant, also satisfy the problem, i.e., the desired pair of functions is not unique;
2. the boundary conditions are of asymptotic character.

The posed problem can be solved by using the following shooting-type method. At the central point $x_C = 0$, we impose the additional condition $p(x_C) = (p_R + p_L)/2$. This ensures the uniqueness of the solution, and there arises one more nonasymptotic condition. Assume that the value $\rho(x_C) = \rho_C$ is the parameter of the problem. For each fixed value of this parameter, we obtain the Cauchy problem for the system of two equations with two initial conditions at the point x_C. We solve the obtained Cauchy problem by using the Runge–Kutta method and find the unknown value of the parameter by selection starting from the necessity to satisfy the asymptotic conditions (C.13) and (C.14).

For the solution of the problem, we use the following algorithm.

1. Choose a sufficiently large closed interval $[x_L, x_R]$ on which the solution is constructed.
2. Choose the initial value ρ_C, the initial step δ, and the admissible error Δ.
3. Solve the problem for the current ρ_C.
4. If the solution with $\rho(x_R) < \rho_R - \Delta$ is obtained, then enlarge ρ_C by δ and pass to step 3.
5. If the solution with $\rho(x_R) > \rho_R + \Delta$ is obtained, then diminish ρ_C by $\delta/2$, diminish δ by 2 times, and pass to step 3.

At step 3, the solution is constructed by the Rosenbrock formulas of the second order in the Matlab environment (see [177]). The error Δ is chosen to be equal to 0.001, the initial step is $\delta = (\rho_R - \rho_L)/3$, and the initial value of the parameter is $\rho_C = \rho_L$.

As an example, in Fig. C.4, we present the sequential iterations of the solution $\rho(x)$ for argon ($\gamma = 5/3$, $\mathrm{Pr} = 2/3$, $\omega = 0.81$) when $\mathrm{Ma} = 4$ for different values of the parameter ρ_C. All the obtained $\rho(x)$ satisfy the asymptotic condition on the left boundary, and only the desired solution satisfies the condition on the right boundary.

Also, it is seen from the graph that the solutions obtained are very sensitive to a variation of the parameter ρ_C, and their deviations from the desired solution are attained at a right boundary point. The exact value of the parameter can be found by the dichotomy method. In this case, the fulfillment of the right boundary conditions with the relative accuracy 10^{-2} is attained for the parameter error $\rho_C < 10^{-7}$.

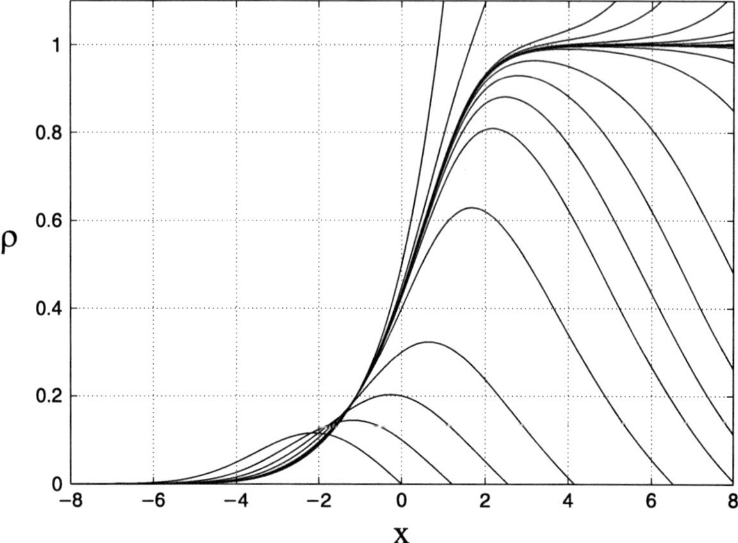

Fig. C.4 Family of normalized $\rho(x)$ obtained as a result of numerical solution of the problem for various values of ρ_C; the *bold line* is the desired solution of the shock wave type

Therefore, we obtain the dependence of $\rho(x)$ that differs from the desired dependence at the point x_C lesser than by 10^{-7}. It is seen from Eqs. (C.11) and (C.12) that the error of computation of the derivatives $d\rho/dx$ and dp/dx at the center of integration closed interval is also very small.

With a higher accuracy, the results obtained for argon correspond to the data obtained for the Navier–Stokes equations by using the time-evolution method in the previous section. This additionally justifies the accuracy of the obtained solutions.

For the QGD equations, the posed problem can be reduced to a system of three ordinary differential equations whose solution by the shooting method described above is of essential difficulty.

C.5 Results of Computations for Nitrogen

Nitrogen is a two-atom gas that has two rotational degrees of freedom. According to [28], the values of the molecular parameters are $\gamma = 7/5$, $\omega = 0.74$, $\mathrm{Pr} = 14/19$, and $\mathrm{Sc} = 0.746$.

To improve the description of gas flows with inner degrees of freedom, in calculating the viscous stress tensor, we need to take into account the second or volume viscosity. In this case, the viscous stress tensor is represented in the form

$$\Pi = \frac{4}{3}\mu\frac{\partial u}{\partial x} + \zeta\frac{\partial u}{\partial x}. \tag{C.15}$$

As was shown in Chap. 3, the second viscosity coefficient can be written in the form

$$\zeta = \mu \left(\frac{5}{3} - \gamma \right) B. \tag{C.16}$$

According to the results of Sect. 3.4, the dimensionless coefficient B is given by

$$B = \frac{3}{2}(\gamma - 1)A\sqrt{\frac{\pi}{8}} \cdot Z_{\text{rot}}, \tag{C.17}$$

$$A = \frac{2(7 - 2\omega)(5 - 2\omega)}{15\sqrt{2\pi}}, \tag{C.18}$$

$$Z_{\text{rot}} = \frac{Z_\infty}{1 + (\pi^{3/2}/2)(T^*/T)^{1/2} + (\pi + \pi^2/4)(T^*/T)}, \tag{C.19}$$

where $Z_\infty = 23$ and $T^* = 91.5\,\text{K}$ for nitrogen (see [28]). In the dimensionless form, under the gas temperature in the unperturbed flow equal to 273 K, we have $T^* = 91.5/273$. The quantity $1/Z_{\text{rot}}$ is the relative frequency of nonelastic collisions.

In what follows, we present the results of computation of the shock wave structure in nitrogen obtained on the basis of the Navier–Stokes system and the QGD equations with account for the volume viscosity ζ of the form (C.16).

The Navier–Stokes system of equations is solved by reducing it to the Cauchy problem for a system of two differential equations of the first order according to the method described in the previous section. In this case, Eqs. (C.9) and (C.10) are written with account for the volume viscosity coefficient in the form

$$\frac{d\rho}{dx} = \frac{\rho^2}{\left(\frac{4}{3}\mu + \zeta\right)c_0}\left(c_1 - p - \frac{c_0^2}{\rho}\right),$$

$$\frac{dp}{dx} = \frac{d\rho}{dx}\cdot\frac{p}{\rho} - \frac{R\rho}{\varkappa}\left(c_2 - \left(\frac{4}{3}\mu + \zeta\right)\frac{c_0^2}{\rho^3}\frac{d\rho}{dx} - \frac{c_0^3}{2\rho^2} - \frac{c_0 p}{\rho}\cdot\frac{\gamma}{\gamma - 1}\right).$$

The other aspects of the procedure do not change.

The QGD system is solved by the time-evolution method in which we use the mesh with number of nodes $N = 1200$, the grid step is $h = 0.25$, and the step in time is $\Delta t = \beta h / \max(\sqrt{T} + |u|)$, where $\beta = 0.001$.

In Fig. C.5, we present the data for the reciprocal thickness of the shock wave in nitrogen in comparison with the experimental data [6] denoted by markers. The experimental data are normalized by the free path of the upstream flow for argon computed according to Eq. (C.5) and are equal to $\lambda_{\text{Ar}} = 1.098\,\text{mm}$ for $p_\infty = 50\,\text{mTorr}$, $T_\infty = 300\,\text{K}$. Under these conditions, $\lambda_{\text{Ar}}/\lambda_{N_2} = \mu_{\text{Ar}}/\mu_{N_2}\sqrt{m_{\text{Ar}}/m_{N_2}} = 1.060$.

We present three variants of computation: without account for the second viscosity $B = 0$ (line 1), with account for the second viscosity in the simplified form $B = 1$ (2), and for B in the form (C.17) (line 3). As follows from the dependence of

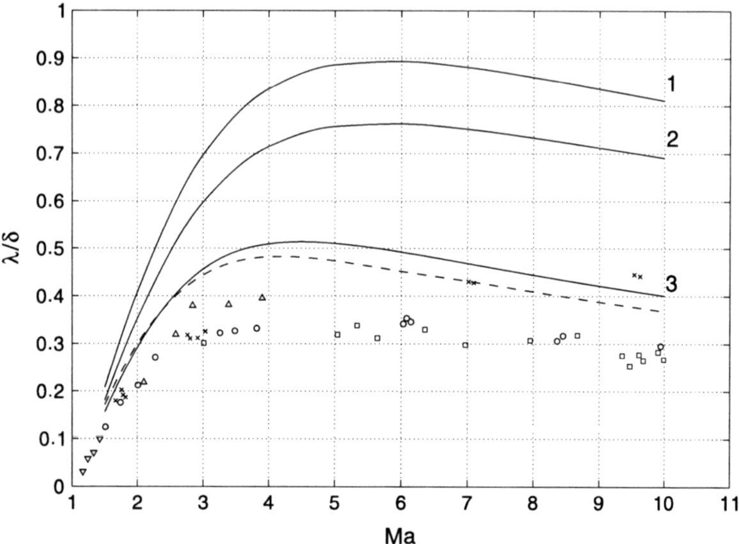

Fig. C.5 Reciprocal thickness of the shock wave λ_∞/δ in comparison with the experimental data for nitrogen (*markers*) vs Mach number. *Continuous lines* are the computations by the Navier–Stokes equations for $B = 0$ (1), $B = 1$ (2), and $B(Z_{\mathrm{rot}}, T)$ (3). The *doted line* is the computation by the QGD model for $B(Z_{\mathrm{rot}}, T)$

$B(T)$ (Sect. 3.4), under the change of the Mach number, the coefficient B varies in the limits 0.5–3. The thickness of the density profile without account for the volume viscosity ($B = 0$) for Ma > 3 turns out to be less than that in the experiment by almost 3 times. The account for the influence of the inner degrees of freedom by introducing the volume viscosity coefficient improves the accuracy of description of the density profile, and for the choice of B in the form (C.17), it essentially approximates the numerical solution to the experimental data with accuracy of order 30% (see Fig. C.5).

To study the temperature nonequilibrium in this problem, we can use the QGDR model for a gas with two rotational degrees of freedom. The corresponding system of equations and some computational results are presented in Chap. 8.

In Fig. C.6, we show the normalized density profiles in the shock wave for Ma $= 6.1$ in comparison with the experimental data. We see a good agreement of the measured and computed values with account for the volume viscosity coefficient.

Therefore, for the shock wave profile in nitrogen, the Navier–Stokes model and the QGD equations yield close results. When introducing the volume viscosity coefficient onto the model, the thickness of the density profile in the computation differs moderately from the experimental data even for Ma > 2. In conclusion, in Table C.1, we present the results of computing the inverse thickness of the shock wave for argon and nitrogen.

The problem presented above makes it clear that there are advantages and disadvantages for both used mathematical models. So, it is appropriate to perform the

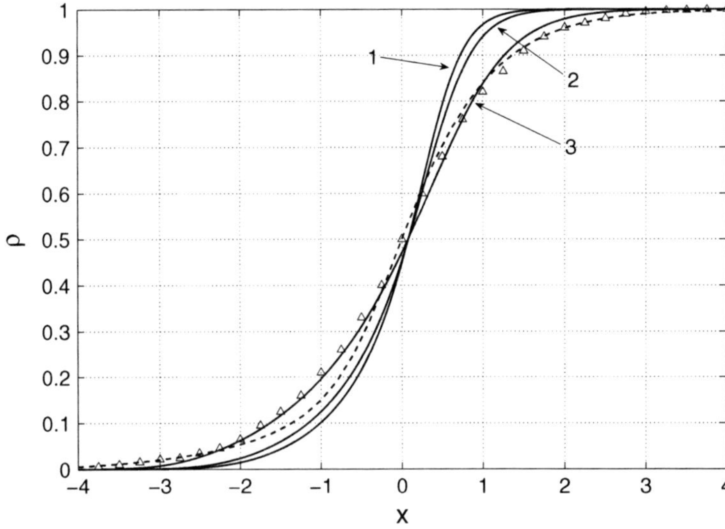

Fig. C.6 Density profiles in nitrogen for Ma $= 6.1$ in comparison with the experimental data (*markers*). The *continuous lines* are the computations by the Navier–Stokes equations for $B = 0$ (1), $B = 1$ (2), and $B(Z_{\mathrm{rot}}, T)$ (3). The *dotted line* is the computation by the QGD model for $B(Z_{\mathrm{rot}}, T)$

Table C.1 Results of numerical computation of the inverse thickness of the shock wave; the Navier–Stokes model

		Nitrogen		
Ma	Argon	$B = 0$	$B = 1$	$B(Z_{rot}, T)$
1.5	0.15877	0.20759	0.18088	0.15591
2	0.27232	0.40876	0.35330	0.29220
3	0.37882	0.69726	0.59801	0.45643
4	0.40028	0.83537	0.71425	0.50934
5	0.39137	0.88571	0.75611	0.51047
6	0.37313	0.89316	0.76189	0.49234
7	0.35307	0.88095	0.75098	0.46854
8	0.33375	0.86001	0.73292	0.44434
9	0.31600	0.83572	0.71209	0.42165
10	0.29996	0.81073	0.69069	0.40100

computation by the Navier–Stokes model for a one-dimensional stationary flow by using the shooting method, which rapidly converges and allows us to find the solution with a higher accuracy. The time-evolution method in this case converges very slowly. For the QGD equations, on the contrary, because of a more complicated and nonlinear form of equations, the shooting method is of great difficulty, whereas the time-evolution method converges sufficiently rapidly.

Appendix D
Backward-Facing Step Flow in a Channel: Laminar–Turbulent Transition

In this appendix, we present the results of numerical simulation of a subsonic compressible gas flow in a channel with a sudden expansion. The calculations are carried out in the range of Reynolds numbers, including laminar and turbulent flow regimes. We obtain that with the increasing of Reynolds numbers, there is a bifurcation of the solution, and the stationary laminar flow regime is changed by a nonstationary turbulent regime. We show that the QGD model allows us to describe both these flow regimes in the framework of a uniform algorithm.

The presented results are based on [75, 76, 82, 83].

D.1 Introduction

The scheme of calculation domain is presented in Fig. D.1. Here, h and l are the height and the length of the step, and H and L are the height and the length of the channel, respectively. The flow is characterized by the Reynolds and Mach numbers

$$\mathrm{Re} = \frac{\rho_0 U_0 h}{\mu_0}, \quad \mathrm{Ma} = \frac{U_0}{c_0}, \tag{D.1}$$

where U_0 is the gas mean velocity in the entrance section of the channel, ρ_0, μ_0, and c_0 are the density, the viscosity coefficient, and the sound speed in the same section, respectively. We consider subsonic flows corresponding to $0.01 < \mathrm{Ma} < 1$. The length L_s of the reverse flow domain, or the separation zone, which is formed in such a flow after the step, is a representative and sensitive characteristic of the flow regime.

According to [14], for the Reynolds numbers satisfying $\mathrm{Re} < 600$, the flow is laminar[1]. In this case, L_s increases almost linearly as Re grows.

For Reynolds numbers satisfying $\mathrm{Re} > 600$, we see the transition from the laminar flow to a turbulent flow. In this case, the length of the separation zone diminishes when Re grows. In the regime of the developed turbulent flow, the length of the

---

[1] The Reynolds number in [14] is calculated according to Eq. (D.1) for $D = 2h$.

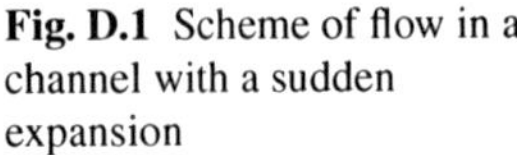

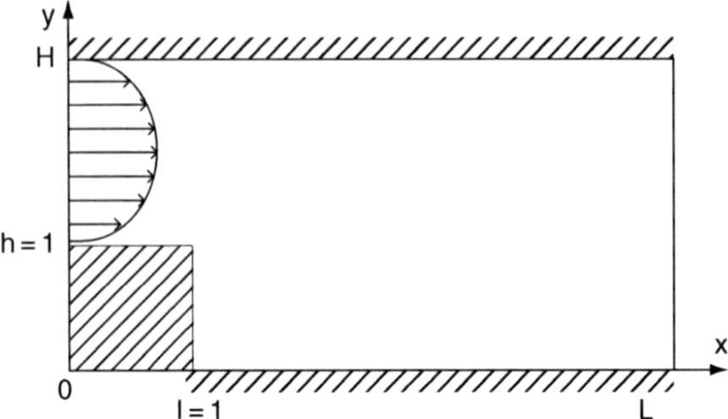

Fig. D.1 Scheme of flow in a channel with a sudden expansion

separation zone referred to the step height, L_s/h, is practically independent of the Reynolds number and, according to different estimates, is varied from 5 up to 8 depending on the problem geometry. An especial difficulty for numerical modelling are the flows with the Reynolds number corresponding to the process of transition from the laminar to the turbulent flow.

The numerical models for computing turbulent flows, except for the method of direct numerical modelling (DNS), are based on the introduction of an additional dissipation to the gas-dynamic equations. In this case, the molecular transition coefficients in the Navier–Stokes equations are replaced by some effective values, where the effective viscosity and heat conduction are calculated as the sum of the molecular viscosity, the heat conductivity, and some additions. Such a method for introducing the turbulent dissipation does not change the general structure of hydrodynamic equations. In the framework of Reynolds-averaged Navier–Stokes equations (RANS model), these additions are calculated on the basis of algebraic or differential models. The computations of two-dimensional flows after the step on the basis of these approaches are contained in [139,150,200,206] and the bibliographies therein. In the framework of large eddy simulation models (LES), the addition to the coefficients is found by using the subgrid viscosity models and turns out to be related to the spatial grid step. Here the step of the spatial grid plays the role of the spatial filter. There exist models in which the role of the turbulent dissipation on scales less than the spatial grid step is played by artificial regularizes [23]. The dissipative mechanism inherent to the numerical regularizers plays the role of a filter smoothing the subgrid pulses of the gas-dynamic quantities.

All the listed models include constants not known in advance, and, as a rule, the equations used in these models require the statement of special conditions on the boundary, which are also not sufficiently theoretically elaborated.

In this section, for the numerical modelling of the laminar–turbulent transition, we use the QGD equations without using the traditional turbulence models. In this case, the role of the turbulent dissipation is played by nonlinear dissipative summands proportional to the coefficient τ, which enter all the equations of the system and are participated in calculations of laminar, as well as turbulent, regimes. This computational model can be referred to the class of filter-type models in which the size of filter cells is determined by the step of the finite-difference grid.

For slowly varying laminar flows, τ-terms play the role of effective regularizers of the numerical algorithm. Their value is proportional to $O(\tau^2)$, and they have a small influence on the flow structure. For rapidly varying turbulent flows, the value of the QGD τ-additions is of order $O(\tau)$, and they can essentially influence the flow character. The coefficient τ is related to the characteristic time for which the perturbation crosses the computational cell. Such a method for numerical modelling allows us to uniformly describe the laminar flow regime and the passage from the laminar flow to the turbulent flow under increasing Reynolds number.

The results of the QGD computations are compared with the data of experiment of [14], which are classical and, up to the present, are used for testing numerical algorithms (see, e.g., [140, 206] and the bibliographies therein). In the mentioned experiment, the conditions under which the flow after the step can be considered as two-dimensional with a sufficient accuracy are reproduced.

D.2 Statement of the Problem

In accordance with the experiment of [14], we assume that in the input of the channel, the flow velocity is the Poiseuille parabola:

$$u_x(0, y) = -\frac{\mathrm{Re}}{2}\frac{\partial p}{\partial x}(H - y)(h - y), \quad u_y(0, y) = 0. \tag{D.2}$$

From the condition that the mean dimension-free velocity in the input of the channel is equal to 1,

$$U_0 = U_x(0) = \int_h^H u_x(y)dy = 1,$$

we calculate the gradient of the pressure in the input of the channel. In accordance with [14], we have $H/h = 2$ and

$$u_x(y) = -6(2 - y)(1 - y), \quad \frac{\partial p}{\partial x} = -\frac{12}{\mathrm{Re}}. \tag{D.3}$$

We complement the boundary conditions (D.3) on the input by the additional condition $\rho(y) = 1$ for the density and the condition $u_y(y) = 0$ for the vertical component of the velocity.

On the output boundary, for the subsonic flow, we use the "soft" boundary conditions for the density and the velocity components and assume that the pressure is constant (see Sect. (5.10.2)):

$$\frac{\partial \rho}{\partial x} = 0, \quad \frac{\partial u_x}{\partial x} = 0, \quad \frac{\partial u_y}{\partial x} = 0, \quad p = \frac{1}{\gamma\,\mathrm{Ma}^2}. \tag{D.4}$$

On the rigid walls, we pose the "adhesion" and "non-leakage" conditions for the velocity, together with the adiabatic condition for the temperature. On the upper and lower walls, these conditions have the form

$$u_x = u_y = 0, \quad \frac{\partial p}{\partial y} = \frac{\partial T}{\partial y} = 0.$$

As the initial conditions, we choose the conditions of the input unperturbed flow in the zone over the step and the state of unmovable gas in the zone after the step.

The numerical algorithm being a finite-difference scheme explicit in time with approximation of the second-order accuracy for all spatial derivatives and also the method for reducing the equations to dimension-free form and introducing an artificial dissipation are described in Chap. 5 in detail. In computations, we use the spatial grid uniform in both directions with grid step $h_x = h_y = h_{xy}$. The smoothing parameter τ is found in the form

$$\tau = \frac{\mu}{p\,\mathrm{Sc}} + \alpha\frac{h_{xy}}{c}, \tag{D.5}$$

where $\mu = \mu_0(T/T_0)^\omega$ is the dynamical viscosity coefficient, Sc is the Schmidt number, $c = \sqrt{\gamma RT}$ is the sound speed, and $0 < \alpha < 1$ is a numerical coefficient. The quantity h_{xy}/c corresponds to time for which the perturbation crosses the grid cell. The step in time is chosen from the stability condition $\Delta t = \beta h_{xy}/c$, where the coefficient $0 < \beta < 1$ is selected in the computation process from the stability condition.

The computations are carried out for subsonic air flows under the normal pressure. The molecular parameters are $\gamma = 1.4$, $\mathrm{Pr} = 0.737$, $\omega = 0.74$, and $\mathrm{Sc} = 0.746$.

D.3 Numerical Computations and Discussion of the Results

D.3.1 Laminar Flows

This series of computation is carried out for the Reynolds numbers Re $= 100$, 200, 300, and 400 and the Mach numbers from Ma $= 0.5$ up to 0.01 on uniform spatial grids with steps $h_{xy} = 0.1$ and 0.05. The steps Δt in time in the dimensional-free units are varied in the interval 10^{-3}–10^{-4}. The stationary solution is found by the time-evolution method. The computation is terminated when the accuracy $\varepsilon \sim 0.01$–0.001 is attained.

For the considered flows, the gradients of the density are proportional to $1/\mathrm{Ma}^2$, which allow us to compare the solution with the computations carried out in the approximation of viscous incompressible flow (Chap. 7) and the data of experiments [14] carried out for the Mach numbers Ma ~ 0.01–0.001. As numerical computations show, the length of the separation zone is practically independent of

Table D.1 Reduced length L_s/h of the separation zone for different numbers Re

Re	Experiment [14]	Fluid (Chap. 7) $h_{xy} = 0.05$	Gas, Ma $= 0.1$ $h_{xy} = 0.1$	$h_{xy} = 0.05$
100	5	5	5.1	5.00
200	8.3	8.2	8.5	8.35
300	11.3	10.1	10.1	10.3
400	14.2	14.8	11.3	12.70

the Mach number for $0.01 <$ Ma < 0.5. For a further decrease (Ma < 0.01), the rate of convergence of solution is considerably decelerated, since the dimension-free ascertainment time of solution increases as $\sim 1/$ Ma.

In Table D.1, we present the main characteristics of computations of laminar flows for Ma $= 0.1$, $\alpha = 0.5$. The corresponding pictures of the steady-state solutions are presented in Fig. D.2.

For all variants of computations, the length L_s of the separation zone corresponds to the standard results for Ma $= 100, 200$, and 300. When the spatial grid step diminishes, the length of the separation zone for Re $= 400$ becomes close to the experimental value (see Table D.1).

In Fig. D.3, we depict the process of solution ascertainment for the variant Re $= 100$, $\alpha = 0.5$, and Ma $= 0.1$. On the background of distribution of the vertical velocity component u_y, we present the streamlines for the instants of time $t = 0.5$, $1, 5, 6$, and 81. In the process of ascertainment, the perturbations arising in the flow field freely cross the input and output boundaries of the computational domain.

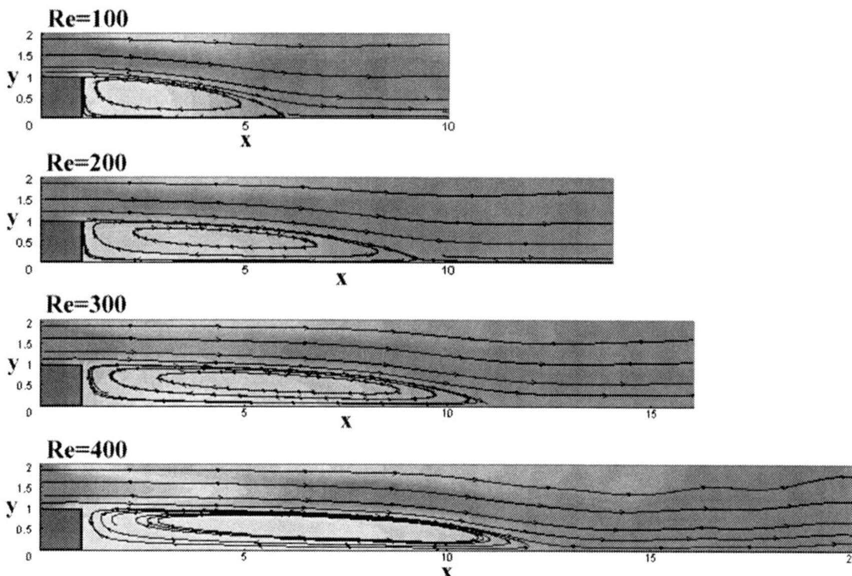

Fig. D.2 Distributions of density and the streamlines for Re $= 100, 200, 300$, and 400; Ma $= 0.1$

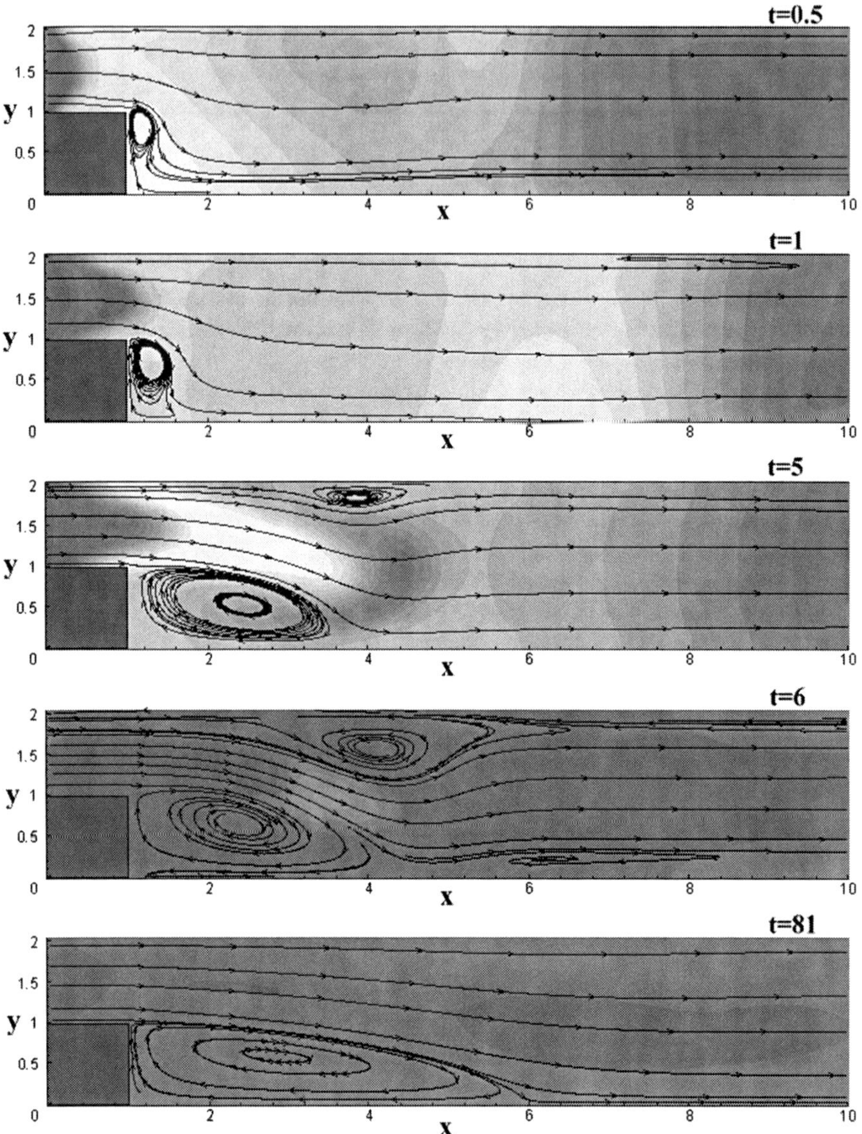

Fig. D.3 Time evolution of the subsonic flow in a channel with a sudden expansion, Re = 100, Ma = 0.1

D.3.2 Laminar–Turbulent Transition

For laminar flows (Re < 600), the backward-facing step flow is stationary, and the length of the separation zone is in a good agreement with the experimental data and the known computations. For Re > 600, oscillations arise in the flow field, and it

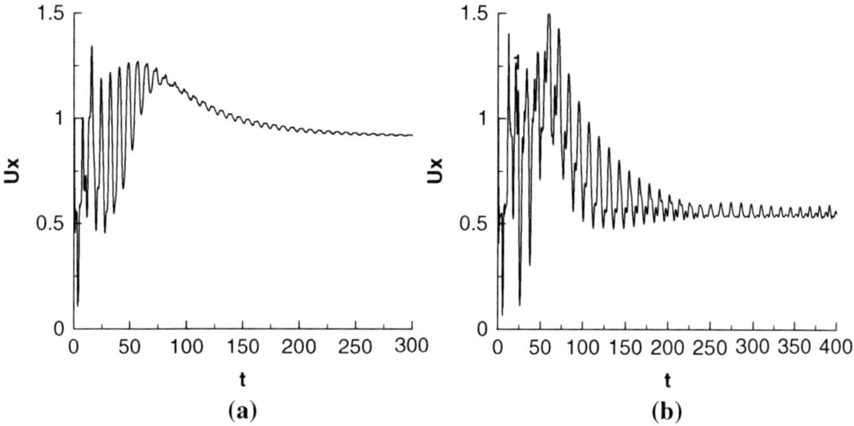

Fig. D.4 Time–velocity (u_x) evolution for Re $=$ 300 (**a**) and Re $=$ 600 (**b**)

becomes nonstationary and turbulent. The length of the separation zone in this case is found by the velocity field averaged in time.

In Fig. D.4, we present the evolutions in time of the longitudinal velocity for Re $=$ 300 (to the left) and Re $=$ 600 (to the right) at the point with coordinates $(8, 1)$. For the laminar flow for Re $=$ 300, the ascertainment process goes to the stationary regime. The value Re $=$ 600 corresponds to the transition regime for which the birth of velocity oscillations is seen.

In Fig. D.5, we present a fragment of time evolution of the longitudinal velocity for the turbulent flow regime for Re $=$ 1000 (to the left) and the energy spectra $E(k)$ in the logarithmic scale (to the right). Here, we also present the dependence $E(k) \sim k^{-5/3}$ (the Kolmogorov–Obukhov law in the spectral form [23]). This law of kinetic energy dissipation is characteristic for the developed turbulent flow.

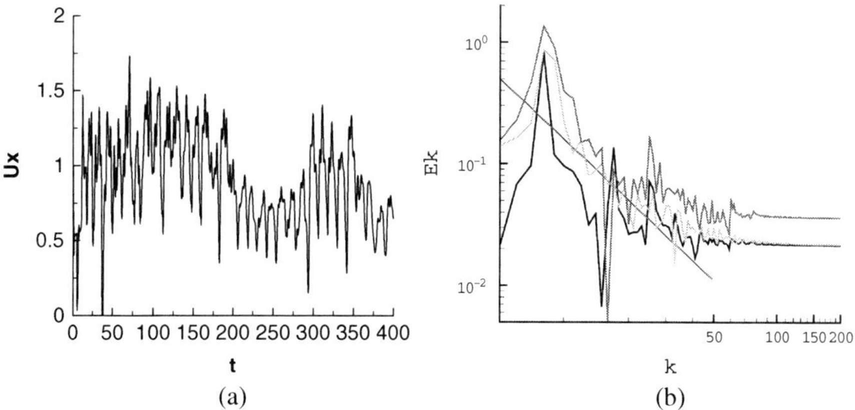

Fig. D.5 Fragments of velocity–time evolution Re $=$ 1000 (**a**) and the energy spectrum of the x-component of the flow velocity in the logarithmic scale (**b**) for Re $=$ 1000, 3500, and 4667

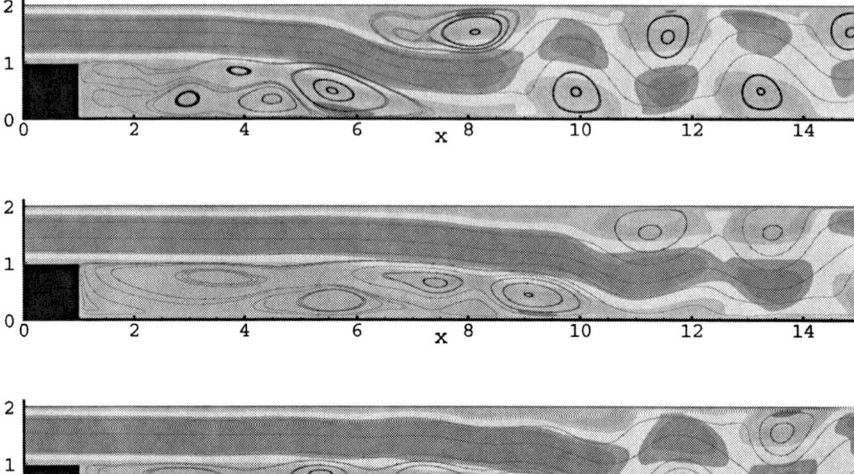

Fig. D.6 Momentary streamlines, Re $= 1000$, $t = 80$, $t = 360$, $t = 400$, $\alpha = 0.3$. By saturation of gray, we label the value of the component u_x of the velocity

The developed flow regime here corresponds to Re > 6000. The energy spectrum obtained in the numerical experiment in the frequency range corresponding to the resolution of the spatial grid corresponds as a whole to the classical decay law of the kinetic energy E_k as the wave number k grows.

The turbulent flows obtained in computations turn out to be essentially nonstationary. To compare the length of the separation zone with experimental data, we perform the averaging of the velocity field on a certain interval of time. The averaged flow weakly depends on the length of the averaging interval if this interval is chosen sufficiently distant from the point $t = 0$ and its value is essentially greater than the characteristic period Δt^* of oscillations. In this case, $\Delta t^* \sim 5$, and the averaging period is $t_1 = 300 < t < t_2 = 400$. In Figs. D.6 and D.7, we present the momentary and averaged pictures of flow for Re $= 1000$. Both figures qualitatively correspond to the results of [206] obtained in the two-dimensional computation for Re $= 3700$ by the method of direct numerical modelling of a viscous incompressible fluid flow on the basis of the Navier–Stokes equations. Here the finite-difference approximations of the third and fourth order of accuracy in time and space were used.

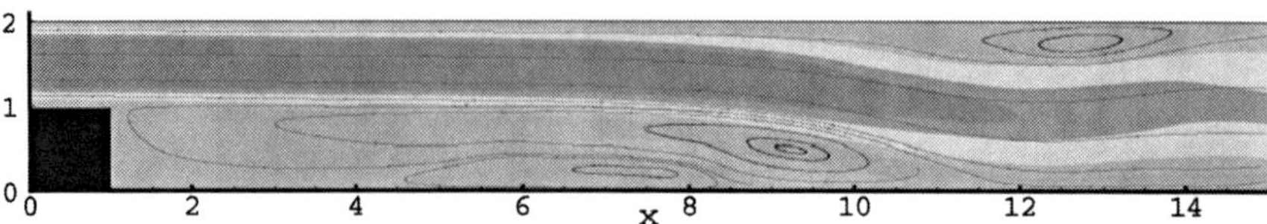

Fig. D.7 Streamlines for the averaged flow, Re $= 1000$; a fragment

Table D.2 Parameters and results of computations

Re	h_{xy}	α	β	Δt	L	L_s [14]	L_s	T_f
100	0.05	0.5	0.2	0.0010	10	5.0	5.0	81
200	0.05	0.5	0.2	0.0010	14	8.5	8.4	204
300	0.05	0.5	0.3	0.0015	20	11.3	10.4	300
400	0.05	0.5	0.2	0.0010	20	14.2	12.7	626
600	0.05	0.3	0.5	0.0025	30	17.0	16.0	400
		0.5	0.3	0.0015			15.5	
700	0.05	0.3	0.5	0.0025	30	15.5	15.7	400
		0.5	0.3	0.0015			17.0	
1000	0.03	0.3	0.5	0.0015	30	13.8	10.6	400
		0.5	0.3	0.0009			11.4	
2000	0.05	0.3	0.5	0.0025	30	9.0	9.1	400
		0.5	0.3	0.0015			10.9	
2800	0.05	0.3	0.5	0.0025	30	6.1	7.2	400
		0.5	0.3	0.0015			8.8	
3500	0.05	0.3	0.5	0.0025	30	8.0	7.8	400
		0.5	0.3	0.0015			9.0	

In Table D.2, we collect the main computational data carried out for $Ma = 0.1$ in the range of the Reynolds numbers including the laminar regime and the transition from the laminar regime to the developed turbulent flow. Here, T_f is the dimension-free time of stopping the computation. The length L_s of the separation zone weakly depends on the spatial grid step for $h_{xy} < 0.1$. For the transition domain Re ~ 600, such a dependence is observed.

In Fig. D.8, the computed flow pictures are compared with the experimental data of [14] in the length L_s of the separation zone for two values of the parameter: $\alpha = 0.3$ and 0.5.

We see a reasonable agreement between the computation and the experimental data in the whole range of Reynolds numbers: the enlargement of the separation zone size as Re grows for laminar regimes, the diminishing of L_s for a further growth of Re, and the formation of the maximum of $L_s(Re)$, which determines the transition domain for these two regimes. The location of the curve maximum is in concordance with the experiment of [14] with good accuracy. The non-monotone character of diminishing of L_s in the experiment for $700 < Re < 3500$ can be related with the appearance of three-dimensional vortex structures in the flow whose appearance in the two-dimensional numerical computation carried out is impossible.

In contrast to most traditional turbulence models, in the QGD approaches, when computing the smoothing parameter τ (see Eq. (D.5)), there exists only one free coefficient, the numerical coefficient $0 < \alpha < 1$. In some sense, this coefficient can be considered as an analog of the numerical coefficient in Smagorinsky's LES model. The study of the character of flow and its stability depending on the value of the coefficient α is in particular carried out for $Re = 1000$ in the range $0.05 < \alpha < 0.5$ and is presented in Fig. 5.13 (see Sect. 5.11). The maximal step in time ensuring the stability of the difference algorithm is attained for $0.2 < \alpha < 0.5$.

According to the computation practice, for laminar flows, the separation zone size weakly depends on α. In computing the laminar–turbulent transition, we

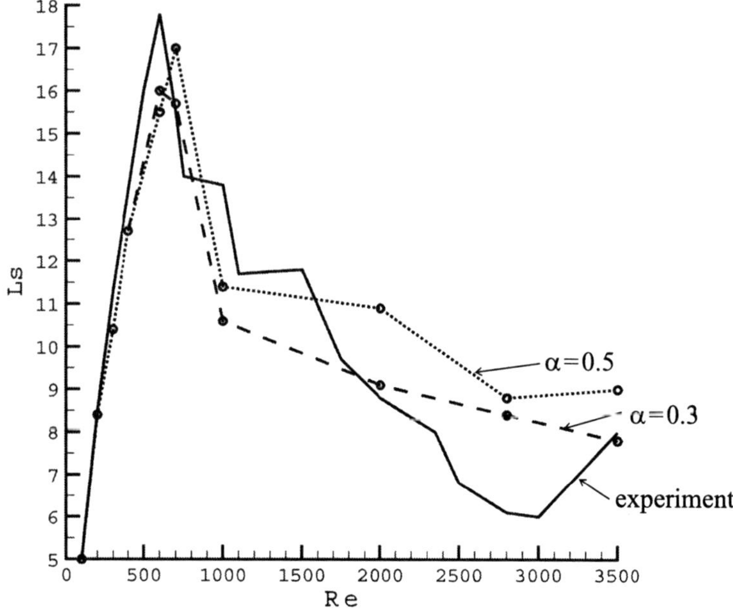

Fig. D.8 Comparison of the computed separation zone length for $\alpha = 0.5$ and 0.3 (*dotted lines*; the markers denote the computations carried out) with the data of experiment in [14] (*continuous line*)

observe a certain dependence on α. For $\alpha > 0.5$, the extra dissipation prevents the appearance of a nonstationary flow. For small values of this coefficient, the step in time ensuring the stability of the method becomes unjustifiably small. Therefore, $\alpha = 0.3$–0.5 is the optimal value. We can expect that when the grid step diminishes, the influence of the parameter α on the averaged flow decreases, since the part of the energy spectrum and vortex structures being greater and greater is resolved by the spatial grid.

Close results in comparing the separation zone structure with the experiment are obtained in modelling laminar and turbulent viscous incompressible flows after the backward step in the framework of Sheretov's quasi-hydrodynamic (QHD) equations [65].

The example considered here shows that the QGD model allows us to perform the numerical modelling of a laminar flow and its transition to the turbulent regime in the framework of the same numerical algorithm without use of additional models. The homogeneity of the computational model seems to be of especial importance for engineering applications where the moment of the transition in flow regime is not known in advance.

The modelling of the laminar–turbulent transition in the framework of a three-dimensional model is also of undoubted interest, since the described algorithm naturally extends to the three-dimensional flows and is effectively adopted for implementing in multi-processor computational systems (see, e.g., [74, 107]).

References

1. I. V. Abalakin, A. V. Zhokhova, and B. N. Chetverushkin, Kinetically consistent finite-difference schemes on irregular grids. *Mat. Model.*, **9**, No. 7, 44–53 (1997).
2. I. V. Abalakin, A. V. Zhokhova, and B. N. Chetverushkin, Kinetically consistent algorithm for simulation of gas-dynamic flows on triangular grids. *Mat. Model.*, **10**, No. 4, 51–60 (1998).
3. G. N. Abramovich, *Applied Gas Dynamics* [in Russian], Vol. 2, Nauka, Moscow (1992).
4. B. V. Alekseev, Generalized Boltzmann physical kinetics. *Thermal Phys. High Temperatures*, **35**, No. 1, 129–146 (1997).
5. B. V. Alekseev, *Generalized Boltzmann Physical Kinetics*, Elsevier, Amsterdam (2004).
6. H. Alsmeyer, Density profiles in argon and nitrogen shock waves measured by the absorption of an electron beam. *J. Fluid Mech.*, **74**, No. 3, 479–513 (1976).
7. D. A. Anderson, J. C. Tannehill, and R. H. Pletcher, *Computational Fluid Mechanics and Heat Transfer*, Hemisphere Publishing Corporation, New York (1984).
8. A. N. Antonov, T. G. Elizarova, A. N. Pavlov, and B. N. Chetverushkin, Mathematical modeling of oscillating regimes during the flow of a body with a spike. *Math. Model.*, **1**, No. 2, 13–23, (1989).
9. A. N. Antonov, T. G. Elizarova, B. N. Chetverushkin, and Yu. V. Sheretov, Numerical modeling of pulsating regimes accompanying supersonic flow round a hollow cylinder. *USSR Comput. Maths. Math. Phys.*, **30**, No. 2, 139–144 (1990).
10. M. A. Antonov, I. A. Graur, L. V. Kosarev, and B. N. Chetverushkin, Numerical modeling of pressure pulsing in three-dimensional cavities. *Mat. Model.*, **8**, No. 5, 76–90 (1996).
11. V. V. Aristov, *Direct Methods for Solving the Boltzmann Equation and Study of Nonequilibrium Flows*. Kluwer Academic Publisher. Dordrecht (2001).
12. V. V. Aristov and F. G. Cheremisin, Solution of the Euler and Navier–Stokes equations on the basis of an operator splitting of the kinetic equation. *Dokl. Akad. Nauk SSSR*, **272**, No. 3, 555–559 (1983).
13. E. B. Arkilic, K. S. Bruer, and M. A. Schmidt, Massflow and tangential momentum accommodation in silicon micromachined channels. *J. Fluid Mech.*, **437**, 29–43 (2001).
14. B. F. Armaly, F. Durst, J. C. F. Pereira, and B. Schonung, Experimental and theoretical investigation of backward-facing step flow. *J. Fluid Mech.*, **127**, 473–496 (1983).
15. C. Aubert and S. Colin, High-order boundary conditions for gaseous flows in rectangular microchannels. *Microscale Thermophys. Eng.*, **5**, No. 1, 41–54 (2001).
16. P. Bassanini, C. Cercigniani, and F. Sernagiotto, Flow of a rarefied gas in a tube of annular section. *Phys. Fluids*, **9**, No. 6, 1174–1178 (1966).
17. K. J. Bathe, *Finite Element Procedures*. Prentice Hall, 1996.
18. K. J. Bathe, The inf-sup condition and its evaluation for mixed finite element methods. *Comput. Struct.*, **79**, 243–252, 971 (2001).
19. K. J. Bathe and H. Zhang, A flow-condition-based interpolation finite element procedure for incompressible fluid flows. *Comput. Struct.*, **80**, 1267–1277 (2002).

20. M. Behnia, Synthesis of finite difference methods. In: *Workshop "Numerical Simulation of Oscillatory Convection in Low*-Pr *Fluids.* Notes Numer. Fluid Dynam., **27** (1990), pp. 265–272.

21. M. Behnia and G. de Vahl Davis, Fine mesh solutions using streamfunction. Vorticity formulation. In: *Workshop "Numerical Simulation of Oscillatory Convection in Low*-Pr *Fluids.* Notes Numer. Fluid Dynam., **27** (1990), pp. 11–18.

22. O. M. Belotserkovskii and Yu. M. Davydov, *Method of Large Particles in Gas Dynamics* [in Russian], Nauka, Moscow (1982).

23. O. M. Belotserkovskii and A. M. Oparin, *Numerical Experiment in Turbulence: From Order to Chaos* [in Russian], Nauka, Moscow (2004).

24. H. Ben Hadid and B. Roux, Buoyancy-driven oscillatory flows in shallow cavities filled with a low-Prandtl number fluid. In: *Workshop "Numerical Simulation of Oscillatory Convection in Low*-Pr *Fluids.* Notes Numer. Fluid Dynam., **27** (1990), pp. 25–34.

25. P. L. Bhatnagar, E. P. Gross, and M. Krook, Model for collision processes in gases, I. Small amplitude processes in charged and neutral one component systems. *Phys. Rev.*, **94**, 511–524 (1954).

26. G. A. Bird, Breakdown of translational and rotational equilibrium in gaseous expansions. *AIAA J.*, **8**, No. 11, 1988–2003 (1970).

27. G. A. Bird, *Molecular Gas Dynamics*, Oxford University Press, Oxford (1976).

28. G. A. Bird, *Molecular Gas Dynamics and the Direct Simulation of Gas Flows*, Clarendon Press, Oxford (1994).

29. S. Biringen, G. Danabasoglu, and T. K. Eastman, A finite-difference method with direct solvers for thermally-driven cavity problems. In: *Workshop "Numerical Simulation of Oscillatory Convection in Low*-Pr *Fluids.* Notes Numer. Fluid Dynam., **27** (1990), pp. 35–42.

30. H. Brenner, Navier–Stokes revisited. *Phys. A*, **349**, 60–132 (2005).

31. H. Brenner, Fluid mechanics revisited. *Phys. A*, **370**, 190–224. (2006).

32. F. Brezzi and K. J. Bathe, A discourse on the stability conditions for mixed finite element formulations. *J. Comput. Meth. Appl. Mech. Eng.*, **82**, 27–57 (1990).

33. C. Cercignani, *Mathematical Methods in Kinetic Theory*, Macmillan, London (1969).

34. C. Cercignani, *Theory and Application of the Boltzmann Equation*, Scottish Academic Press, Edinburg–London (1975).

35. C. Cercigniani, *Rarefied Gas Dynamics. From Basic Concepts to Actual Calculations*, Cambridge University Press, Cambridge (2000).

36. C. Cercigniani, M. Lampis, and S. Lorenzani, Variational approach to gas flows in microchannels. *Phys. Fluids*, **16**, No. 9, 3426–3437 (2004).

37. C. Cercigniani and F. Sernagiotto, Cylindrical Poiseuille flow of a rarefied gas. *Phys. Fluids*, **9**, No. 1, 40–44 (1966).

38. P. Chang, *Separation of Flow*. Pergamon Press, Oxford (1970).

39. S. Chapmen and T. G. Cowling, *The Mathematical Theory of Non-Uniform Gases*, Cambridge University Press, Cambridge (1970).

40. B. N. Chetverushkin, *Kinetically Consistent Schemes in Gas Dynamics* [in Russian], Moscow State University, Moscow (1999).

41. B. N. Chetverushkin, *Kinetic Schemes and Quasi-Gas-Dynamic System* [in Russian], Maks Press, Moscow (2004).

42. L. V. Dorodnitsyn, Artificial boundary conditions in the numerical simulation of subsonic gas flows. *Zh. Vychisl. Mat. Mat. Fiz.*, **45**, No. 7, 1251–1278 (2005).

43. A. Chpoun, T. G. Elizarova, I. A. Graur, and J. C. Lengrand, Simulation of the rarefied gas flow around a perpendicular disk. *Eur. J. Mech.* (B: Fluids), **24**, 457–467 (2005).

44. S. Colin, Rarefaction and compressibility effects on steady and transient gas flows in microchannels. In: *Microfluidics and Nanofluidics*, Springer-Verlag (2004).

45. S. Colin, ed. Microfluidique. Chap. 2. Micro-ecoulements gazeux. EGEM, Lavoisier (2004); www.hermes-science.com.

46. R. G. Deissler, An analysis of second-order slip flow and temperature-jump boundary conditions for rarefied gases. *Int. J. Heat Mass Transfer*, **7**, 681–694 (1964).

47. G. De Vahl Davis and I. P. Jones, Natural convection in a square cavity: a comparison exercise. *Int. J. Numer. Mech. Fluids*, **3**, 227–248 (1983).
48. S. M. Deshpande, Kinetic theory based on new upwind methods for inviscid compressible flows. *AIAA J.*, Paper 86–0275 (1986).
49. B. A. Dubrovin, S. P. Novikov, and A. T. Fomenko, *Modern Geometry. Methods and Applications* [in Russian], Editorial, Moscow (1998).
50. M. van Dyke, *An Album of Fluid Motion*. The parabolic press, Stanford, California (1982).
51. T. G. Elizarova, *Kinetically Compatible Finite-Difference Schemes of Gas Dynamics* [in Russian]. Thesis, Institute of Applied Mathematics, Moscow (1989).
52. T. G. Elizarova, A class of kinetically compatible finite-difference schemes for gas dynamics. *USSR Comput. Math. Math. Phys.*, **27**, No. 6, 98–101 (1987).
53. T. G. Elizarova, Knudsen effect and a unified formula for mass flow-rate in microchannels. In: *Proc. 25 Int. Symp. on Rarefied Gas Dynamics RGD25, St.-Petersburg (Repino), Russia, 22–28 July, 2006* (M. S. Ivanov and A. K. Rebrov, eds.). Sib. Branch of Russian Academy of Science (2007), pp. 1164–1169
54. T. G. Elizarova and B. N. Chetverushkin, On a computational algorithm for calculating gas-dynamic flows. *Dokl. Akad. Nauk SSSR*, **279**, No. 1, 80–83 (1984).
55. T. G. Elizarova and B. N. Chetverushkin, Kinetic algorithms for calculating gas-dynamic flows. *USSR Comput. Math. Math. Phys.*, **25**, No. 5, 164–169 (1985).
56. T. G. Elizarova and B. N. Chetverushkin, The use of kinetic models to calculate gas-dynamic flows. In: *Mathematical Modeling: Processes in Nonlinear Media* [in Russian], Moscow, Nauka (1986), pp. 261–278.
57. T. G. Elizarova, B. N. Chetverushkin, and Yu. V. Sheretov, Quasi-gas-dynamic equations and computer simulation of viscous gas flows. *Lect. Notes Phys.*, **414**. Proc. 13th Int. Conf. on Numerical Methods in Fluid Dynamics, Roma. Springer-Verlag (1992), pp. 421–425.
58. T. G. Elizarova and D. G. Ershov, Mass flow-rate calculations for microchannels. *Moscow Univ. Phys. Bull.*, **62**, No. 1, 21–24 (2007); http://www.phys.msu.ru/eng/
59. T. G. Elizarova, I. A. Graur, and J.-C. Lengrand, Macroscopic equations for a binary gas mixture. In: *Proc. 22nd Int. Symp. on Rarefied Gas Dynamics* (T. G. Bartel and M. A. Gallis, eds.), AIP Conf. Proc., **585**, Melville, New York (2001), pp. 297–304.
60. T. G. Elizarova, I. A. Graur, and J. C. Lengrand, Two-fluid computational model for a binary gas mixture. *Eur. J. Mech.* (B: Fluids), **3**, 351–369 (2001).
61. T. G. Elizarova, I. A. Graur, J. C. Lengrand, and A. Chpoun, Rarefied gas flow simulation based on quasi-gas-dynamic equations. *AIAA J.*, **33**, No. 12, 2316–2324 (1995).
62. T. G. Elizarova, I. A. Graur, and Yu. V. Sheretov, Quasi-gas-dynamic equations and computer simulation of rarefied gas flows. *Proc. 19th Int. Symp. on Shock Waves*, Marseille, France, July 26–30, **4**. Springer-Verlag (1993), pp. 45–50.
63. T. G. Elizarova, I. S. Kalachinskaya, A. V. Kluchnikova, and Yu. V. Sheretov, Application of quasi-hydrodynamic equations in the modeling of low-Prandtl thermal convection. *Comput. Math. Math. Phys.*, **38**, No. 10, 1662–1671 (1998).
64. T. G. Elizarova, I. S. Kalachinskaya, A. V. Klyuchnikova, and Yu. V. Sheretov, Computation of convective flows using quasihydrodynamic equations. *Comput. Math. Model.*, **10**, No. 2, 160–171 (1999).
65. T. G. Elizarova, I. S. Kalachinskaya, Yu. V. Sheretov, and E. V. Shil'nikov, Simulation of separating flows over a backward-facing step. *Comput. Math. Model.*, **15**, No. 2, 167–193 (2004).
66. T. G. Elizarova, I. S. Kalachinskaya, and Yu. V. Sheretov, Separating flow behind a back-step, I. Quasi-hydrodynamic equations and computation of a laminar flow. http://arXiv.org/abs/math-ph/0407053.
67. T. G. Elizarova, I. S. Kalachinskaya, Yu. V. Sheretov, and I. A. Shirokov, Numerical simulation of electrically conducting liquid flows in an external magnetic field. *J. Commun. Technology Electronics*, **50**, No. 2, 227–233 (2005).

68. T. G. Elizarova and A. A. Khokhlov, Numerical simulation of shock-wave structure by solution of steady-state Navier–Stokes equations. *Vestn. Mosk. Univ. Ser. Fiz. Astron.*, **3**, 28–32 (2006).

69. T. G. Elizarova and A. A. Khokhlov, Quasi-gas-dynamic equations for flows with external heat sources. *Moscow Univ. Phys. Bull.*, **62**, No. 3, 138–142 (2007).

70. T. G. Elizarova, A. A. Khokhlov, and S. Montero, Numerical simulation of shock wave structure in nitrogen. *Phys. Fluids*, **19**, No. 6, 068102 (2007).

71. T. G. Elizarova, A. A. Khokhlov, and Yu. V. Sheretov, Quasi-gas-dynamic numerical algorithm for gas flow simulations. *Int. J. Numer. Meth. Fluids*, **56**, No. 8, 1209–1215 (2008).

72. T. G. Elizarova, J. C. Lengrand, and I. A. Graur, Gradient expansions for distribution functions and derivation of moment equations. *Proc. 21st Int. Symp. on RGD* (R. Brun, ed.), Marseille, France, July 26–30, 1998. Toulouse, France, Cepadues (1999), pp. 119–126.

73. T. G. Elizarova and O. Yu. Milyukova, Parallel algorithm for numerical simulation of 3D incompressible flows. In: *Parallel Computational Fluid Dynamics. Advanced Numerical Methods, Software, and Applications* (B. Chetverushkin et al., eds.) Proc. of the Parallel CFD 2003 Conf. Moscow, Russia, May 13–15, 2003. Elsevier (2004), pp. 65–72.

74. T. G. Elizarova and O. Yu. Milyukova, Numerical simulation of viscous incompressible flow in a cubic cavity. *Comput. Math. Math. Phys.*, **43**, No. 3, 433–445 (2003).

75. T. G. Elizarova and P. N. Nikolskiy, Numerical simulation of laminar-turbulent transition in backward-step flow. *Moscow Univ. Phys. Bull.*, **62**, No. 4, 216–220 (2007).

76. T. G. Elizarova, P. N. Nikolsky, and J. C. Lengrand, A new variant of subgrid dissipation for LES method and simulation of laminar-turbulent transition in subsonic gas flows. In: *Advances in Hybrid RANS-LES Modelling* (S.-H. Peng and W. Haase, eds.). Notes on Numerical Fluid Mechanics and Multidisciplinary Design, **97**, Springer-Verlag, Berlin–Heidelberg (2008), pp. 289–298.

77. T. G. Elizarova, A. N. Pavlov, and B. N. Chetverushkin, *Obtaining the equations that describe gas flows by using the kinetic model* [in Russian]. Preprint No. 144, Institute of Applied Mathematics, Moscow (1983).

78. T. G. Elizarova and V. V. Seregin, Approximation of quasi-gas-dynamic equations on triangle grids. *Vestn. Mosk. Univ. Ser. Fiz. Astron.*, **4**, 15–18 (2005).

79. T. G. Elizarova and V. V. Seregin, Quasi-gas-dynamic equations and approximation for bulk viscosity coefficient. *Vestn. Mosk. Univ. Ser. Fiz. Astron.*, **1**, 15–18 (2006).

80. T. G. Elizarova and M. E. Sokolova, Dissipative terms in quasi-gas-dynamic equations and its influence on the shock-wave flow field. *Vestn. Mosk. Univ. Ser. Fiz. Astron.*, **5**, 19–22 (2001).

81. T. G. Elizarova and M. E. Sokolova, Numerical algorithm for simulation supersonic gas flows based on quasi-gas-dynamic equations. *Vestn. Mosk. Univ. Ser. Fiz. Astron.*, **1**, 10–15 (2004).

82. T. G. Elizarova and M. E. Sokolova, Algorithm for numerical simulation of subsonic viscous gas flows. *Vestn. Mosk. Univ. Ser. Fiz. Astron.*, **3**, 6–10 (2005).

83. T. G. Elizarova, M. E. Sokolova, and Yu. V. Sheretov, Quasi-gas-dynamic equations and numerical simulation of viscous gas flows. *Comput. Math. Math. Phys.*, **45**, No. 3, 524–534 (2005).

84. T. G. Elizarova and Yu. V. Sheretov, Theoretical and numerical analysis of quasi-gas-dynamic and quasi-hydrodynamic equations. *Comput. Math. Math. Phys.*, **41**, No. 2, 219–234 (2001).

85. T. G. Elizarova and Yu. V. Sheretov, Analyse du probleme de l'ecoulement gazeux dans les microcanaux par les equations quasi hydrodynamiques. *J. La Houille Blanche. Revue Internationale de l'Eau*, **5**, 66–72 (2003); see also *Congr. Soc. Hydrotechnique de France (SHF) "Microfluidique,"* Toulouse, 3–5 Decembre (2002), pp. 309–318.

86. T. G. Elizarova, E. V. Shilnikov, R. Weber, and J. Hureau, Separated flow behind a backward-facing step, II. Experimental and numerical investigation of a turbulent flow. http://arXiv.org/abs/math-ph/0410023.

87. T. G. Elizarova, E. V. Shilnikov, R. Weber, J. Hureau, and J. C. Lengrand, Experimental and numerical investigation of the turbulent flow behind a backward-facing step. In: *Proc. Int. Conf. on Boundary and Interior Layes (BAIL)*, France, Toulouse, 5–7 July (2004).

88. T. G. Elizarova and I. A. Shirokov, Macroscopic model for gas with translational-rotational nonequilibrium. *Comput. Math. Math. Phys.*, **39**, No. 1, 135–146 (1999).

89. T. G. Elizarova and I. A. Shirokov, Simulating shock waves in argon, helium, and nitrogen. *Comput. Math. Model.*, **16**, No. 4, 336–348 (2005).

90. T. G. Elizarova, I. A. Shirokov, and S. Montero, Numerical simulation of shock wave structure for argon and helium. *Phys. Fluids*, **17**, No. 6, 068101 (2005).

91. T. G. Elizarova, A. V. Zherikov, I. S. Kalachinskaya, and Yu. V. Sheretov, Simulation of convective flow of an electrically conducting fluid in a cavity. *Comput. Math. Model.*, **15**, No. 1, 52–68 (2004).

92. T. G. Elizarova, A. V. Zherikov, and I. S. Kalachinskaya, Numerical solution of quasi-hydrodynamic equations in a domain of complex shape. *Differ. Equations*, **43**, No. 9, 1255–1262 (2007).

93. S. M. Ermakov, *Monte-Carlo Method and Adjacent Problems* [in Russian]. Nauka, Moscow (1975).

94. P. G. Frey and P.-L. George, *Mesh Generation. Application of Finite Elements*, HERMES Science Publ. (2000).

95. V. E. Fortov, V. K. Levin, G. I. Savin, A. V. Zabrodin, V. V. Karatanov, G. S. Elizarov, V. V. Korneev, and B. M. Shabanov, Supercomputer MBC–1000M and its perspectives. *Nauka Promyshl. Ross.*, **55**, 49–52 (2001).

96. A. N. Gilmanov, *Method of Adaptive Grids in Problems of Gas Dynamics* [in Russian]. Fizmatlit, Moscow (2000).

97. A. S. Gmurczyk, M. Tarczyński, and Z. A. Walenta, Shock wave structure in the binary mixtures of gases with disparate molecular masses. In: *Rarefied Gas Dynamics* (R. Compargue, ed.), **1**. CEA, Paris (1979), pp. 333–341.

98. E. Goldman and L. Sirovich, Equations for gas mixtures. *Phys. Fluids*, **10**, No. 9, 1928–1940 (1967).

99. Yu. P. Golovachev, *Numerical Simulation of Viscous Gas Flows in Shock Layers* [in Russian], Fizmatlit, Moscow (1996).

100. I. A. Graur, Algorithms for numerical solution of quasi-gas-dynamic equations. *J. Comput. Math. Math. Phys.*, **39**, No. 8, 1300–1315 (1999).

101. I. A. Graur, Method of quasi-gas-dynamic splitting for solving the Euler equations. *J. Comput. Math. Math. Phys.*, **41**, No. 10, 1506–1518 (2001).

102. I. A. Graur, Second-order algorithms for numerical solution of quasi-gas-dynamic equations. *J. Comput. Math. Math. Phys.*, **42**, No. 5, 668–678 (2002).

103. I. A. Graur, Splitting finite-difference schemes for the Euler equations based on quasi-gas-dynamic equations. *J. Comput. Math. Math. Phys.*, **44**, No. 1, 155–167 (2004).

104. I. A. Graur, T. G. Elizarova, and B. N. Chetverushkin, Simulation of complex gas-dynamic flows on the basis of kinetic algorithms. *Diff. Uravn.*, **22**, No. 7, 1112–1118 (1986).

105. I. A. Graur, T. G. Elizarova, and B. N. Chetverushkin, Numerical simulation of flow of cavities by a supersonic streams of a viscous compressible gas. *Inzh. Fiz. Zh.*, **61**, No. 4, 570–577 (1991).

106. I. A. Graur, T. G. Elizarova, T. A. Kudryashova, S. V. Polyakov, In: *A parallel implementation of an implicit scheme for underexpanded jet problems. Parallel Computational Fluid Dynamics – New Frontiers and Multy-Disciplinary Applications* (K. Matsund et al., eds.), Japan, Elsevier (2003), pp. 273–280.

107. I. A. Graur, T. G. Elizarova, T. A. Kudryashova, S. V. Polyakov, and S. Montero, Implementation of underexpanded jet problems on multiprocessor computer systems. In: *Parallel Computational Fluid Dynamics. Practice and Theory* (P. Wilders et al., eds.), North Holland, Elsevier (2002), pp. 151–158.

108. I. A. Graur, T. G. Elizarova, and J. C. Lengrand, *Quasi-gas-dynamic equations with multiple translational temperatures*. Report R97-1, Laboratoire d'A'erothermique du CNRS, Meudon, France (1997).

109. I. A. Graur, T. G. Elizarova, A. Ramos, G. Tejeda, J. M. Fernandez, and S. Montero, A study of shock waves in expanding flows on the basis of spectroscopic experiments and quasi-gas-dynamic equations. *J. Fluid Mech.*, **504**, 239–270 (2004).

110. I. A. Graur, M. S. Ivanov, G. N. Markelov, Y. Burtschell, E. Valerio, and D. Zeitoun, Comparison of kinetic and continuum approaches for simulation of shock wave/boundary layer interaction. *Shock Waves*, **12**, 343–350 (2003).

111. D. B. Gurov, T. G. Elizarova, and Yu. V. Sheretov, Numerical simulation of fluid flows in a cavity by the quasi-hydrodynamic system. *Mat. Model.*, **8**, No. 7, 33–44 (1996).

112. L. P. Hackman, G. D. Raithby, A. B. Strong, Numerical predictions of flows over backward-facing steps. *Int. J. Numerical Methods Fluids*, **4**, No. 8, 711–724 (1984).

113. J. O. Hirschfelder, C. F. Curtiss, and R. B. Bird, *The Molecular Theory of Gases and Liquids.* Wiley, New York (1954).

114. M. A. Ilgamov and A. N. Gilmanov, *Non-Reflecting Boundary Conditions for Computational Domains* [in Russian]. Fizmatlit, Moscow (2003).

115. S. A. Isaev, P. A. Baranov, N. A. Kudryavtsev, D. A. Lysenko, and A. E. Usachov, Comparative analysis of the software vp2/3 and fluent for the simulation of a nonstationary flow of a cylinder by using the Spalart–Allmaras and Menter turbulency models. *Inzh. Fiz. Zh.*, **78**, No. 6, 148–162 (2005).

116. G. Kardiadakis, A. Beskok, and N. Aluru, *Microflows and Nanoflows. Fundamentals and Simulation*, Springer-Verlag (2005).

117. J. Kim and P. Moin, Application of a fractional-step method to incompressible Navier–Stokes equations. *J. Computer Phys.*, **59**, 308–323 (1985).

118. Yu. L. Klimontovich, *Statistical Physics* [in Russian]. Nauka, Moscow (1982).

119. Yu. L. Klimontovich, *Turbulent Flows and Structure of Chaos* [in Russian]. Nauka, Moscow (1990).

120. Yu. L. Klimontovich, On the necessity and possibility of unified description of kinetical and hydrodynamical processes. *Teor. Mat. Fiz.*, **92**, 312–330 (1992).

121. Yu. L. Klimontovich, *Statistical Theory of Open Systems* [in Russian]. Yanus, Moscow (1995).

122. M. N. Kogan, *Rarefied Gas Dynamics*, Plenum Press, New York (1969).

123. H. Kohno and K. J. Bathe, A flow-condition-based interpolation finite element procedure for triangular grids. *Int. J. Num. Meth. Fluids*, **51**, 673–699 (2006).

124. H. Kohno and K. J. Bathe, Insight into the flow-condition-based interpolation finite element approach: Solution of steady-state advection-diffusion problems. *Int. J. Num. Meth. Eng.*, **63**, 197–217 (2005).

125. G. Korn and T. Korn, *Mathematical Handbook.* McGraw-Hill, New York etc. (1968).

126. J. R. Koseff and R. L. Street, On the influence of butt walls on the flow in a cavity with moving lid. *J. Fluids Eng.*, **106**, No. 4, 285 (1984).

127. Yu. A. Koshmarov and Yu. A. Ryzhov, *Applied Dynamics of Rarefied Gases* [in Russian]. Mashinostroenie, Moscow (1977).

128. S. Kosuge, K. Aoki, and S. Takata, *Shock-wave structure for a binary gas mixture: Finite-difference analysis of the Boltzmann equation for hard-sphere molecules.* Preprint: Kyoto Univ. (1999).

129. K. Koura, Monte Carlo direct simulation of rotational relaxation of diatomic molecules using classical trajectory calculations: Nitrogen shock wave. *Phys. Fluids*, **9**, No. 11, 3543–3549 (1997).

130. L. D. Kudryavtsev, *A Course of Analysis* [in Russian], Vysshaya Shkola. Moscow (1981).

131. A. G. Kulikovskii, N. V. Pogorelov, and A. Yu. Semenov, *Mathematical Problems of Numerical Solution of Hyperbolic Systems* [in Russian]. Nauka, Moscow (2001).

132. K. Xu and Z. Li, Microchannel flow in the slip regime: gas-kinetic BGK-Burnett solutions. *J. Fluid Mech.*, **513**, 87–110 (2004).

133. L. D. Landau and E. M. Lifshits, *Mechanics* [in Russian]. Nauka, Moscow (1965).

134. L. D. Landau and E. M. Lifshits, *Hydrodynamics* [in Russian]. Nauka, Moscow (1986).

135. Yu. V. Lapin, *Turbulent Boundary Layers in Supersonic Flows* [in Russian]. Nauka, Moscow (1982).

136. Yu. V. Lapin and M. Kh. Strelets, *Internal Flows of Gas Mixtures* [in Russian]. Nauka, Moscow (1989).

137. I. N. Larina and V. A. Rykov, Similarity of hypersonic flows of a rarefied gas near obtuse bodies. *Mekh. Zhidk. Gas.*, **2**, 130–135 (1981).

138. I. N. Larina and V. A. Rykov, Calculation of the flow of rarefied gas around a circular cylinder at low Knudsen numbers. *Comput. Math. Math. Phys.*, **45**, No. 7, 1259–1275 (2005).

139. W. C. Lasher and D. B. Taulbee, On the computation of turbulent backstep flow. *Int. J. Heat Fluid Flow*, **13**, No. 1, 30–40 (1992).

140. H. Le, P. Moin, and J. Kim, Direct numerical simulation of turbulent flow over a backward-facing step. *J. Fluid Mech.*, **330**, 349–374 (1997).

141. J. L. Lebowitz and E. W. Montroll, eds. Nonequilibrium phenomena, I. The Boltzmann equation. North-Holland, Amsterdam–New York–Oxford (1983).

142. E. M. Lifshits and L. P. Pitaevsky, *Physical Kinetics*. Pergamon Press, London (1963).

143. M. Linzer and D. F. Hornig, Structure of shock fronts in argon and nitrogen. *Phys. Fluids*, **6**, No. 12, 1661–1668 (1963).

144. L. G. Loitsyansky, *Mechanics of Liquids and Gases*. Pergamon Press, London (1966).

145. D. A. Lockerby, J. M. Reese, D. R. Emerson, and R. W. Barber, Velocity boundary condition at the solid walls in rarefied gas calculations. *Phys. Rev.*, **E70**, 017303 (2004).

146. B. Mate, I. A. Graur, T. G. Elizarova, I. A. Shirokov, G. Tejeda, J. M. Fernandez, and S. Montero, Experimental and numerical investigation of an axisymmetric supersonic jet. *J. Fluid Mech.*, **426**, 177–197 (2001).

147. J. Maurer, P. Tabeling, P. Joseph, and H. Willaime, Second order slip laws in microchannels for helium and nitrogen. *Phys. Fluids*, **15**, No. 9, 2613–2621 (2003).

148. J. C. Maxwell, On stresses in rarefied gases arising from inequalities of temperature. *Phil. Trans. Roy. Soc. London*, **70**, No. 6, 231–256 (1879).

149. T. F. Morse, Kinetic model equation for gas mixture. *Phys. Fluids*, **7**, No. 12, 2012–2013 (1964).

150. M. Nallasamy, Turbulence models and their applications to the prediction of internal flows: a review. *Computers Fluids*, **15**, No. 2, 151–194 (1987).

151. E. S. Nikolaev, ed., *Program Library for Solving Grid Equations* [in Russian] Moscow State University, Moscow (1984).

152. M. Ohnishi, H. Azuma, and T. Doi, Computer simulation of oscillatory Marangoni flow. *Acta Astronautica*, **26**, Nos. 8–10, 685–696 (1992).

153. H. Ohshima and H. Ninokata, Numerical simulation of oscillatory convection in low Prandtl number fluids using Aqua code. In: *Workshop "Numerical Simulation of Oscillatory Convection in Low-Pr Fluids*. Notes Numer. Fluid Dynam., **27** (1990), pp. 90–97.

154. H. C. Ottinger, *Beyond Equilibrium Thermodynamics*. John Wiley, New Jersey (2005).

155. W. Pauli, *Relativitätstheorie*, Enzycl. Math. Wiss. (1921).

156. W. G. Polard and R. D. Present, On gaseous self-diffusion in long capillary tubes. *Phys. Rev.*, **73**, No. 7, 762–774 (1948).

157. I. V. Popov and S. V. Polyakov, Construction of adaptive irregular triangular grids for two-dimensional, multiply connected nonconvex domains. *Mat. Model.*, **14**, No. 6, 25 (2002).

158. E. G. Poznyak and E. V. Shikin, *Differential Geometry* [in Russian], Moscow (1990).

159. F. P. Preparata and M. I. Shamos. *Computational Geometry: An Introduction*. Springer-Verlag, New York (1985).

160. R. D. Present, *Kinetic Theory of Gases*, McGraw-Hill, New York–Toronto–London (1958).

161. I. G. Pushkina and V. F. Tishkin, Adaptive computational grids of Dirichlet cells for solution of problems of mathematical physics. *Mat. Model.*, **12**, No. 3, 97 (2000).

162. R. C. Reid, J. M. Prausnitz, and T. K. Sherwood, *The Properties of Gases and Liquids*. McGraw Hill, New York (1977).

163. P. J. Roache, *Computational Hydrodynamics* [Russian translation], Mir Moscow (1970).

164. B. V. Rogov and I. A. Sokolova, A survey of models of viscous internal flows. *Mat. Model.*, **14**, No. 1, 41–72 (2002).

165. P. Roushan and X. L. Wu, Universal wake structures of Karman vortex streets in two-dimensional flows. *Phys. Fluids*, **17**, 073601 (2005).

166. B. Roux, H. Ben Hadid, P. Laure, Hydrodynamical regimes in metallic melts subject to a horizontal temperature gradient. *Eur. J. Mech.* (B: Fluids), **8**, No. 5, 375–396 (1989).

167. B. L. Rozhdestvensky and N. N. Yanenko, *Systems of Quasilinear Equations* [in Russian], Nauka, Moscow (1978).

168. V. A. Rykov, A model kinetic equation for a gas with rotational degrees of freedom. *Mekh. Zhidk. Gaza*, **6**, 107–115 (1975).

169. V. A. Rykov and V. N. Skobelkin, On a macroscopic description of flows of a gas with rotational degrees of freedom. *Mekh. Zhidk. Gaza*, **1**, 180–183 (1978).

170. A. A. Samarsky, *Theory of Finite-Difference Schemes* [in Russian], Nauka, Moscow (1989).

171. A. A. Samarsky and Yu. P. Popov, *Finite-Difference Methods in Gas Dynamics* [in Russian], Nauka, Moscow (1980).

172. L. M. Schwartz and D. F. Hornig, Navier–Stokes calculations of argon shock wave structure. *Phys. Fluids*, **6**, No. 12, 1669–1675 (1963).

173. L. I. Sedov, *A Course in Continuum Mechanics*, Vols. 1, 2. Wolters-Noordhoff, Groningen (1971–1972).

174. M. V. Semenov and Yu. V. Sheretov, Numerical simulation of liquid flows in a neighborhood of a sphere. In: *Applications of Functional Analysis in Approximation Theory* [in Russian], Tver (2005), pp. 107–123.

175. M. V. Semenov and Yu. V. Sheretov, Numerical simulation of subsonic axially symmetric flows of a gas in a neighborhood of a sphere. *Vestn. Tver. Univ.*, **4**, 78–97 (2006).

176. E. M. Shakhov, *Methods of the Study of Flows of Rarified Gases* [in Russian], Nauka, Moscow (1974).

177. L. F. Shampine and M. W. Reichelt, The Matlab code suite. *SIAM J. Sci. Comput.*, **18**, 1–22 (1997).

178. F. Sharipov and V. Seleznev, Data on internal rarefied gas flows. *J. Phys. Chem. Reference Data*, **27**, No. 3, 657–706 (1988).

179. V. G. Sheretov and S. Yu. Shcherbakova, *300 Years of Russian Mathematics* [in Russian]. Tver (2000).

180. Yu. V. Sheretov, On a new mathematical model in hydrodynamics. In: *Applications of Functional Analysis in Approximation Theory* [in Russian], Tver (1996), pp. 124–134.

181. Yu. V. Sheretov, Quasi-hydrodynamic equations as a model of flows of compressible viscous, heat-conducting medium. In: *Applications of Functional Analysis in Approximation Theory* [in Russian], Tver (1997), pp. 127–155.

182. Yu. V. Sheretov, On exact solutions of quasi-hydrodynamic equations. In: *Applications of Functional Analysis in Approximation Theory* [in Russian], Tver (1998), pp. 213–241.

183. Yu. V. Sheretov, Finite-difference schemes of hydrodynamics in the Euler and Lagrange coordinates based on quasi-gas-dynamic and quasi-hydrodynamic equations. In: *Applications of Functional Analysis in Approximation Theory* [in Russian], Tver (1999), pp. 184–208.

184. Yu. V. Sheretov, *Mathematical Modeling of Flows of Liquids and Gases Based on Quasi-Gas-Dynamic and Quasi-Hydrodynamic Equations* [in Russian], Tver State Univ., Tver (2000).

185. Yu. V. Sheretov, Some properties of quasi-gas-dynamic equations. In: *Applications of Functional Analysis in Approximation Theory* [in Russian], Tver (2000), pp. 134–149.

186. Yu. V. Sheretov, On finite-difference approximations of quasi-gas-dynamic equations for axially symmetric flows. In: *Applications of Functional Analysis in Approximation Theory* [in Russian], Tver (2001), pp. 191–207.

187. Yu. V. Sheretov, Parabolized quasi-hydrodynamic equations. *Vestn. Tver. Univ.*, **2**, 79–83 (2003).

188. Yu. V. Sheretov, Equations of hydrodynamics and the Galileo transforms. In: *Applications of Functional Analysis in Approximation Theory* [in Russian], Tver (2003), pp. 187–198.

189. Yu. V. Sheretov, Analysis of the stability of a modified, kinetically consistent scheme in the acoustic approximation. In: *Applications of Functional Analysis in Approximation Theory* [in Russian], Tver (2004), pp. 147–160.

190. Yu. V. Sheretov, *Mathematical Models of Hydrodynamics* [in Russian]. Tver State Univ., Tver (2004).

191. I. A. Shirokov, Solving Poisson's equation on a multiprocessor system in problems of incompressible fluid flow simulation. *Differ. Equations*, **39**, No. 7, 1050–1057 (2003).

192. I. A. Shirokov, T. G. Elizarova, and J. C. Lengrand, Numerical study of shock wave structure based on quasi-gas-dynamic equations with rotational nonequilibrium. In: *21th Int. Symp. on Rarefied Gas Dynamics*, Marseille, France, July 26–30, 1998 (R. Brun et al., eds.). Cepadues, Toulouse (1999), pp. 175–182.

193. L. Sirovich, Kinetic modeling of gas mixtures. *Phys. Fluids*, **5**, No. 8, 908–918 (1962).

194. N. A. Slezkin, On differential equations of motion of gases, *Dokl. Akad. Nauk SSSR*, **77**, No. 2, 205–208 (1951).

195. Y. Sone, Continuum gas dynamics in the light of kinetic theory and new features of rarefied gas flows. In: *Proc. 20th Int. Symp. on Rarefied Gas Dynamics* (C. Shen, ed.). Peking Univ. Press, Beijing, China (1997), pp. 3–24.

196. E. M. Sparrow and W. Chuck, PC solutions for heat transfer and fluid flow downstream of an abrupt, asymmetric enlargement in a channel. *Numerical Heat Transfer*, **12**, 19–40 (1987).

197. D. J. Struik, *A Concise History of Mathematics*. Dover Publ. (1987).

198. V. P. Stulov, *Lectures in Gas Dynamics* [in Russian]. Fizmatlit, Moscow (2004).

199. E. V. Stupochenko, S. A. Losev, and A. I. Osipov, *Relaxation Processes in Shock Waves* [in Russian]. Nauka, Moscow (1965).

200. S. Thangam, C. G. Speziale, Turbulent flow past a backward-facing step: A critical evaluation of two-equation models. *AIAA J.*, **30**, No. 5, 1314–1320 (1992).

201. D. V. Tikhomirov, A tensor representation of the quasi-gas-dynamic system and finite-difference approximation. *Vestn. Mosk. Univ. Ser. Fiz. Astron.* **1**, 31–35 (2006).

202. M. Torrilhon and H. Struchtrup, Regularized 13-moment equations: shock structure calculations and comparison to Burnett models. *J. Fluid Mech.*, **513**, 171–198 (2004).

203. P. N. Vabishchevich, M. M. Makarov, V. V. Chudanov, and A. G. Churbanov, *Numerical simulation of convective flows in the variables "stream function, velocity curl, and temperature"* [in Russian]. Preprint No. 28, Institute of Applied Mathematics, Moscow (1993).

204. S. V. Vallander, Equation of motion of viscous gas. *Dokl. Akad. Nauk SSSR*, **58**, No. 1, 25–27 (1951).

205. W. G. Vincenti and C. H. Kruger Jr., *Introduction to Physical Gas Dynamics*. Wiley (1965).

206. D. Wee, Yi. Tongxun, A. Annaswamy, and A. Ghoniem, Self-sustained oscillations and vortex shedding in backward-facing step flows: Simulation and linear instability analysis. *Phys. Fluids*, **16**, No. 9, 3361–3373 (2004).

207. C. R. Wilke, A viscosity equation for gas mixtures. *J. Chem. Phys.*, **18**, No. 4, 517–522 (1950).

208. P. Woodward and P. J. Collela, The numerical simulation of two-dimensional fluid flow with strong shock. *Comput. Phys.*, **54**, 115–173 (1984).

209. M.-H. Wu, Ch.-Yu. Wen, R.-H. Yen, M.-Ch. Weng, and M.-Ch. Wang, Experimental and numerical study of the separation angle for flow around a circular cylinder at low Reynolds number. *J. Fluid Mech.*, **515**, 233–260 (2004).

210. Y. Wu and C. H. Lee, Kinetic theory of shock tube problems for binary mixtures. *Phys. Fluids*, **14**, No. 2, 313–322 (1971).

211. V. M. Zhdanov, *Transport and Relaxation Processes in Many-Component Plasma* [in Russian]. Energoizdat, Moscow (1982).

212. V. M. Zhdanov and M. Ya. Alievsky, *Transport and Relaxation Processes in Molecular Gases* [in Russian]. Nauka, Moscow (1989).

213. A. I. Zhmakin and A. A. Fursenko, On a monotonic difference scheme of direct calculations. *Zh. Vychisl. Mat. Mat Fiz.*, **20**, No. 4, 1021–1031 (1980).

Index

Printed in the United States
148239LV00001BA/41/P